In this new volume, John Hedley Brooke offers an introduction and critical guide to one of the most fascinating and enduring issues in the development of the modern world: the relationship between scientific thought and religious belief. It is common knowledge that in western societies there have been periods of crisis when new science has threatened established authority. The trial of Galileo in 1633 and the uproar caused by Darwin's *Origin of Species* after its publication in 1859 are two of the most famous examples. Taking account of recent scholarship in the history of science, Brooke takes a fresh look at these and similar episodes, showing that science and religion have been mutually relevant in so rich a variety of ways that no simple generalizations are possible.

A special feature of the book is that Brooke stands back from general theses affirming "conflict" or "harmony," which have so often served partisan interests. His object is to reveal the subtlety, complexity, and diversity of the interaction as it has taken place in the past and in the twentieth century. Instead of treating science and religion as discrete definable entities, a historical approach requires sensitivity to shifting boundaries and a willingness to consider the contexts in which particular forms of "science" could be used both for religious and secular ends. Without assuming specialist knowledge, Brooke provides a wide-ranging study from the Copernican innovation to in vitro fertilization.

Science and Religion

CAMBRIDGE HISTORY OF SCIENCE

Editor
GEORGE BASALLA
University of Delaware

Physical Science in the Middle Ages
EDWARD GRANT

Man and Nature in the Renaissance
ALLEN G. DEBUS

The Construction of Modern Science:
Mechanisms and Mechanics
RICHARD S. WESTFALL

Science and the Enlightenment
THOMAS L. HANKINS

Biology in the Nineteenth Century:
Problems of Form, Function, and Transformation
WILLIAM COLEMAN

Energy, Force, and Matter: The Conceptual Development of
Nineteenth-Century Physics
P. M. HARMAN

Life Science in the Twentieth Century
GARLAND E. ALLEN

The Evolution of Technology
GEORGE BASALLA

Science and Religion: Some Historical Perspectives
JOHN HEDLEY BROOKE

Science and Religion

Some Historical Perspectives

JOHN HEDLEY BROOKE

Department of History, University of Lancaster

CAMBRIDGE
UNIVERSITY PRESS

Published by the Press Syndicate of the University of Cambridge
The Pitt Building, Trumpington Street, Cambridge CB2 1RP
40 West 20th Street, New York, NY 10011-4211, USA
10 Stamford Road, Oakleigh, Melbourne 3166, Australia

First published 1991
Reprinted 1991, 1992, 1993

Printed in the United States of America

Library of Congress Cataloging-in-Publication Data
Brooke, John Hedley.
Science and religion: some historical perspectives / John Hedley
Brooke.
p. cm. – (Cambridge history of science)
ISBN 0-521-23961-3. – ISBN 0-521-28374-4 (pbk.)
1. Religion and science – History.   I. Title   II. Series.
BL245.B77   1991
291.1'75–dc20   90-48909
                                                        CIP

A catalogue record for this book is available from the British Library.

ISBN 0-521-23961-3 hardback
ISBN 0-521-28374-4 paperback

For Janice

Contents

Acknowledgments

My first acknowledgment is to the several generations of students at the University of Lancaster with whom I have explored the issues raised in this book, and from whom I have received unfailing stimulus. It is with great pleasure that I acknowledge a further debt to Lancaster University: to the Humanities Research Committee, which awarded a special year of study-leave to assist the completion of the text. My original research, particularly concerning the British natural theology tradition, on which parts of the discussion are based, has also been generously supported by a research grant from the Royal Society.

In keeping with the style of this series, references in the text to other secondary sources have been kept to a minimum. A bibliographic essay can make but partial amends. If I have misrepresented the sources on which I have drawn, the responsibility is of course entirely mine. To many friends and colleagues in the history of science I have a debt that is impossible to articulate fully. I am especially grateful to colleagues who have made detailed comments on earlier drafts: Professor Michael J. Crowe of the University of Notre Dame; Dr. Geoffrey Cantor of the University of Leeds; the editor of the series, Professor George Basalla of the University of Delaware; and my colleague at Lancaster, Dr. Roger Smith. Their encouragement and advice have been invaluable.

Finally, my greatest debt is to the person at whose invitation this volume was first conceived and under whose guidance (as former coeditor of the series) it assumed its present shape: the late William Coleman of the University of Wisconsin. Many have rightly said that his untimely death deprived the history of science of one of its true masters. The many respects in which this book is the better for his recommendations constitute a further tribute which, had he lived to see it, would not, I hope, have been unwelcome.

Introduction

In a classic discussion of the origins of modern science, the historian Herbert Butterfield drew a much-quoted parallel. Such was the impact of the seventeenth-century Scientific Revolution that the only landmark with which it could be compared was the rise of Christianity. In shaping the values of Western societies, science and the Christian religion had each played a preeminent part and made a lasting impression. Exaggerated or not, such comparisons raise an obvious question. What was the relationship between these powerful cultural forces? Were they complementary in their effects, or were they antagonistic? Did religious movements assist the emergence of the scientific movement, or was there a power struggle from the start? Were scientific and religious beliefs constantly at variance, or were they perhaps more commonly integrated, both by clergy and by practicing men of science? How has the relationship changed over time?

Such questions are easier to formulate than to answer. Since the seventeenth century every generation has taken a view on their importance without, however, reaching any consensus as to how they should be answered. Writing some sixty years ago, the philosopher A. N. Whitehead considered that the future course of history would depend on the decision of his generation as to the proper relations between science and religion – so powerful were the religious symbols through which men and women conferred meaning on their lives, and so powerful the scientific models through which they could manipulate their environment. Because every generation has reappraised the issues, if not always with the same sense of

1

urgency, there has been no shortage of opinion as to what that proper relationship should be.

In popular literature three positions are commonly found, which, though not equally unsatisfactory, turn out to be problematic. One often encounters the view that there is an underlying conflict between scientific and religious mentalities, the one dealing in testable facts, the other deserting reason for faith; the one relishing change as scientific understanding advances, the other finding solace in eternal verities. Where such a view holds sway, it is assumed that historical analysis provides supporting evidence – of territorial squabbles in which cosmologies constructed in the name of religion have been forced into retreat by more sophisticated theories coming from science. The nineteenth-century scholars J. W. Draper and A. D. White constructed catalogs of this kind, in which scientific explanations repeatedly challenged religious sensibilities, in which ecclesiastics invariably protested at the presumption, and in which the scientists would have the last laugh.

Typical was White's account of the reluctance of the clergy to fix lightning rods to their churches. In 1745 the bell tower of St. Mark's in Venice had once again been shattered in a storm. Within ten years, Benjamin Franklin had mastered the electrical nature of lightning. His conducting rod could have saved many a church from that divine voice of rebuke, which thunder had often been supposed to be. But White reported that such meddling with providence, such presumption in controlling the artillery of heaven, was opposed so long by clerical authorities that the tower of St. Mark's was smitten again in 1761 and 1762. Not until 1766 was the conductor fixed – after which the monument was spared. White's picture of religious scruples and shattered towers symbolizes the popular notion of an intrinsic and perennial conflict. An ounce of scientific knowledge could be more effective in controlling the forces of nature than any amount of supplication.

A second, quite different view also appeals to history for its vindication. Science and religion are sometimes presented not as contending forces but as essentially complementary – each answering a different set of human needs. On this view, scientific and theological language have to be related to different spheres of practice. Discourse about God, which is inappropriate in the context of laboratory practice, may be appropriate in the context of worship, or of self-examination.

Figure Int. 1. Francesco Guardi (1712–93), *Venice Piazza San Marco*. Dating from c. 1760, this painting shows St. Mark's Cathedral and its bell tower. Reproduced by courtesy of the Trustees, The National Gallery, London.

Historical analysis is often invoked to support this case for separation because it can always be argued that the conflicts of the past were the result of misunderstanding. If only the clergy had not pontificated about the workings of nature, and if only the scientists had not been so arrogant as to imagine that scientific information could meet the deepest human needs, all would have been sweetness and light.

It has been argued, for example, that much of the heat could have been taken out of the Darwinian debates if only the Christian doctrine of creation had been properly formulated. That doctrine, it is said, refers to the ultimate dependence of everything that exists on a Creator. It need not entail the separate creation of every species. Some twentieth-century theologians, notably Rudolph Bultmann, have gone so far as to say that the doctrine of creation has nothing to do with the physical world. Its correct application is to the creation within men and women of an authentic stance toward their earthly predicament. By such means the spheres of science and religion are insulated one from the other.

A third view, which can also be overstated, expresses a more intimate relationship between scientific and religious concerns. Contrary to the first – the conflict model – it is asserted that certain religious beliefs may be conducive to scientific activity. And contrary to the second – the separationist position – it is argued that interaction between religion and science, far from being detrimental, can work to the advantage of both. This more open position clearly appealed to Whitehead, for he raised the question whether the assumption of seventeenth-century natural philosophers, that there was an order imposed on nature, might not have been an unconscious derivative of medieval theology. And he also argued that interaction between religion and science could purge the former of superfluous and obsolete imagery. Once again, the appeal to history is essential to the enterprise. The thesis of the American sociologist, R. K. Merton, that puritan values assisted the expansion of science in seventeenth-century England, would be a good example of historical scholarship in which the mutual relevance of science and religion is affirmed, rather than constant conflict or complete separation.

There are, of course, many variants of these positions. But in their presentation it is almost always assumed that there are lessons to be learned from history. The object of this book is not to deny that assumption but to show that the lessons are far from simple. The chapters that follow do not pretend to tell a complete or definitive story. They should be read as a historically based commentary rather than as a conventional historical narrative. The principal aim is to assist in the creation of critical perspectives, not to describe a continuous series of seemingly decisive transformations.

Serious scholarship in the history of science has revealed so extraordinarily rich and complex a relationship between science and religion in the past that general theses are difficult to sustain. The real lesson turns out to be the complexity. Members of the Christian churches have not all been obscurantists; many scientists of stature have professed a religious faith, even if their theology was sometimes suspect. Conflicts allegedly between science and religion may turn out to be between rival scientific interests, or conversely between rival theological factions. Issues of political power, social prestige, and intellectual authority have repeatedly been at stake. And the histories written by protagonists have reflected their own preoccupations. In his efforts to boost the profile of a rapidly professionalizing scientific community, at the expense of the cultural and educational leadership of the clergy, Darwin's champion, T. H. Huxley, found a conflict model congenial. Extinguished theologians, he declared, lie about the cradle of every science as the strangled snakes beside that of Hercules.

The purpose of this book is not to recover the corpses. It is to display the diversity, the subtlety, and ingenuity of the methods employed, both by apologists for science and for religion, as they have wrestled with fundamental questions concerning their relationship with nature and with God. Such is the richness of the subject that it is well to set aside one's preconceptions. There are surprises in store. The same Franklin who devised the lightning conductor was not ashamed to say that, as for the nature of electricity, he was still in the dark. He *was* ashamed about the confidence with which he had earlier thought the subject mastered. As he reflected on the succession of his theories, he observed that one use of electricity had been to make a vain man humble. Franklin had recognized, as Francis Bacon had before him, a congruence between the virtue of humility and the demands of an experimental method. He had recognized that the majestic towers of scientific theory could crumble as spectacularly as the towers of great cathedrals.

It is just such a succession of incompletely successful theories that the history of science reveals, the survivors having some advantage over their predecessors, but rarely in a manner that made evaluation at their inception a straightforward matter. The popular antithesis between science, conceived as a body of unassailable facts, and religion, conceived as a set of unverifiable beliefs, is assuredly simplistic. Theoretical innovations have usually been controversial, often divisive, within scientific communities. Consequently, when they have impinged on the sacred, there has usually been considerable room for debate. To portray the relations between science and religion as a continuous retreat of theological dogma before a cumulative and infallible science is to overlook the fine structure of scientific controversy, in which religious interests certainly intruded, but often in subtle rather than overtly obstructive ways.

An obvious difficulty arises at this stage. How can one speak about the relationship between science and religion, either as practices or as systems of belief, without first defining terms? It is possible to go only so far in meeting this objection. *Religion* has been defined in terms of belief in supernatural beings or in terms of a commitment to some transcendent "other," which serves to integrate one's life. It may refer to organized institutions that, through creed and ritual, claim to give coherent answers to questions of human destiny. Or it may simply refer to any deeply held convictions that find expression in moral imperatives. Although there is often overlap between such definitions, there need not be. In some of the world's religions, Buddhism for example, belief in a transcendent Creator is not affirmed. Because this book is concerned with the relationship between science and religion in the West, most of the contexts in which the word religion is used will be those in which some variant or some critique of the Christian faith was at stake. Too restrictive a definition can, however, be counterproductive because it may exclude too many questions before they have been asked. If the study of history is to be instructive, it is important not to establish foregone conclusions through the rigidity of definitions.

The same difficulty arises with the word *science*. There have been so many definitions offered by philosophers, and by scientists themselves, that it would require another book to consider them. Many refer to some unique "scientific method" to which exemplary science is supposed to conform. But, as the Cambridge philosopher William Whewell observed, almost a hundred fifty years ago, the *history* of science already showed that each new branch of scientific inquiry had required its own distinctive methodology. And that very process of increasing differentiation reflected a more fundamental change in the meaning of science – from when it had referred to all knowledge and when theology was "queen of the sciences," to its more modern connotations of empirical investigation and high specialization.

There are at least three reasons why the historian might recoil from the demand that "science" and "religion" be rigorously defined before the exercise may begin. The first can be illustrated by a celebrated remark of Isaac Newton. His most famous book, in which planetary orbits were explained by his gravitational theory, was entitled *Mathematical principles of natural philosophy* (1687). It was not entitled *Mathematical principles of natural science.* When seventeenth-century students of nature called themselves natural philosophers, they were identifying themselves with intellectual traditions in which broader issues than immediate scientific technicalities were discussed. Newton himself remarked that it was part of the business of natural philosophy to discuss such questions as the attributes of God and His relationship to the physical world. Very few physicists today would conceive their role in such terms. The point is that if we prejudge what we mean by science and religion, we might be in no position to appreciate the distinctiveness of Newton's vision. There would be a degree of artificiality in asking how Newton reconciled his "science" and his "religion," if he saw himself pursuing a form of "natural philosophy," in which the two interests were integrated.

The second reason for resisting definitions that might prove too constrictive can also be illustrated from the late seventeenth century when Thomas Burnet wrote his *Sacred theory of the earth* (1684). In it he assumed the role of a Christian apologist, using a knowledge of history to identify certain mistakes that should not be made when theologizing about nature. Thus he applauded St. Augustine for his warning that science and religion should not be too tightly interlocked, that it was dangerous to invoke the authority of Scripture in disputes about the natural world. The danger as Burnet saw it was this: As scientific understanding advanced, propositions that Scripture had been made to affirm would be proved false. Its authority would then be jeopardized on far more important matters. But, says Burnet with evident condescension, Augustine had fallen into the very trap he had identified. He had used the Bible in his dismissal of inhabitants at the Antipodes. Burnet, so much wiser in the late seventeenth century, is even more aware of the danger and knows how to avoid it.

And yet, anyone reading Burnet's *Sacred theory* today would be struck by the fact that he falls headlong into the selfsame trap. Instead of keeping the spheres of science and the Bible apart, he brings them together. He offers a mechanistic account of how the Genesis flood had come about, and he defines the main epochs of earth history with reference to information gleaned from his Bible. His picture of a submerged earth, in which Noah's ark is conspicuous, shows how the flood was made constitutive of the earth's physical history. The point of the example is not to score points against Burnet but to raise the more sympathetic question: How was it possible for Augustine to behave in a manner that, to a later generation, looked inconsistent? And similarly for Burnet. Part of the answer is that the domains of science and religion were separated by different boundaries in Augustine's day from those in Burnet's day, and in Burnet's day from our own. Precisely because the boundaries have shifted with time, it would be artificial to ask about the relationship between "science" and "religion" as if modern definitions of their provenance had some timeless validity.

A third reason for tolerance on matters of definition is that it can be artificial in a quite different sense to ask about the relationship between science and religion in the past. Because both are rooted in human concerns and human endeavor, it would be a profound mistake to treat them as if they were entities in themselves – as if they could be completely abstracted from the social contexts in which those concerns and endeavors took their distinctive forms. To understand the predicament of Galileo in his relations with the Roman Catholic Church, it is not enough to say that science was in conflict with religion. The political ramifications of the Counter-Reformation were such that Galileo's science (which was not self-evidently correct) acquired meanings and implications that it might otherwise not have carried. The metaphor of a changing boundary or interface is itself too superficial to cope with this particular problem. The way in which

Figure Int. 2. Frontispiece from Thomas Burnet's *Theory of the earth* (1684). Burnet identified seven phases in the earth's physical history. In clockwise sequence, the chaos of the primeval earth as described in Genesis 1 was followed by the smooth-surfaced globe corresponding to a state of paradise. In this second phase the earth's axis of rotation was still vertical, with the consequence that the Garden of Eden, occupying a middle latitude in the Southern Hemisphere, enjoyed a perpetual spring – until the third phase, that of Noah's flood, when the earth's crust collapsed and its axis tilted. The fourth phase is represented and depicted by the earth's current surface. Following a global conflagration, there is the promise of a new heaven and a new earth, with paradise restored before the final consummation. Reproduced by permission of the Syndics of Cambridge University Library.

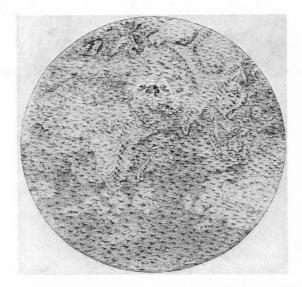

Figure Int. 3. Illustration from page 101 of Thomas Burnet's *Theory of the earth* (1684). According to Burnet, the deluge had covered the whole of the earth, though he discounted the divine creation of new water. There had been a subterranean aquatic layer concentric with the earth's original crust. The floodwaters had been released when the crust had cracked. The synchronization of these physical events with a crisis in the moral history of humanity, as recounted in Genesis, was, for Burnet, a powerful argument for divine providence. Reproduced by permission of the Syndics of Cambridge University Library.

the relationship between scientific and religious claims has been perceived in the past has depended on social and political circumstances that the historian cannot ignore. The evolutionary speculations of Charles Darwin's grandfather, Erasmus Darwin, occasioned little hostility in England during the 1780s. But in the conservative backlash, following the French Revolution, they were noisily condemned as atheistic.

The existence of a political dimension to many of the debates in which scientific and religious interests were involved means that to abstract both the "science" and the "religion" and then try to establish their mutual relationship can be highly artificial. Indeed, it is tempting to say that we should be more concerned with the use to which scientific and religious ideas have been put in different societies than with some notional relationship between them. Satisfying that concern would require a larger and more detailed book than this is designed to be. But it is important to appreciate that scientific conclusions, however provisional, have often been used by religious apologists in pursuing their own designs. The pressures to protect their authority have sometimes been felt so keenly that they have appropriated the latest science to demonstrate the vitality of their position. In some cases they have run into more trouble by adopting scientific innovations than if they had opposed them.

The fact that science has been used as a resource both by Christians and their critics may call into question another common assumption – that modern science has been largely responsible for the secularization of society. Have the sciences not progressively diminished that sense of awe and mystery that once induced deference to the gods? It is an argument with a long history and a fine pedigree. But it has not passed unchallenged. Critics point out that it may be an unquestioned survival from nineteenth-century positivism when a sense of liberation through science was at its height. They add that it mistakenly assumes religion to depend more on the physical environment than on the quality of social relations. And it ignores those respects in which scientific knowledge may magnify rather than diminish a sense of awe. The place of science in the process of secularization may also require reevaluation in the light of religious resurgence in polities where science-based technologies are not conspicuous by their absence. In the pages that follow, we shall try to keep an open mind on issues of this kind.

In Chapter I we consider a series of examples in which statements about God and statements about nature are closely interrelated. They are designed to show how theological and scientific concerns have been mutually relevant in the past. Because statements of an ostensibly religious character fulfilled many different roles in the context of interpreting nature, these examples help to establish the richness and diversity of the interaction. But precisely because they show religious beliefs functioning, as it were, *within* science, they also illustrate the artificiality of discussing the relationship between science and religion as if the province of each had already been established. The implications of these examples for the "conflict thesis" will then be assessed, for they exclude the reductionist view that representatives of science and representatives of the churches have invariably been locked in battle. In seeking to attain what in matters of this kind is ultimately unattainable – a balanced view – it is also necessary to consider whether revisionist histories, structured around a critique of the conflict thesis, have not gone too far in the opposite direction. The apologetic intentions of both secularists and religious thinkers have so colored the literature that a fresh approach is required.

In Chapter II we address a specific historical problem: The interpretation of those shifts in the understanding of nature that, during the sixteenth and seventeenth centuries, added up to what has traditionally been called the Scientific Revolution. A common view is that by the end of the seventeenth century, a recognizably *modern* science had emerged, separated at last from a preoccupation with matters of philosophy and religion. The object of the chapter is to identify the difficulties that arise in sustaining that view. While it is true that investigations into nature were often subordinate to religious concerns in the late medieval period, it would be misleading to imply that they were bound together in an indissoluble complex until they were prized apart in the seventeenth century. It is also argued that, although there were certain levels on which the problems of science and those of theology were increasingly differentiated during the seventeenth century, the assertion of an absolute separation would be too extreme. Scientific innovations continued to be presented in theological terms and divine attributes continued to be given physical meanings – as when Newton insisted that space was constituted by God's omnipresence.

In Chapter III we raise the question whether parallels can be drawn between the reform of learning through experimental science and the reform of religion that occurred through the Protestant Reformation. The reaction to the sun-centered cosmos of Copernicus provides an appropriate case study because opposition to Galileo, its most famous popularizer, has symbolized the suppression of scientific freedom by the Roman Catholic Church. In taking the Copernican innovation as a test case, however, it soon becomes clear that complications arise when assessing receptivity to new hypotheses. While

there is circumstantial evidence to suggest that certain Protestant societies were more tolerant toward new scientific
learning, the difficulties that arise in testing such generalizations can be formidable. Just how formidable will be shown
in the context of evaluating Merton's thesis that puritan values were propitious for science in seventeenth-century England.

The theme of Chapter IV is the mechanization of the natural world — that seventeenth-century development which
has often been seen as a decisive advance on organic models
of the cosmos. The reconstruction of nature through mechanical metaphors has also been seen as a crucial step in the
secularization of knowledge; for, if nature ran like clockwork, what room was there for God's direct activity or special providence? Once again, the issues turn out to be far
more subtle than that simple question suggests. In the case
of Robert Boyle, mechanical images of nature were enlisted
in the defense of Christianity. Not only did they reinforce
the view that nature was a designed system; they could also
be used to emphasize God's absolute sovereignty. If physical
phenomena were to be explained exclusively in terms of matter and motion, and if, as both Boyle and Newton insisted,
matter itself was inert, then God could be made directly responsible for the motion. But the clockwork image could be
read in other ways. One of the many ironies in our story is
that a model for the universe, which in the seventeenth century was used to affirm God's sovereignty, was used by the
deists of the eighteenth century in their attacks on established religion.

Chapter V takes us into the eighteenth century and into
that period of "Enlightenment" when the sciences were hailed
as instruments of progress and when institutionalized religion, especially in Catholic countries, was vilified for its superstition and priestcraft. In an age when unprecedented
confidence was placed in the power of human reason, the
methods and achievements of the sciences were a powerful
resource for those who, with a variety of motives, launched
their assault on established Christianity. But to reduce the
relations between science and religion to a polarity between
reason and superstition is inadmissible, even for that period
when it had such rhetorical force. It was often not the natural
philosophers themselves, but thinkers with a social or political ax to grind, who transformed the sciences into a secularizing force. Although certain scientific discoveries could be

invoked to support a materialist philosophy, they were usually susceptible to less radical interpretations. And in the confrontation between skepticism and Christianity, science could still be on the side of the angels — especially in England where arguments for design retained a strategic role in the defense of the faith. Whereas in France, the materialist La Mettrie would claim that the study of nature made only unbelievers, the contrary claim of Robert Boyle, that one could only be an atheist if one had *not* studied nature, was the more common sentiment in the English-speaking world.

The diverse functions of natural theology, including its role in the popularization of science, is the subject of Chapter VI. The claim that from scientific study one could learn something about God falls strangely on twentieth-century ears. It was, however, a commonplace in the writing of British scientists until the generation that saw the publication of Darwin's *Origin of species* (1859). The object of the chapter is to uncover some of the reasons why this integration of science and religion proved so viable, despite the existence of trenchant critiques. We shall also consider the extent to which a commitment to natural theology affected the scientific enterprise and the extent to which advances in science affected the plausibility of arguments from design.

For much of the eighteenth century it had been commonly assumed that human history and the history of the earth were coextensive. But during the late eighteenth and early nineteenth centuries, new visions of earth history emerged, presenting a challenge to popular religious belief. As evolutionary models came to the fore in astronomy, geology, and biology, traditional beliefs about humanity's place in nature were increasingly difficult to defend. The emergence of these historical sciences, culminating in Darwin's theory of evolution, is the subject of Chapter VII. The assumptions made in reconstructing the past were often highly controversial even among naturalists themselves. We shall therefore stress the competition between rival scenarios, in which political and religious preferences sometimes constituted a hidden agenda. Although there were countless attempts to harmonize these disturbing vistas with biblical texts, they were eventually abandoned — at least among academic theologians — as the methods of historical research were brought to bear on questions of biblical authorship.

In Chapter VIII we try to take as broad a view as possible

of the Darwinian challenge to popular Christianity, in the knowledge that there are circles in which the issues are far from dead. To focus on the post-Darwinian debates is not to deny that the physical sciences also raised new and exciting issues. During the nineteenth century the science of thermodynamics, for example, created new vistas for the ultimate fate of the universe. The emergence of statistical models in explaining the behavior of gases reopened questions about the nature of scientific "laws." But few would dispute that the impact of evolutionary theory had the more penetrating and enduring effects. We shall therefore survey the many uses to which the Darwinian theory (and corruptions of it) were put. Because the popularization of evolutionary science was so intimately associated with the promotion of social and political ideologies, this chapter provides the most telling illustration of the point made earlier – that it can be highly artificial to abstract the science and the religion from the ensuing debates with a view to determining their mutual bearing. We shall, however, examine the capital that secularists could make out of Darwin's theory and the response of religious thinkers who looked for conciliation rather than confrontation.

In the pluralistic and largely secular societies of the West, a preoccupation with the demands of traditional religion has given way to a humanism that has been described as *beyond* atheism. In helping twentieth-century humanists to rationalize their unbelief, few thinkers have been as influential as Freud, whose attitude toward belief in God is the starting point for a concluding postscript. Despite a prevailing ethos, in which science and secularization are seen as linked together in the constitution of modern culture, the twentieth century has witnessed certain developments in the sciences that have given solace to the religious apologist. Innovations in subatomic physics have been seen as a license for more organic and less deterministic models of reality. Moreover, new scientific techniques have raised ethical questions of such gravity that the general public has become aware of an interface along which science and human values meet. As long as the world's religions continue to stake a claim in the articulation of those values, it is unlikely that the two spheres of science and religion will be completely divorced.

CHAPTER I

Interaction between Science and Religion: Some Preliminary Considerations

Introduction

During their history, the natural sciences have been invested with religious meaning, with antireligious implications and, in many contexts, with no religious significance at all. The object of this book is to offer some insight into the connections that *have* been made between statements about nature and statements about God. As we noted in the introduction, however, problems arise as soon as one enquires about the relationship between "science" and "religion" in the past. Not only have the boundaries between them shifted with time, but to abstract them from their historical contexts can lead to artificiality as well as anachronism.

How, for example, do we deal with that late nineteenth-century evolutionist Henry Drummond, who insisted that it was wrong to speak of *reconciling* Christianity with evolution since the two were one? In a glowing vision of *The ascent of man* (1894), Drummond acknowledged the Darwinian struggle for life, but he also referred to a struggle for the life of others. The former was essential to the evolutionary process, as individuals competed for resources. But, Drummond argued, so too was the latter: Once the human mind had evolved, self-sacrifice, cooperation, and maternal love would each contribute to the survival of societies in which those virtues were encouraged. Christianity and evolution were ultimately one because both denoted a method of creation; both had as their object the making of more perfect beings. Because altruistic love was germane to both, he could argue for a perfect union. Evolution embraced progress in spirit as well as in matter.

Many of Drummond's contemporaries did know how to deal with him. His secular critics saw the misguided attempt of an evangelical Christian to adapt his theology to an insuperable threat from science. Nor, despite a close friendship with the evangelist D. L. Moody, could he pacify the more conservative evangelicals in his audience. "Many fell upon me and rent me," he complained, after addressing a Northfield conference in 1893.

It is not clear, however, that the historian need follow their example. From a late twentieth-century perspective are we not more likely to disregard than dissect him? It would be easy to do so on the basis that his fusion of science and religion was really a confusion. Our own century has seen so many movements against the conflation of scientific and religious language that we may, on purely philosophical criteria, judge him to have been misguided. By those who treat scientific and religious discourse as two distinct language games, or who detach doctrines of creation from statements about the physical world, Drummond's attempt to integrate his faith with evolutionary science would be dismissed as an unfortunate error.

The problem for the historian, however, is that sophisticated twentieth-century distinctions may not always be the most sensitive instruments for understanding the issues as they were formulated in the past. There were Henry Drummonds in every generation who did make connections between their theology and their science. If we begin with too cavalier an attitude toward them, we may miss a certain richness in our intellectual heritage and one that had a profound bearing on how the word *science* was understood, by both practitioners and their public.

Consider, for example, three contrasts commonly drawn by scholars when comparing our modern scientific age with earlier times when magic prevailed. Science, it is said, operates within a world-view that regards natural phenomena as the product of impersonal forces. By contrast, religious and magical systems involve personalized gods, spirits, or demons. Whereas the scientific enterprise is legitimated by agreed testing procedures, the theological enterprise has been characterized by dogmatism. Whereas religions have required worship, ceremony, and sacrifice, these are forms of activity alien to Western science.

On such grounds science and religion are commonly differentiated. Closer inspection of the history of science, however, suggests a more complex picture. Science may be concerned with impersonal forces, religion with personalized gods; but the very word *force* carried religious meanings, even for Isaac Newton (1642–1727) who, in describing the operation of a gravitational force in mathematical terms, also ascribed it to an omnipotent God. By his critic Gottfried Wilhelm Leibniz (1646–1716), he was accused of making the action of gravity a perpetual miracle.

The contrast between self-criticism in science and an uncritical spirit in religion cannot be made absolute. Scientific theories have often been cherished when in the eyes of their critics they ought to have perished. Recalling a symposium held at St. Louis in 1961, the distinguished organic chemist Herbert C. Brown has commented on the reaction when he suggested that many of the proposed nonclassical structures for carbonium ions had weak experimental support. He had hoped for a critical reexamination of the field. Instead, many fell upon him and rent him. His suggestions were treated as "a heresy, triggering what appeared to be a 'holy war' to prove me wrong." Conversely, a spirit of self-criticism and renewal can hardly be said to be absent from the religious sphere when one of the problems faced by institutionalized orthodoxies has been to contain the eruption of reform and sectarian revolt.

There may be no obvious equivalent in science of the call to worship in religion. And yet, there were those in the late seventeenth century, Robert Boyle (1627–91) and John Ray (1627–1705) among them, who envisaged scientific inquiry itself as a form of worship. The image of nature as temple, the scientist as priest, was explicit in Boyle. Just as fine music was best appreciated by the trained musician, so God's craftsmanship in creation could be celebrated by the skilled anatomist. There was even revelation in this temple of nature. Insights that a modern chemist might attribute to serendipity were ascribed by Boyle to "pregnant hints" received from a greater chemist than he.

There is, therefore, a strong case for bringing an open mind to the richness of the subject. The main purpose of this chapter is to identify some of the levels on which statements about nature and statements about God have coexisted. We can then consider the implications of this analysis for one of the general views identified in the introduction – that there has been nothing but conflict between science and religion.

The Diversity of Interaction

The very possibility of a rational science of nature is usually considered to depend on a uniformity in the relations between cause and effect. In the past, religious beliefs have served as a *presupposition* of the scientific enterprise insofar as they have underwritten that uniformity. Natural philosophers of the seventeenth century would present their work as the search for order in a universe regulated by an intelligent Creator. A created universe, unlike one that had always existed, was one in which the Creator had been free to exercise His will in devising the laws that nature should obey. A doctrine of creation could give coherence to scientific endeavor insofar as it implied a dependable order behind the flux of nature.

To say that religious belief could function as a presupposition of science need not entail the strong claim that, without a prior theology, science would never have taken off. But it does mean that the particular conceptions of science held by its pioneers were often informed by theological and metaphysical beliefs. When natural philosophers referred to *laws* of nature, they were not glibly choosing that metaphor. Laws were the result of legislation by an intelligent deity. Thus the philosopher René Descartes (1596–1650) insisted that he was discovering the "laws that God has put into nature." Later, Newton would declare that the regulation of the solar system presupposed the "counsel and dominion of an intelligent and powerful Being."

A doctrine of creation could underwrite the scientific enterprise in a second respect. If the human mind had been created in such a way that it was matched to the intelligibility of nature, then the possibility of secure scientific knowledge could be affirmed. Some two hundred years after Descartes had formulated his concept of mechanical laws, a proven capacity to discover the laws of nature and to express them mathematically was taken by the first systematic historian of science, William Whewell, as evidence of an affinity between the human and the divine mind. As the astronomer Johannes Kepler (1571–1630) had put it, in exposing the geometry of

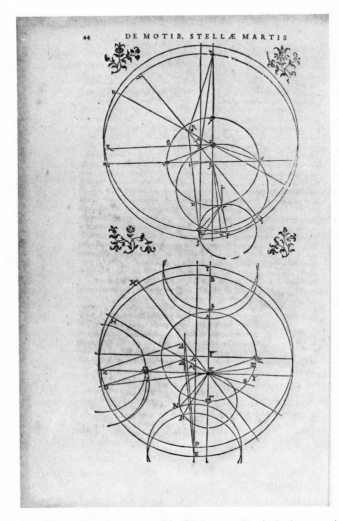

Figure I. 1. Illustration from page 44 of Johannes Kepler's *Astronomia nova* (1609). The imposition of order on a wayward planet. Kepler considers the orbit of Mars in the context of differentiating between three competing models of the world system: the Copernican (in which all the planets — earth included — orbit the sun), the Ptolemaic (in which sun and planets orbit a stationary earth), and the alternative geostatic system of Tycho Brahe (see Fig. I.2), in which the planets — but not the earth — revolve around the sun, which in turn orbits the earth. Kepler was to argue that the divinely ordained order behind the appearances was best revealed by supposing Mars to have an elliptical orbit, with the sun stationary at one focus of the ellipse. Reproduced by permission of the Syndics of Cambridge University Library.

Prius .n. erat ut δ β ad δ λ minorem quam eſt dimidia δ β, ſic β κ ad λ μ. Ptolemæo vero eſſet, ut κ δ terra ad κ δ dimidiam ſic δ δ æqualis ipſi ſ δ κ ad M δ.

DENIQVE EADEM & in TYCHONICA hypotheſi deducam.

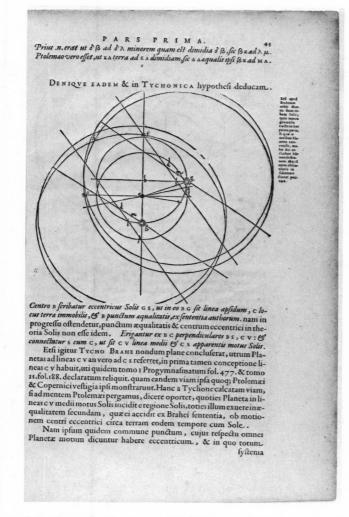

Centro B *ſcribatur eccentricus Solis* G S, *ut in eo* B G *ſit linea apſidum,* C *locus terræ immobilis,* & B *punctum æqualitatis, ex ſententia authorum.* nam in progreſſu oſtendetur, punctum æqualitatis & centrum eccentrici in theoria Solis non eſſe idem. *Erigantur ex* B C *perpendiculares* B S, C V: & *connectatur* S *cum* C, *ut ſit* C V *linea medii* & C S *apparentis motus Solis.*

Etſi igitur TYCHO BRAHE nondum plane concluſerat, utrum Planetas ad lineas C V an vero ad C S referret, in prima tamen conceptione lineas C V habuit, uti quidem tomo I Progymnaſmatum fol. 477. & tomo II. fol. 188. declaratum reliquit. quam eandem viam ipſa quoq; Ptolemæi & Copernici veſtigia ipſi monſtrarunt. Hanc a Tychone calcatam viam, ſi ad mentem Ptolemæi pergamus, dicere oportet, quoties Planeta in lineas C V medii motus Solis incidit e regione Solis, toties illum exuere inæqualitatem ſecundam, quæ ei accidit ex Brahei ſententia, ob motionem centri eccentrici circa terram eodem tempore cum Sole.

Nam ipſum quidem commune punctum, cujus reſpectu omnes Planetæ motum dicuntur habere eccentricum, & in quo totum ſyſtema

Figure I. 2. Illustration from page 45 of Johannes Kepler's *Astronomia nova* (1609). Reproduced by permission of the Syndics of Cambridge University Library.

creation one was thinking God's thoughts after Him. The idea of a First Cause, Whewell suggested, was not extracted from natural phenomena. Rather it had been *assumed* in order that those phenomena could become intelligible to the mind.

In addition to providing presuppositions for science, religious doctrines have also offered *sanction* or justification. This has been a recurring function as scientists have repeatedly had to justify the place of science in their culture. Proponents of scientific inquiry would often argue that God had revealed Himself in two books – the book of His words (the Bible) and the book of His works (nature). As one was under obligation to study the former, so too there was an obligation to study the latter. According to a manuscript of that great diplomat for science, Francis Bacon (1561–1626), the rise of experimental science was sanctioned not merely by religion, but by God Himself. For, in the book of Daniel (12:4) there is a prophecy that seemed to speak of a time when many would pass to and fro and knowledge would be increased. Recent improvements in navigation and the expansion of commerce persuaded Bacon that the time had come. He knew that the Genesis account of man's fall from grace could be read as an indictment of a thirst for knowledge. But he was able to anticipate that objection by suggesting that the prohibition applied only to knowledge sought in self-aggrandizement, not to that which was of service to humanity.

A religious sanction of a different kind was offered by Thomas Sprat in his *History of the Royal Society* (1667). Sprat suggested that, of all pursuits, the study of experimental philosophy was most likely to engender a spirit of piety, perseverance, and humility – the hallmark of Christian virtue. The contemplation of God's works could be a constructive diversion from the unhappy doctrinal disputes that had torn Christianity apart since the Reformation. Science could offer both sanctuary and its own brand of sanctification. Many years later when another scientific institution was founded in Britain, the British Association for the Advancement of Science, a similar refrain was still to be heard. Speaking in 1833, the Cambridge geologist and cleric Adam Sedgwick referred both to the wisdom of God as manifested in creation and to the wisdom of the association in seeking to transcend political and religious divisions. Otherwise the "foul demon of discord" would find his way into "our Eden of philosophy." Nat-

Figure I. 3. Cartoon by George Cruickshank from page 209 of *Bentley's Miscellany*, vol. 4, 1838. The necessity for scientists to find a justification for their endeavors is thrown into relief by this cartoon, which refers to a model exhibited before section B of the "Mudfog Association," a satirical creation of Charles Dickens who lampooned the British Association for the Advancement of Science with his "Full Report of the Second Meeting of the Mudfog Association for the Advancement of Everything." In his satire of the pretensions of mechanical science Dickens proclaims the value to society of automatons filling the role of policemen. Hence the images in the cartoon; for, in Dickens's account, the robot policemen were to be kept on shelves in the police office until required. Reproduced by permission of the Syndics of Cambridge University Library.

ural philosophy had its own garden of Eden, the arguments of natural theology providing a sanction for its preservation.

If religious beliefs have supplied presuppositions and sanctions for science, they have also supplied *motives*. The analysis of human motives is, of course, a precarious undertaking; but some of the connections made in the past between scientific and religious ideals were strong enough to bear reconstruction in such terms. It has certainly been claimed that values associated with ascetic Protestantism provided new motivation for scientific inquiry, particularly in Holland and England during the seventeenth century. A Protestant emphasis on improving the world, under the aegis of providence, could confer dignity on scientific activity that promised both glory to God and the relief of human suffering. Conceptions of a better world over which Christ would reign for a thousand years (the millennium) were later secularized to yield visions of a purely earthly utopia, in which a perfect human society might be possible in the absence of coercive measures. Notwithstanding the unbridled optimism of many such visions, it cannot be denied that one source of the modern idea of progress was this millenarian theology of puritan reformers anxious to transform the world in readiness for Christ's second coming.

The possibility of religious motivation behind scientific inquiry is perhaps most visible in systems of natural theology in which scientific knowledge was used to establish the existence and attributes of God. In polemics against infidels and skeptics, who were commonly thought subversive of a stable society, science could be an impressive ally. Noting that the lens and pupil of the eye were "so finely shaped and fitted for vision that no artist can mend them," Newton asked whether blind chance could have known sufficient of light and its refraction to have effected the design. To one who had made a special study of refraction, the answer was obvious. There was a Being who had made all things and who was to be feared.

A motive for the study of nature could therefore exist in the desire to confirm a God to be feared. But also a God to be praised. In John Ray's *The wisdom of God manifested in the works of creation* (1691), there was a sense of exultation in the wonders of nature. So marvelous was the migrating instinct of birds that he could only ascribe it to the superior intelligence of their Creator. Nor did Ray merely exploit the inexplicable. It had been scientific ignorance, he suggested, that had permitted the clumsy models of the universe characteristic of Ptolemy's astronomy. With the advent of the Copernican system, the universe, according to Ray, had acquired a new elegance, more in keeping with what might be expected of a divine architect.

Claims for religious motivation behind the pursuit of science are difficult to test, but there have been circumstances in which they seem appropriate. Where a protagonist may have political reasons for wishing to differentiate one theological position from another, he may place a high value on scientific results that assist the differentiation. In late eighteenth-century England, much of Joseph Priestley's motivation as a religious dissenter came from a desire to establish a form of Christianity that could withstand rational criticism. European philosophers who had rejected the Christian religion had, he believed, rejected a corrupt form. His aim was to reconvert them to a Unitarian Christianity devoid of superstition. The doctrine of the Trinity − that God is three persons in one − had infected the early Christian church through contact with Plato's philosophy. Because science could help to eliminate superstition, there undoubtedly were connections in Priestley's mind between scientific activity and the promotion of a religion shorn of Platonist corruption. The rapid progress of scientific knowledge, he declared, would be "the means under God of extirpating all error and prejudice, and of putting an end to all undue and usurped authority in the business of religion as well as of science."

What Priestley called the business of religion could enter scientific debate on yet another level. It could reinforce prescriptions for an appropriate scientific method. Each science in its infancy has had to establish the assumptions and procedures by which it could claim to extend our knowledge of nature. These have been the subject of intense debate, with religious preferences sometimes intruding. To say that religious beliefs have infiltrated discussions of scientific method is not to say that they have directly affected scientific practice, for statements about methodology have often been rationalizations, used to justify a research program already in existence. Such rationalization is, however, of great interest to the historian because it reveals something of the social processes involved in gaining respect for scientific work both inside and outside a scientific community.

In the 1830s, the British geologist Charles Lyell argued that to establish geology as a rigorous science, it had to be assumed that the forces that had sculpted the earth's surface in the past were identical, both in kind and intensity, to those acting now. But there were contemporary geologists who had misgivings. Was it not too restrictive to preclude the possibility that forces of greater intensity had acted in the past? In expressing that objection, other contemporaries were undoubtedly swayed by the realization that Lyell's axiom would so greatly increase the age of the earth that it might threaten even a generous reading of Genesis. When the Scottish physicist David Brewster protested against Lyell's principle of uniformity, he invoked biblical considerations to justify an alternative methodology.

In earlier periods, this role of religious belief in *regulating scientific methodology* was extremely common. The colorful chemical and medical reformer Paracelsus (c. 1493–1541) suggested that, in creating the world, God had left a magic sign on every herb – a clue to its efficacy. In some cases the divine signature was perfectly legible: The thistle would relieve a prickly cough. In others a little more imagination was required:

Behold the satyrion root, is it not formed like the male privy parts? No one can deny this. Accordingly magic discovered it and revealed that it can restore a man's virility.[1]

This may not sound much like science, but it represented a significant departure from the book learning of his day. For Paracelsus it implied a departure into the hills with a stout pair of boots to explore, decode, and harness the magic power. It implied empiricism of a sort.

Empiricism of a different sort was practiced by Kepler, who welcomed the astronomical data of Tycho Brahe (1546–1601) as a means of confirming preconceived beliefs about the geometry of the universe. The data were vital, but so was the preconception that the planetary orbits could be inscribed within, and circumscribed without, the five regular Greek solids. It was empiricism of a sort, but regulated by Kepler's fusion of a Pythagorean harmony with a Christian doctrine of creation. It was a brand of empiricism that his critic Marin Mersenne (1588–1648) renounced as not empirical enough. Mersenne argued that, because the structure of the solar system is only one of infinitely numerous possibilities (and therefore ultimately dependent on the choice of the deity), it would be wrong to cherish a preconceived pattern. A more open mind was required to discover which of the many possible patterns God had actually chosen to instantiate. Such references to the freedom of the divine will were often used in the seventeenth century to justify attacks on rationalist theories of nature, whose authors presumed to know how God *must* have shaped the world.

The term *modern science* usually connotes a complete openness to empirical testing. Scientists in the past, however, experienced difficulty (and there are contexts in which they still do) when experimental criteria failed to discriminate deci-

In Chymicis versanti Natura, Ratio, Experientia & lectio,
sint Dux, scipio, perspicilia & lampas.

EPIGRAMMA XLII.

DUx Natura tibi, tuque arte pedissequus illi
Esto lubens, erras, ni comes ipsa viæ est.
Det ratio scipionis opem, Experientia firmet
Lumina, quò possit cernere posta procul.
Lectio sit lampas tenebris dilucida, rerum
Verborúmque strues providus ut caveas.   Z   CAS-

Figure I. 4. Plate from page 177 of Michael Maier's *Atalanta fugiens* (1618).
To learn nature's secrets, the chemist follows in her footsteps. In the ac-
companying text, Nature is the guide, reason the staff, experience the
spectacles, and reading the lamp, all prerequisites of successful inquiry.
Reproduced by courtesy of The Bodleian Library, Oxford; Shelfmark Vet.
D2. e. 18.

sively between two or more theories. Recent work in the philosophy and sociology of science has emphasized that experimental results are rarely adequate for making a definitive choice between competing theories. Not uncommonly there have been problems in replicating both experimental procedures and the results claimed for them. Certain problems have long been acknowledged even in popular accounts of scientific practice where emphasis is placed on the role of aesthetic criteria in theory selection. How often one hears that when two theories look equally plausible, one opts for the simpler. In the process of theory selection, religious (and antireligious) preferences have again intruded. The criterion of simplicity could itself be referred, as it was by Newton, to a God who had ensured that nature did nothing in vain or, as it was by Michael Faraday in the nineteenth century, to a God who had ensured that the book of his works would be as simple to comprehend as the book of his words.

An example from the astronomical debates of the early seventeenth century may illuminate this *selective* role of religious belief. In 1600, the renegade monk Giordano Bruno was burned at the stake for numerous heresies, including the proposition that the universe is infinite and that it contains an infinite number of worlds. One of the reasons why this proposition was distasteful to many of Bruno's contemporaries was that it deprived humanity of a privileged place in the cosmos. In resisting Bruno's vision, Kepler did all he could to protect the cosmic identity of his own solar system and (in accordance with his religious beliefs) of humanity's special place within it. Despite the Copernican transformation, Kepler could still argue for a unique earth. It was the planet with the central *orbit* in a system that had as its focus a symbol of the living God: the most resplendent sun in the universe. Bruno's vista of an infinite plurality of star systems, scattered through an infinite universe, was unacceptable because of its relativistic implications. Accordingly, when Kepler first heard of Galileo's telescopic observations, he was alarmed lest the satellites reported should be planets orbiting another sun. How great his relief when he learned they were moons of Jupiter! His insistence that our sun should be preeminent also predisposed him toward the belief that previously invisible stars, now revealed by Galileo's telescope, had been too small rather than too distant to be seen. In such a way have religious beliefs sometimes shaped the interpretation of scientific data.

They have operated, too, in less subtle ways, sometimes fulfilling an explanatory role before being overtaken by more sophisticated scientific developments. We can speak here of the *constitutive* role of religious beliefs — in the sense that they have constituted an explanation for phenomena that have subsequently proved explicable without theological reference. For Newton there was no satisfactory account of why the planets should orbit the sun in the same direction and in roughly the same plane: This aesthetically pleasing scheme could only be explained by appealing to God's initial design. When Laplace, late in the eighteenth century, applied his nebular hypothesis to the problem, deliberately dispensing with concepts of design, it could look as if Newton's God had been made redundant.

Propositions deduced from Scripture have also fulfilled an explanatory function. The legend of Noah's ark once purported to say something about the geographical distribution of animals. For the Oxford geologist William Buckland the flood was still constitutive of the earth's physical history in the 1820s. The concept of separate creation, which Darwin found inadequate, had often been presented as a biblical view. It is this constitutive role of religious belief that so often comes to mind when the words science and religion are juxtaposed; for the explanatory pretensions of the world's religions have made them vulnerable to scientific advance. This is one reason why it has become routine among Christian writers to warn against a god-of-the-gaps. Past precedents indicate that conflict will assuredly arise, if statements about God are used to fill the gaps in scientific explanation.

It should not be assumed, however, that when religious beliefs have functioned as primitive science, it has always been to the detriment of further inquiry. They could stimulate investigation before being discarded. Take the case of Noah's ark. A conventional view would be that the proliferation of new species, discovered between 1650 and 1750, sank the ark by the sheer weight of their numbers. Certainly, by the 1740s, the Swedish taxonomist Linnaeus found it incredible that the fifty-six hundred species he had named could have been crammed into the elaborate vessels designed by biblical literalists. And yet, in the preceding century, the legend had played its part in stimulating inquiry into the geographical distribution of species. It had focused discussion on the plausibility of explanations based on dispersal from a single point.

Figure I. 5. Illustration following page 122 of Athanasius Kircher's *Arca Noë* (1675). *Preparing to board an ark in which there was room for every species.* Courtesy of Dr. Janet Browne and reproduced by courtesy of the Wellcome Institute Library, London.

The desire of biblical literalists to show that the story could have been history acted as a spur to the study of zoogeography, even though the eventual, and ironic, outcome was that they sank their own ship. And even when the ark was abandoned, it still left its mark on the science. Linnaeus himself retained a dispersal of all species from a single mountain, formerly surrounded by sea.

If religious concepts have sometimes functioned as primitive science, the converse has also occurred. As religious beliefs have constituted science, so scientific creeds have constituted an alternative religion. Scientists themselves have often drawn parallels between the experience of a scientific vocation and certain forms of religious experience. In late eighteenth- and early nineteenth-century Europe, it was not unusual to hear naturalists speaking of their scientific awakening in terms that might be used of a religious conversion. One could experience a sense of ecstasy at the disclosure of nature's secrets and a sense of the sublime in contemplating its works. Whenever scientific rhetoric assumed the form of a counterreligion, it could easily become a substitute religion, as it began to do in France during the revolutionary era. Later champions of a scientific naturalism would sometimes refer to the "church scientific," as if to emphasize their usurping the role of the clergy in the arbitration of cultural values. Charles Darwin's advocate T. H. Huxley made science the subject of what he chose to call "lay sermons." When one of the most energetic spokesmen for applied science in nineteenth-century Britain, Lyon Playfair, addressed the members of a mechanics institute in 1853, he unashamedly declared that "science is a religion and its philosophers are the priests of nature."

It would not be difficult to extend this analysis and to identify other types of interaction. A concern with theological questions has sometimes prompted new lines of scientific inquiry, as when Richard Bentley (1662–1742), wishing to use the latest science to defend his Christian theism, asked Newton whether he thought it possible that the frame of the world could have been produced, from a uniform distribution of matter, by mechanical principles alone. His question prompted a fresh line of reasoning, Newton acknowledging that he had barely considered the point before receiving his letter.

The point we need to consider is whether, in the light of such diversity of interaction, it is appropriate to focus exclu-

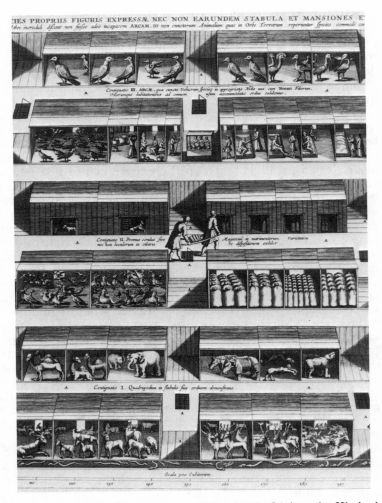

Figure I. 6. Illustration between pages 108 and 109 of Athanasius Kircher's *Arca Noë* (1675). Some of the carefully calculated compartments in Kircher's ark. Reproduced by permission of the Syndics of Cambridge University Library.

sively on the impact *of* science *on* religion. Standard treatments of the subject are often preoccupied with that formulation, as if the streams of relevance and implication could flow in one direction only. But if religious beliefs have provided presupposition, sanction, even motivation for science; if they have regulated discussions of method and played a selective role in the evaluation of rival theories, the possibility of a more wide-ranging, and I hope rewarding, inquiry opens up. This is not to deny that the custodians of institutionalized religion have often done their best to censure what they perceive to be damaging scientific conclusions. But it is to suggest that an image of perennial conflict between science and religion is inappropriate as a guiding principle. We shall now consider further reasons why this is so.

Conflict between Science and Religion

With reference to the effect of scientific innovation on religious belief, two propositions are often juxtaposed. First, when there has been confrontation, it has been religion that has yielded in the end. Second, despite recurrent concession, the religion itself has survived. This may indicate that the conflict has been over peripheral matters, that a core belief in a transcendent power could retain its plausibility, untouched by changing conceptions of the physical world. Or it might suggest that the religious symbols, by which men and women have conferred meaning on their lives, meet such pressing psychological needs that they remain impervious to scientific frames of meaning, which, in the last analysis, only dictate how things are and how they came to be so, without pretending to answer the why and the wherefore. Or it might suggest that, despite a total lack of plausibility in their claims for knowledge, religious movements have survived either through institutional inertia, or insofar as they assist in conferring a sense of identity on national or local communities. For one of the founders of the sociology of religion, Emile Durkheim, history showed beyond doubt that religion had come to embrace a progressively smaller sector of social life. Political, economic, and scientific functions had gradually freed themselves from religious control. At the same time, there was something eternal in religion, destined to survive all the particular symbols in which religious thought had successively enveloped itself. A fully secularized society was a contradiction in terms, for there could be no society that would not feel the need to uphold the collective sentiments that gave it its unity and individuality.

One reason why there can be no simple answer to the question whether there has been irresolvable conflict between science and religion is that the answer may well depend on which of these types of explanation is favored. So much depends on the prior plausibility ascribed to core religious beliefs. And in religious communities, perhaps more than any other, the experiences and perceptions of the insider will contrast with those of the outsider. What for the insider might be the purification of a defensible religion by its exposure to scientific criticism, would for the outsider be one more step along the path to destruction as an inherently implausible account of human destiny was shown up for what it is. Despite the impossibility of an objective assessment, there are, however, considerations that suggest that the conflict has been exaggerated in the interests of scientism and secularism. Because it has also been underplayed in the interests of religious apologetics, a degree of critical detachment is required.

In his *History of the conflict between religion and science* (1875), J. W. Draper put forward a principle of interpretation that still enjoys popular support. The history of science, he wrote, is a narrative of the conflict of two contending powers, the expansive force of the human intellect on one side and the compression arising from traditional faith, and human interests, on the other. Draper was an English scientist who became the first president of the American Chemical Society. Living through the post-Darwinian debates, he invariably took up the cudgels on behalf of scientific rationalism. He had even spoken – interminably – at the famous meeting of the British Association for the Advancement of Science, held at Oxford in 1860, when Huxley allegedly scored his victory over Bishop Wilberforce with the retort that he would rather have an ape for an ancestor than a man who used his privileged position to pronounce on matters he knew nothing about. For Draper, the Darwinian debates had focused attention on a crucial issue – whether the government of the world is by incessant divine intervention or by the operation of unchangeable laws.

One reason why his historical narrative, and others like it, must not be read uncritically stems from the fact that he projected current issues backward in time. Far from being impartial, he had an obvious target. Whereas science had been unstained by cruelty, hands in the Vatican had been steeped in blood. Draper's history was a diatribe against the Roman Catholic Church. It reflected in part his reaction to the encyclical *Quanta cura* of 1864. The accompanying "Syllabus of errors" had deemed erroneous the belief that public institutions devoted to teaching literature and science should be exempt from the Church's authority. Liberal protests notwithstanding, it had been announced in 1870 that the pope, speaking ex cathedra, is gifted with infallibility when defining doctrines to do with faith and morals. Such developments were, for Draper, red rag to a bull – hence his recourse to history for the counterattack. A parade of martyred scientists would show by whose hands the real errors had been committed.

Although he differentiated his position from that of Draper, suggesting that the struggle had been between science and dogmatic theology rather than between science and religion, A. D. White insisted that there had been a theological and a scientific view of every question, invariably at odds. He, too, had lived with the Darwinian controversies and indulged in a similar backward projection. As with Draper, White had a personal investment. He was smarting from the clerical opposition that had been marshaled against his nonsectarian charter for Cornell University. The prominence he had given to instruction in science had evidently made matters worse. He was forced to conclude that there was "antagonism between the theological and scientific view of the universe and of education in relation to it." The result was *A history of the warfare of science with theology in Christendom* (1895), suffused with foolish remarks by uninformed priests, but hardly a model of discrimination.

To understand why Draper and White wrote as they did is not of itself sufficient to impugn their conclusions. Their arguments have to be judged on their merits. On closer inspection, however, they turn out to be deeply flawed. They share a defect in common with all historical reconstruction that is only concerned with extreme positions. They neglect the efforts of those who have regarded scientific and religious discourse as complementary rather than mutually exclusive. Their preconception that, as science has advanced, phenomena once considered supernatural have yielded to naturalistic explanation, is not without support. But it assumes a dichotomy between nature and supernature that oversimplifies the theologies of the past. If a supernatural power was envisaged as working *through,* as distinct from *interfering with,* nature, the antithesis would partially collapse. Or, in the vocabulary often employed by earlier natural philosophers, an explanation in terms of secondary causes need not exclude ultimate reference to a primary cause.

Anthropologists who have studied witchcraft among African tribes have commented on the reaction to suggestions that the disease of an individual may be the result of a virus, not of witchcraft. A common reaction is, Who sent the virus? This is instructive because it shows how an explanation in terms of secondary causes is not perceived to exclude a customary reference to the will of a person. The significance given to explanations in terms of natural causes depends on higher-level assumptions embedded in a broader cultural framework. In the history of Western culture, it has not simply been a case of nature swallowing supernature. Something had to happen to change the higher-level assumptions if conflict between science and religion was to achieve that self-evident status proclaimed by Draper, White, and their successors.

There is, here, a philosophical as well as an anthropological point. It was made by T. H. Huxley who, for all his anticlericalism, nevertheless acknowledged that there were limits to the conflict he had himself been stirring. There was a wider teleology not touched by the doctrine of evolution. For it was always possible to argue that a primordial molecular arrangement had been designed in such a way that the universe, as we know it, would be the outcome. In that sense, Huxley suggested, evolution had no more to do with theism than had the first book of Euclid. A complementarity between scientific and theological interpretation was still possible, even if he himself preferred to remain agnostic.

The conflict histories were flawed in a further respect. The scientific achievements of the past were crudely evaluated according to their contribution to later knowledge. A more sensitive approach requires that scientific innovations be judged against the background of prevailing knowledge at the time they were announced. The kind of history in which later knowledge is made the yardstick by which to judge earlier theories is now widely recognized as profoundly unhistorical. In the minds of those who practice it (and it is still quite common in scientific textbooks), the scientists of the past are effectively divided into heroes and villains. The former anticipate later developments, the latter fail to see the light. Given a cast of heroes and villains, there was no question but that clerical opponents of fruitful innovation would be placed among the latter. And if some brilliant anticipation of later theory failed to make its mark, there was a convenient explanation to hand: the opposition of "the Church."

The problem is that such accounts overlook the dialogue between established and innovative science. In criticizing scientific novelty, existing theories are usually available as a resource. Consequently, a case could usually be made out on scientific grounds for resisting hypotheses that appeared premature. It would, for example, be quite wrong to imagine that opposition to the Copernican theory derived only from religious prejudice. In 1543 an earth-centered cosmos was the physical orthodoxy of the day, supported by philosophical arguments that, at the time, were peculiarly compelling. Until an effective principle of inertia had been formulated, the earth's motion was contradicted by common sense. Surely, if the earth were moving, an object dropped from a tower would no longer hit the ground immediately beneath the point of release. Prevailing Aristotelian philosophy also affirmed a fundamental division between two regions in the cosmos. Beyond the moon all was perfect and immutable. Beneath the moon, all was corruption and change. To wrench the earth from the sublunar region and to place it among the planets was to violate the entire cosmos. Certainly the Catholic Church had a vested interest in Aristotelian philosophy, but much of the conflict ostensibly between science and religion turns out to have been between new science and the sanctified science of the previous generation.

Appeal to ecclesiastical censure as a way of explaining the misfortunes of scientific theories is a card that can be overplayed. Galileo seems to have felt that his difficulties with the Catholic Church had their origin in the resentment of academic philosophers who had put pressure on ecclesiastical authorities to denounce him. A conventional conflict thesis can also conceal vital distinctions between different religious traditions and between liberal and conservative representatives of those traditions. One has to ask whether different pressure groups within the Roman Church became equally hostile to Galileo, or whether, as some believed at the time, he was a victim of a Jesuit plot – of an act of revenge for insults he had meted out to prominent members of that order. The pope under whose jurisdiction he eventually fell, Urban VIII, had once been a friend and ally. Even after he had been summoned to the Holy Office in April 1633, Galileo could still arouse sympathy in high places. The commissary general in charge of the prosecution, Firenzuola, apparently admitted that he did not consider the Copernican system unacceptable. The pope's nephew, Cardinal Barberini, appears to have shared with Firenzuola the suspicion that the trial had more to do with personal revenge than doctrinal necessity. The issues in this famous case are extremely involved and we shall consider them further in Chapter III. But the need for discrimination is plain enough. During the 1620s the opinion had been expressed in Catholic circles that the Copernican system ought not to be condemned. To do so would constitute a stumbling block to the reconversion of Protestants who favored the new astronomy.

In assessing the attitudes of religious thinkers to the natural sciences, discrimination is required in another respect. For there is a difference between hostility and indifference. Many spokesmen for the Christian religion have echoed the conviction that St. Basil expressed with his statement that a life of meekness and piety knew higher concerns than whether the earth was a sphere, cylinder, or disk. Sometimes the investigation of nature has sunk to the bottom in a stratification of priorities. Another of the early church fathers, Tertullian, observed that it had "served Thales of Miletus quite right, when, star gazing as he walked . . . , he had the mortification of falling into a well."

It was easy for Draper and White to mock the church fathers for their naïveté on matters of natural philosophy, forgetting that they did have higher concerns – forgetting, too, that their naïveté often reflected the pagan wisdom of their day. According to Draper, no one had done more than St. Augustine to bring science and religion into antagonism. It was he who had made the Bible an arbiter of human knowledge. Yet Draper failed to mention that, in his exegesis of Scripture, Augustine had not been a hide-bound literalist. In fact he had specifically warned against taking the "days" of Genesis literally. Because eggs and seeds took time to develop, reflecting a "wonderful constancy in an established order," the creation narrative had to be read with care. Drawing on a Stoic concept of seeds, which he applied even to the origin of Adam and Eve, he had a sense of natural order that could hardly be considered hostile to further inquiry. He may have shown indifference with his remark that Christians need not be ashamed if they knew nothing of the number and properties of the elements. But he could also be dismayed if Christians were heard by pagans to be talking nonsense about nature.

To reduce the relationship between science and religion to one of conflict may also obscure the possibility that, when students of nature were persecuted by ecclesiastical authorities, it was for theological heresies rather than for scientific heterodoxy. Two examples have frequently been pressed into the mold of religion versus science: the burning of Michael Servetus (c. 1511–53) in Protestant Geneva and of Giordano Bruno (1548–1600) by the Roman Inquisition. Servetus is remembered for his description of the lesser, pulmonary, circulation of the blood. Rejecting Galen's theory, that blood passes directly from right to left ventricle, he proposed a vital role for the lungs, where a vital spirit emerged from the mixture of air and blood. His insight, however, was achieved in a theological work discussing the relationship between "spirit" and air. His concern was to understand the dispensation of God's Spirit to mankind. But he was burned in 1553 not so much for attacking Galen, as for attacking Calvin. Geneva was still technically under the Holy Roman Empire where the Justinian code allowed burning for those who denied either infant baptism or the divinity of Christ. Servetus had denied both. Calvin himself had no doubt that Servetus should be executed, but he was actually unsure whether burning, as recommended by the town council, was the appropriate sentence.

An advocate of Copernican astronomy, an infinite universe, and a plurality of worlds, Bruno has often been seen as the archetypal scientific martyr. Although his proposal of plural worlds *was* considered heretical, it is difficult to believe it was that which determined his fate. A renegade monk, he made no secret of his unorthodox Christology. It was rumoured that he had declared Christ a rogue, all monks asses, and Catholic doctrines asinine. Behind his hostility lay a conviction that the Roman Church represented a corruption of an earlier, undefiled religion that he associated with the Egyptians. Bruno was familiar with the collection of texts known as the *Corpus Hermeticum,* then attributed to an Egyptian philosopher, Hermes Trismegistus. Where some saw in the Hermetic texts an anticipation of Christianity, Bruno saw an alternative. Indeed, he hoped they would provide the basis of a religion that could unite the warring factions of the Church. His world-picture was colored by a magical philosophy that almost became his religion. He described Moses as a magus who, learning his magic from the Egyptians, had outconjured the magicians of Pharaoh. The true cross, for Bruno, was the Egyptian cross — full of magic power for tapping astral influence. The Christian cross was a weak derivative. The evidence suggests, not surprisingly, that his interrogators were more concerned by his theology, by matters of church discipline, and by his contacts with other known heretics, than by his Copernicanism.

The dependence of the conflict thesis on legends that, on closer examination, prove misleading is a more general defect than isolated examples might suggest. Consider the case of Charles Lyell, who lectured on geology at King's College London in 1832 and 1833. Appointed to the chair of geology in April 1831, he had resigned by October 1833. The anecdotal view has been that his brevity of tenure reflected animosity from an Anglican establishment, disturbed by Lyell's rejection of a recent, universal flood. It was not, however, that simple. Even the one bishop who had been apprehensive, Edward Copleston, acknowledged that the progress of geology could be accommodated by a readjustment of biblical interpretation. His principal concern had been pastoral — that Lyell might abuse his position of responsibility by drawing conclusions of general human significance, not strictly deducible from his science. Lyell resigned not because of some inherent conflict between science and religion, but because he enjoyed neither the prestige nor the remuneration he had expected from the post. His lectures would have been more lucrative had they attracted more women. But it was during his tenure that the College Council debarred them. With the prospect of a depleted income, Lyell considered that his energies would be better spent in completing his *Principles of geology.* His resignation arose not so much because an audience was to be debarred from geology, as because women were to be debarred from the audience.

Women *were* in the audience when T. H. Huxley and Bishop

Wilberforce had their encounter at Oxford in 1860. One fainted, according to some reports, and had to be carried out – so tremendous had been the spectacle of a man of God slain by a man of science. The bishop had asked whether Huxley would prefer to think of himself descended from an ape on his grandfather's or his grandmother's side, thereby touching the sensitive nerve of female ancestry. Huxley's retort that he would rather have an ape for an ancestor than a bishop – or words to that effect – has come to symbolize not merely the conflict between Darwinism and the Bible but the victory of science over religion. "The Lord hath delivered him into mine hands" was a phrase that came into Huxley's mind, as he reminisced about the incident later. But it was thirty-one years later and recent scholarship has questioned whether the anecdote might not have been a retrospective invention. Contemporary accounts were certainly not unanimous in recording a triumphant Huxley and a humiliated bishop. Huxley had turned the tables admirably, according to Sir Joseph Hooker, but he had been unable to throw his voice over so large an assembly. He had not dealt with the bishop's weak points nor, according to Hooker, had he carried the audience with him. One convert to Darwin's theory, Henry Baker Tristram, who had applied the concept of natural selection to larks and chats in the Sahara desert, was actually deconverted as he witnessed the debate! There was, undeniably, an issue: whether the nonspecialist had any right to expatiate on scientific matters. But that is precisely why the legend, once created, would become part of the folklore of scientific professionalism.

From popular texts on the Darwinian revolution, one would never suspect that another clergyman spoke out during the Oxford meeting of the British Association. Frederick Temple, headmaster of Rugby School, and future archbishop of Canterbury, preached the official sermon on 1 July 1860. Taking a more liberal line than the bishop, Temple argued that the finger of God was to be discerned in the laws of nature, not in the current limits of scientific knowledge. Tacitly, he made room for Darwin's science and was even said by one observer to have espoused Darwin's ideas fully. The point is that memorable anecdotes can create false perspectives. Wilberforce was not the only representative of Anglican opinion, and even he was not as obscurantist as is sometimes

implied. His printed review of the *Origin of species* attracted
Darwin's comment that "it picks out with skill all the most
conjectural parts, and brings forward well all the difficulties."

The fundamental weakness of the conflict thesis is its tendency to portray science and religion as hypostatized forces,
as entities in themselves. They should rather be seen as complex social activities involving different expressions of human
concern, the same individuals often participating in both. In
its traditional forms, the thesis has been largely discredited.
It is, in fact, so easy a target that scholars reacting against it
have constructed a revised view that has also been driven
to excess. If past conflicts can be spirited away, it is tempting for the religious apologist to step in and to paint a more
harmonious picture. After all, if religious beliefs did fur-
nish presupposition, sanction, and motivation for science,
they might be looked upon more favorably in an otherwise
secular society. The argument sometimes takes the form of a
diagnosis — that much of the perplexity of our modern age is
due to the severance of science from the religious values that
once shaped it. Christians, both Catholic and Protestant, have
argued thus, as have Muslim scholars looking back to a golden
age when Islamic thinkers were at the forefront of the physical sciences. This apologetic use of revisionist history is,
however, so full of pitfalls that it, too, deserves critical attention.

Harmony between Science and Religion

Religious apologists have often argued that when science and
religion are properly understood, they can be in perfect harmony. The implication is that such conflicts as there have
been in the past were merely the result of misunderstanding.
Historical analysis showing that at least some of the conflict
has been exaggerated can be used to support the argument.
The problem is, however, that claims for inherent harmony
are vulnerable to the same kinds of objection as claims for an
inherent conflict. The idea that there is some correct and
timeless view, against which historical controversies can be
judged, can prove an insensitive guide to the issues as they
were perceived at the time.

When the history of science is hijacked for apologetic purposes, it is often marred by a cultural chauvinism. The realization that religious beliefs were relevant to the rise of sci-

ence is transformed into the more parochial claim that a particular religion, or religious tradition, was uniquely propitious. Thus one often encounters the claim that a large number of Quakers (and other dissenters) contributed significantly to eighteenth- and nineteenth-century physical science and that Jews have been preeminent in mathematics, physics, and psychiatry in the twentieth century. It is wise for the historian to be suspicious if claims for a special relationship are superimposed upon such alleged correlations. This is not to deny that there may be good reasons why members of religious minorities should have sought a career in science and why they excelled in particular areas. But these reasons may have little to do with intrinsic features of their religion. In eighteenth-century England, for example, the practical sciences may have appealed to some dissenters because they were denied access to other professions.

Cultural chauvinism can have insidious effects on the historiography of science. Even when they have not been beating a religious drum, Western historians have been censured for their myopia in treating modern science as if it were an exclusively Western phenomenon. Significant contributions by Arab scholars are easily overlooked if there is the preconception that science was uniquely the product of a Hebraic–Christian culture. In the centuries preceding the scientific renaissance in the West, Muslim scholars did far more than merely reproduce their heritage from the Greeks. In algebra they developed a concept of polynomials and pioneered an algebraic geometry that has traditionally been ascribed to Descartes. Different types of experimentation were introduced as a conventional barrier between theoretical science and the practical arts was weakened. A fruitful marriage of mathematics with physics was achieved by Alhazen (c. 965–1040) for whom optics became a study of the geometry of vision. By the early fourteenth century, al-Farisi was constructing experimental models – glass spheres filled with water to simulate rain drops – for the mathematical analysis of the rainbow. Not surprisingly, Muslim scholars see in sixteenth-century Europe not a scientific renaissance but a reactivation.

Because the history of science has become embroiled in religious polemics, it is important to appreciate that the chauvinism appears not only between religions but within them. In claims that one tradition rather than another led to scientific innovation, one often detects an apologetic intention.

One of the most prolific writers on Islamic science has insisted that most, if not all, the seminal contributions of Arab scholars are to be ascribed to a Shi'ite gnostic tradition. The rival theology of Sunni Muslims, particularly that of the Ash'arites, then suffers by comparison. Similarly, the thesis of Western historians that Protestant Christianity was more conducive to the expansion of science than Catholicism has often carried a hidden agenda. The great defect in such writing is that it is often structured by questions of the form Who discovered such and such a fact first? or Who first anticipated such and such a concept? Although the underlying message is quite different from the conflict thesis of White and Draper, there is that same streamlining of history into heroes and villains.

Religious apologists have also fallen into the same trap as White and Draper in projecting backward a model – in this case of harmony – that, although consonant with their own reconstructed religion, may not fit the religious beliefs of the past. The danger consists in imagining some essence of Christianity, for example, that, because it may be shown to be immune to scientific criticism today, is assumed always to have existed and therefore, properly understood, is always impervious to criticism. Apologists wishing to stress the harmony between science and religion may gloss over those facets of Christianity *as it was* that distinguished it from Christianity *as they now wish it to be*. They may wish to say that their religion no longer needs a geocentric universe, a physical location for heaven and hell, a personal devil, or even divine intervention in the physical world. The danger is that they may minimize, or even overlook, the importance of such beliefs for the Christian societies of the past. However condescending contemporary apologists may be to archaic conceptions of divine intervention, it is almost impossible to exaggerate the extent to which belief in such intervention once permeated European societies, creating popular images of the disruption of nature that could hardly have been congenial to a critical science of nature.

Apologists wishing to exploit a revisionist history of science invariably stress the profoundly religious orientation of many prominent scientists. While it is true that many of the great names of the past have been theists rather than atheists, their orientation has often been unorthodox when judged against the norms of their day. Idiosyncrasy has been at least as common as conformity. The originality of a critical mind, expressing itself in creative and sometimes idiosyncratic science, has often manifested itself in theological deviation. The eighteenth-century chemist and Unitarian minister Joseph Priestley (1733–1804) was as keen to remove spirits from Christianity as he was from chemistry. A mind used to questioning scientific hypotheses might experience discomfort with contentious elements in a classical creed – just as Newton, before Priestley, dismissed the doctrine of the Trinity as a Platonist corruption of a biblical Christianity. One of the points to emerge in later chapters will be that eminent scientific figures have rarely been typical representatives of the religious traditions in which they were nurtured.

The apologist has a problem even where expressions of religious belief have an orthodox ring. For the cynic will always say that the scientists of the past simply feigned their belief in order to escape persecution. As a general rule, this cynical objection is far too crude, not least because it often rests on a false assumption – that it is impossible for theological statements to be both placatory and genuine. Nevertheless, such statements *were* sometimes placatory and social pressures toward conformity could be strong. As late as 1865, the botanist J. D. Hooker wrote a revealing letter to Darwin in which he complained of the stance taken by their contemporary, Alfred Russel Wallace, who had wondered that scientists should be so afraid to say what they think. "Had he as many kind and good relations as I have," Hooker wrote, "who would be grieved and pained to hear me say what I think, and had he children who would be placed in predicaments most detrimental to children's minds . . . he would not wonder so much."

The point is not that the theological pronouncements of scientists are to be discounted but rather that they do sometimes have to be seen as efforts at mediation. What may look like a straightforward affirmation of harmony between Christianity and science may turn out to have been a counter in a dialogue between new and entrenched religious positions. The apologist who extracts the affirmation for his own purposes may overlook the dynamics of the dialogue. In the writings of Galileo, for example, he would find a clear statement of compatibility between Copernican astronomy and Catholic Christianity. In his *Letter to the Grand Duchess Christina* (1615), Galileo argued that the language of the Bible had been accommodated to the minds of the uneducated, with the consequence that texts that superficially implied a stationary earth and a moving sun were not to be treated as literal scientific descriptions. Beneath the surface Scripture had deeper meanings to which a knowledge of nature might help to give access. There was room for both science and religion since the Bible taught how to go to heaven, not how the heavens go. Its jurisdiction was principally on moral and spiritual matters. But this affirmation has to be seen in context. When Galileo wrote his letter, he was already on the defense. There were already rumors that the new astronomy was incompatible with Scripture, and he had already been denounced from the pulpit. Statements of harmony, as with statements of conflict, have to be placed in their context. They may define positions that have often been repeated, but they cannot be given a timeless quality.

The Galileo affair illustrates a further difficulty for the apologetic use of history. In criticizing the conflict thesis, we noted that a misleading sketch of opposition to science can be traced if the sample is confined to religious extremists. Everything depends on where the circle is drawn. But it is possible to draw so tight a circle around an enlightened few that the picture becomes distorted in the opposite direction. There was indeed a range of enlightened opinion within the Roman hierarchy at the time of Galileo's trial; but the trial itself was not averted. The point is that it is not just a case of drawing circles around thinkers whose stance one may happen to approve. To understand the events in question, it is also necessary to consider in whose hands political power was concentrated. It was not always the more enlightened who got their way. Galileo was fighting against a religious bureaucracy, which, through an elaborate mechanism of censorship, sought to control the boundaries of legitimate opinion. As we shall see in Chapter III, the threat from the Protestant churches was an integral part of the political complex.

The use of history to identify a privileged standpoint from which science and religion could always be harmonized runs into trouble for yet another reason. The diversification of the sciences and the theoretical changes within them make it extremely difficult to locate a unique set of principles by which harmony could be guaranteed. Among the physical scientists of the seventeenth century it was achieved by stressing the harmony of the universe itself – a divinely conceived harmony, expressible in mathematical terms, and which, Copernicus had declared, it was the duty of the astronomer to display. By the nineteenth century, however, and especially in the life sciences, the metaphysics that had underpinned the work of Copernicus, Kepler, and Newton had become obstructive. The idea that living organisms had been structured according to a divine plan, transparent to the human mind, still featured in the work of eminent naturalists. It was defended in Britain by Richard Owen and in America by the Harvard professor, Louis Agassiz. Although it was not incompatible with the view that new species had arisen during the earth's long history, it was incompatible with evolutionary theory as conceived by Darwin. Agassiz expressed his sadness that Darwin's theory had so quickly won acclaim, while Darwin could find in Agassiz's idealism nothing but "empty sounds." A metaphysics stressing the invariance of divine ideas created dissonance where it had once created harmony.

The strong claim that some one religious tradition was uniquely propitious for science also has its difficulties. The problem is that, although certain doctrines may have licensed scientific inquiry, others within the same tradition could stifle it. This ambivalence certainly existed in seventeenth-century Christian theology. Although a doctrine of creation could countenance the search for order behind the flux of natural phenomena, the doctrine of Adam's Fall was sometimes used to suppress it. Fears were expressed by puritan divines that to thirst after natural knowledge was to run the risk of elevating reason at the expense of faith. And if human reason had been impaired by the Fall, what guarantee was there that one *could* think God's thoughts after Him? Everything hinged on how such doctrines were formulated. As we saw earlier, Francis Bacon was able to turn the doctrine of the Fall to advantage: Science would help restore that dominion over nature lost to humanity through Adam's sin. In assessing whether a particular doctrine might have encouraged scientific inquiry, such ambivalence must be taken into account. Even the claim that the natural order reflected the contingency of a divine will could pull in two directions. It could be used, as it was by Bacon and Mersenne, to justify an empirical rather than a rationalist approach to nature. But it could also be used to dismiss the claims, even of empiricists, that they knew how nature works. For if God could have made the world work in any number of ways, would it not always

HARMONIE
VNIVERSELLE.

De antiquo marmore Illustrissimi Marchionis Mathei Romæ.

Nam & ego confitebor tibi in vafis pfalmi veritatē tuam:
Deus pfallam tibi in Cithara, fanctus Ifrael. *Pfalme 70.*

Figure I. 7. Engraved title page from Marin Mersenne's *Harmonie Vniverselle* (1636–7). An intimate relationship between musical harmony and the mathematical representation of physical data was epitomized by the correlation, ascribed to Pythagoras, between string length and pitch. To halve the length of the string, for example, was to raise the pitch by exactly one octave. In the seventeenth century, Mersenne undertook a systematic experimental inquiry, showing that pitch depended on frequency of vibration, which in turn was determined by the length, tension, and thickness of the string. Concepts of universal harmony could be sustained on the basis of verifiable relationships between regular motion and pleasing sounds. Reproduced by courtesy of The Bodleian Library, Oxford; Shelfmark M. 3. 16. Art.

Figure I. 8. Illustration from page 207 of Johannes Kepler's *Harmonices mundi* (1619). In Kepler's construction of the solar system, each planet had its own melodic line associated with a changing speed that increased as it approached its closest point to the sun. The less the variation in speed, the more monotonous the theme. Kepler even conjectured that, by a backward projection from current orbital positions, it might be possible to determine the age of the earth – on the supposition that the most harmonious sound would have corresponded to the original creation. Reproduced by permission of the Syndics of Cambridge University Library.

be presumptuous to pretend that one had actually pinned Him down? That was an objection with which Galileo had to contend.

In revisionist histories constructed for apologetic purposes, it is not usually denied that there were conflicts in the past. But it is often implied that these were the result of particular social and political circumstances. Once these contingencies are removed from the picture, no – or very little – substantive conflict remains. Now, there is no denying that past controversies were rooted in circumstances that were peculiar to the age in which they occurred. But to suggest that the conflict was over political rather than intellectual matters is to introduce a false antithesis. The fact is that ideas about the relationship between science and religion have themselves been weapons in these political confrontations. To explain the intellectual polarities by reference to historical contingencies is not necessarily to explain them away.

The debate over human origins in nineteenth-century Britain provides a helpful example. Historians have correctly pointed out that to understand the kind of exchange that took place between Huxley and Wilberforce it is important to recognize that a major social transformation was taking place in which the clergy were losing their domination of the intellectual life of the nation. One symptom of this was their increasingly marginal involvement in scientific activity – a trend reinforced by the establishment of professional elites within the scientific movement. When the British Association for the Advancement of Science had been founded in the early 1830s, clerics had constituted some thirty percent of its membership. In the period 1831–65 no fewer than forty-one Anglican clergymen had presided over its various sections. Between 1866 and 1900 the number was three. During that latter period, the function of science in British society was increasingly perceived in terms of national prosperity, economic strength, and military security. Formidable writers on science, such as Huxley and Darwin's cousin Francis Galton, still stressed a moral utility for their subject, but the traditional arguments for its religious utility were wearing rather thin.

It would be unwise to interpret the exchange between Wilberforce and Huxley without taking this social transformation into account. Huxley had every reason to bash a bishop. It would be in keeping with his desire to promote the status of the professional scientist and to exclude clerical interference. But because there were social and political reasons for Huxley's aggression, it does not follow that there were no substantive intellectual issues. To suggest that the struggle was not between science and religion but over cultural leadership may expel the conflict through the front door, but it still returns through the back. Earlier in the chapter, reasons were given why the mythology surrounding this event should be questioned. In the words of one historian, the legend was a one-sided effusion from the winning side. But, in such a statement, the fact that were *sides* has crept in round the back. And the apologist who might wish to make Huxley the aggressor should know that one observer, the zoologist Alfred Newton, clearly reported that it was Huxley who had first been chaffed by the bishop.

The purpose of this chapter has been to establish three propositions: that religious beliefs have penetrated scientific discussion on many levels, that to reduce the relationship between science and religion to one of conflict is therefore inadequate, but that to construct a revisionist history for apologetic purposes would be just as problematic. To enhance the standing of a particular religion or religious tradition by making it the mother of science may be short-sighted in at least two other respects. It may overlook the fact that science could become a very rebellious offspring; and, as an apologetic strategy, it could easily backfire if science (as some believe it has been) were to be devalued in public estimation through association with polluting and exploitative technologies.

Much of the writing on science and religion has been structured by a preoccupation either with conflict or with harmony. It is necessary to transcend these constraints if the interaction, in all its richness and fascination, is to be appreciated. In the following chapter, we shall examine this interaction more closely for the period during which the foundations of modern science were laid.

CHAPTER II

Science and Religion in the Scientific Revolution

Introduction

In the creation of our modern world-view, few periods of Western history have been as decisive as the hundred fifty years that followed the publication, in 1543, of Copernicus's sun-centered astronomy. During that period, what had been an earth-centered cosmos exploded into an infinite universe. The achievements in this age of genius have become legendary. In the physical sciences alone, there were momentous changes. When Kepler refined the Copernican system, he broke with centuries of tradition in suggesting that the planets moved in ellipses, not in circles. In his own colorful expression, he had laid a monstrous egg. Through his telescope Galileo observed more things in the heavens than had ever been dreamed of: moons of Jupiter and myriads of stars invisible to the naked eye. At a more technical level, the motion of bodies was subjected to mathematical analysis, concepts of inertia were formulated, and, as a fitting climax, Newton formulated his inverse-square law for gravity, which explained the planetary orbits. In the words of his contemporary Edmond Halley, he had penetrated the secret mansions of the gods.

We continue to speak of a scientific *revolution* because earlier systems of belief were emphatically overthrown. In the philosophy of Aristotle, which had dominated the intellectual life of Europe since the thirteenth century, there had been a clear division in the cosmos between the perfect spheres beyond the moon and the corrupt sublunar sphere at the center of which stood the earth. By the end of the seventeenth

52

century, in Newton's science, the terms of reference had changed. There was now a universal law of gravitation. Because it applied to all bodies everywhere, the universe had at last *become* a universe. To divide it into two parts, heaven and earth, might still be admissible, as it was for Newton, when discussing theological and political symbolism. But the closed, partitioned, and spherical cosmos had had its day.

We speak of a revolution; and yet this intellectual transformation took so long to accomplish that there is no straightforward parallel with a political revolution. During the hundred fifty years in question, there was so great a diversity of views about nature that, on closer inspection, it becomes difficult to achieve a succinct characterization. There is, however, one that has found favor among many commentators. The scientific revolution, it is often said, saw the *separation* of science from religion. The object of this chapter is to consider whether this assessment is accurate. It may be seductive, but it is also deceptive.

The concept of a separation of science from religion during the seventeenth century implies that during preceding centuries there had been a fusion. And it also implies that where there had once been marriage, there was now divorce. But how closely do these inferences match the reality? It is true that, from the thirteenth to the sixteenth century, the pursuit of natural knowledge was often subordinate to theological concerns. But a relationship of subordination is not the same as one of fusion. It is also true that, during the seventeenth century, the domains of science and theology were differentiated in new and challenging ways. But a differentiation is not the same as an ultimate separation. In this chapter, I hope to show that these distinctions are important and not merely a play on words. New patterns of differentiation were often sought with a view to the reintegration of scientific and religious belief. It is even possible to argue that the scientific revolution saw an unprecedented *fusion* of science with theology, resulting in more secular forms of piety.

The Separation of Science from Religion?

The image of a separation between science and religion during the scientific revolution is certainly attractive. It conforms to our modern secular perception of what ought to have happened. The great achievements of the seventeenth century seem to testify to a new independence of scientific inquiry. There were even explicit statements to the effect that science and religion should not be mixed. Francis Bacon took that line when he suggested that, in seeking to understand the immediate causes of physical phenomena, the experimental philosopher should not be diverted by metaphysical considerations concerning the purpose of the phenomenon in question.

To understand this protest, it is important to appreciate that Aristotle had made the quest for final causes, for ends or purposes, the highest aim of the student of nature. To answer the question Why does a stone fall to earth? Aristotle would look for an answer in terms of purpose. The stone falls in order to return to its natural place, made as it is from the element earth. This would be one example of what we call teleological explanation. Aristotle had implied that an explanation was never complete unless a final cause had been specified. As he rather paradoxically put it, the final cause had to be first. It had pride of place. But not for Bacon, who spoke of final causes as barren virgins. And not for Descartes either, who expelled them in order to concentrate on the immediate mechanical causes of natural phenomena.

This rejection of final causes from physical explanation is often identified with a separation of science from religion because one of the traditional links between nature and God was apparently broken. The traditional view was expressed by Sir Thomas Browne in his *Religio medici* (1642). Every created thing had a purpose. There was some positive end both of its essence and its operation. And this, he wrote, "is the cause I grope after in the works of nature; on this hangs the providence of God." To reject final causes from the study of nature would apparently leave providence hanging.

A separation of science from religion has also been seen in a diminished authority for the Bible in matters of natural philosophy. It is encapsulated in Galileo's quip that the Bible teaches how to go to heaven, not how the heavens go. The English Copernican John Wilkins also found it necessary to revise concepts of biblical authority in order to make room for scientific propositions that seemed at odds with the plain meaning of Scripture. As he did so, he fell foul of reactionary figures such as the Aristotelian schoolmaster Alexander Ross. Wilkins had implied that the Bible was not to be taken as a scientific text since the Holy Spirit had used a form of words comprehensible to the common man. What worried Ross was a dangerous bias. It seemed that Wilkins, and other Copernicans, allowed certain astronomical discoveries to confirm the truth of Scripture; but they would not let Scripture pronounce on matters of astronomy. Ross's complaint was that astronomy was more in need of confirmation from Scripture than vice-versa. It is difficult not to see in such debates a growing independence of science from biblical constraint. Wilkins even argued that where Scripture was silent on a given issue – the plurality of worlds, for example – this was license to consider the possibility, not to exclude it.

An interpretation that stresses the separation of science from religion in the seventeenth century is far from implausible. A new intellectual independence can also be correlated with longer-term social trends. Although there were notable exceptions, the most eminent philosophers of nature were not members of the clergy but were lay figures who developed their own theologies. This was a continuation of that trend away from a clerical domination of scholarship that had been accelerated by the appearance of the printing press. Superimposed on this was the organization of scientific activity in societies and academies – institutions with codes of practice that often outlawed political and religious disputes. The Accademia dei Lincei, to which Galileo belonged, was one of the first scientific societies to claim equal status alongside other institutions of learning. Later organizations, such as the Royal Society, founded in England in 1660, could be even more ambitious in their goals. "You will please to remember," its secretary, Henry Oldenburg, told the governor of Connecticut in 1667, "that we have taken to task the whole universe, and that we were obliged to do so by the nature of our design."

In one of the informal groups that preceded the foundation of the Royal Society, one finds that affirmation of independence that gives weight to the separation thesis. Of meetings which had begun in London in 1645, the mathematician John Wallis could say:

Our business was (precluding matters of Theology and State Affairs) to discourse and consider of Philosophical Enquiries, and such as related thereunto; as Physick, Anatomy, Geometry, Astronomy, Navigation, Staticks, Magneticks, Chymicks, Mechanicks, and Natural Experiments.[1]

These remarks indicate the scope that existed by the mid-seventeenth century for differentiation between the sciences.

And the reference to an avoidance of theology may well suggest that a separation from the queen of the sciences had already been achieved.

The need for differentiation was clearly felt by seventeenth-century philosophers. When the French thinker Blaise Pascal joined those who, in revolt against Aristotle, were prepared to admit a vacuum in nature, he drew a distinction between theology and the natural sciences based on their respective sources of authority. A reverence for antiquity, though an appropriate stance for theology, was inappropriate for natural philosophy where reason and the senses held sway. It is also clear that traditional syntheses between science and faith were badly shaken by new conceptions of nature. The disintegration of an older order, at once exhilarating and disturbing, was most famously expressed by the poet John Donne. His *An anatomy of the world* (1611) contained the lines:

> And New Philosophy calls all in doubt,
> The Element of fire is quite put out;
> The Sun is lost, and th'Earth, and no man's wit
> Can well direct him where to look for it.

All was in pieces, "all coherence gone."

If we are looking for a separation of science from religion in the seventeenth century, we shall surely find it. The difficulty is that, however strong in outline, the characterization turns out to be weak in detail. A finer brush is required to depict the intellectual debates of the period, as distinct from merely documenting achievements or contrasting new with old cosmologies. Even some of the century's most notable achievements were presented in theological terms. As we shall show in greater detail later, Descartes justified his principle of linear inertia by ostensibly deducing it from the immutability of God – a God who conserved the simplest kind of motion in the world. A mechanized universe, dear to both Descartes and Boyle, was often valued as a vehicle for emphasizing God's sovereignty over His creation. As long as matter was considered to be inert, its motion could be ascribed to the divine will. A universe that ran like clockwork also evinced design. Although Boyle agreed with Bacon and Descartes that the quest for final causes would be injurious to science if it jeopardized the search for physical causes, the great beauty of a mechanical philosophy was that, properly understood, it allowed one to have both. To ask how a piece of machinery worked was not to deny that each part had been designed with a function in mind. It was Boyle's contention that if William Harvey had not asked himself about the purpose of valves in the veins, he might well not have discovered the circulation of the blood.

For all that he warned against the mixing of science and religion, Bacon remained convinced that scientific conclusions had still to be limited by religion. On issues such as the size and eternity of the universe, his own faith played a selective role, setting the conditions for admissible theories. An eternal universe, for example, was incompatible with a created universe — a universe that Bacon supposed would one day be liquidated by its Creator. And, as we saw in the previous chapter, he gave science a religious sanction, in that it promised the restoration of a dominion over nature that had been God's intention for humanity.

Even on the issue of biblical interpretation, the concept of a total separation would be too crude. Galileo's conception of biblical authority will be examined later in the chapter. But we shall not find a consistent position in which the tasks of biblical exegesis and scientific inquiry were no longer mutually relevant. Galileo was even prepared to give the Bible jurisdiction over scientific hypotheses that had not been rigorously demonstrated. Both he and Wilkins were able to invoke past precedent for their concepts of biblical accommodation. The image of a revolutionary break is, to some degree, a distortion. Long before Wilkins sought to popularize the Copernican system, the Protestant reformer Calvin had declared that those who wished to learn astronomy should not look to the Bible. Nor was Wilkins a layman championing a secular cause. He was a moderate clergyman, with puritan sympathies, destined to become a bishop in the Church of England.

Even if the most eminent natural philosophers of the seventeenth century were laymen, it would still not follow that all connections between science and religion were severed. Robert Boyle once said that one of the reasons why he had chosen not to be ordained was to preempt the criticism that his pious remarks about nature merely reflected a vested interest. To be a layman, even to be anticlerical, is not necessarily to be irreligious. Similarly, a consensus within early scientific societies, that theological and political discussion should be excluded, must be interpreted with care. Clearly, a new independence was being asserted. But the exclusion of theology could be inspired by purely pragmatic considerations. Given the divisiveness of religious opinion, heightened in England by a civil war, it would be expedient to avoid any issues that might jeopardize a union of scientific interests. This would not, however, prevent individual members of these early societies from holding distinctive views on the relationship between their scientific and religious beliefs.

For these reasons we shall drop references to the *separation* of science and religion during the scientific revolution. A change in the status of natural philosophy was certainly achieved, but by differentiation from, and reintegration with, religious belief rather than by complete severance. From the thirteenth to the sixteenth century, the study of nature had, in various ways, been subordinate to theological interests. Rather than identify a point in time when some separation was effected, it is more constructive to locate successive shifts in the degree of subordination and the grounds of differentiation. The sequence was extraordinarily complex because in one shift – toward a magical philosophy of nature – a new integration of scientific and religious motifs actually discouraged differentiation of the kind that Descartes and Boyle, through their mechanical philosophies, were later to effect. There was no linear process either of differentiation or secularization.

The Subordination of Science to Theology

In this section, we shall examine what it means to say that the study of nature was subordinate to theology in the period that separated the two Bacons, Roger (?1214–94) and Francis (1561–1626). Different patterns of subordination will be distinguished with reference to the work of Roger Bacon, to the synthesis of Aristotelian philosophy and Christian theology achieved by Thomas Aquinas (?1225–74), and to a critique of the Aristotelian world-picture by Nicole Oresme (c. 1323–82).

Roger Bacon has enjoyed renown, both for an emphasis on firsthand experience of nature and for an advocacy of mathematics that, for the thirteenth century, was remarkably energetic. His promotion of mathematics illustrates at least two senses in which the interests of the natural sciences could be subordinate to those of theology. First, there was subservience in that Bacon looked toward the pope for patronage. He had much to gain if he could persuade a powerful patron that mathematics, and especially his own, could be of service to Christendom. Second, it was the theological uses of mathematics on which Bacon waxed eloquent. There was subordination, but not a complete fusion; for Bacon was perfectly aware of subject boundaries, differentiating mathematics itself from grammar, perspective, experimental science, and moral philosophy. The special value of mathematics lay in its applicability to all other sciences.

Of what use was mathematics to medieval Christianity? Bacon's contention was that the theologian had to know the things of this world if he wished to understand the Scriptures. Consequently, mathematics could assist that understanding — an urgent matter if, as Bacon sometimes implied, all knowledge was ultimately to be found in the Bible. A knowledge of geometry was to be highly prized because it could help one to visualize the geometrical forms alluded to in Scripture. As the servant of astronomy, mathematics was peculiarly fitting for theological study, since, as Bacon put it, "throughout the Scriptures we are called away from terrestrial . . . and awakened to heavenly things." Mathematical astronomy could assist in correcting the calendar. Through its involvement in astrology it could also contribute to the governance of the Church, proving the truth of Christianity, forecasting the future, and predicting the course of disease. Through astrology, Bacon believed, it was possible to investigate the great sects of human history: Jews, Chaldeans, Egyptians, Saracens, and Christians. More urgently, one could be forearmed against the Antichrist, the powerful, personal opponent of Christ who (as in 1 John 2:18) was expected to appear before the end of the world. Through astrology one could be forewarned of the exact time of his coming.

Since the Antichrist would use mathematical knowledge in exercising malignant power, it was vital for the Church to be prepared. Bacon could argue that Antichrist would invoke stellar influences and magic words having the power to produce physical effects. The ultimate justification for mathematical research was to equip the Church with a counter-magic of its own. Knowing the things of this world was certainly subordinate to theological concerns. Because the very act of knowing involved divine grace and illumination, because it was not the fruit of work alone, the subordination was marked indeed. By taking a particular view of the elements, Bacon even set out to show that human bodies could remain in hell-fire for eternity.

Subordination of another kind may be discerned in the works of Thomas Aquinas. References to his synthesis sometimes give the impression that he achieved a complete fusion of Aristotelian with Christian doctrine, as if the two were rendered fully compatible. It is important to realize, however, that the introduction of the Aristotelian corpus into medieval Christendom had occasioned alarm because it contained doctrines that clashed with Christian belief. Aristotelian philosophy affirmed that the universe is eternal, that objects in the world are not dependent upon God for their existence. As interpreted by the twelfth-century Muslim commentator Averroes, it even denied the freedom of the will. It could also be construed as a naturalistic rival to supernatural religion, in that it presented the world as a closed system in which the deity was little more than a physical hypothesis to explain motion or change – a world in which there was no room for a higher human destiny. Accordingly, three popes – Honorius III, Gregory X, and Urban IV – issued decrees forbidding or restricting the teaching of Aristotelian metaphysics.

Strictly speaking, it was impossible to effect a fusion of Christianity with Aristotle – as Aquinas was well aware. In selecting those facets of Aristotle's teaching that he considered illuminating, he was guided by the demands of his faith. In this respect, Aristotelian metaphysics remained subordinate to Christian doctrine. From his commentary on Aristotle's *Physics*, it is clear that Aquinas could never regard the Greek philosopher as an authority on matters of faith. Discussing Aristotle's argument for the eternity of the world, he protested that motion had not always existed. "According to our faith," he wrote, "nothing has always existed except God alone, Who is altogether immobile."

This subordination is reflected in a change of emphasis that Aquinas brought to the study of metaphysics. Aristotle had been primarily concerned with the problem of motion, broadly understood as a process of becoming. Aquinas took the problem of existence, of being, as the fundamental issue. And the whole of metaphysics was directed toward knowledge of God. If the world had not first existed in the mind of God, it would not exist at all. And perfect knowledge of the world was impossible without knowledge of its cause. Because, for Aquinas, there was a limit to what could be known of God by reason alone, and a point where faith and revelation had to take over, physical knowledge was subordinate to metaphysical knowledge, and ultimately to faith.

Knowledge of nature was, however, granted a degree of autonomy. It was still possible to study natural objects without infringing on the theologians' treatment of those same objects as signs of God. Although there was no clear distinction between philosophy and the particular sciences, the subordination of metaphysics to theology did not necessarily entail an obstruction to the study of nature. Aquinas himself saw no finality in the solutions offered by either Aristotle or Ptolemy to the problems of planetary motion. Empirically based hypotheses were subject to revision; for, as he put it himself, the facts might be explained in another way as yet unknown to men. Because he made that statement when discussing the concentric spheres with which Aristotle had built the cosmos, he was evidently aware that authoritative world-pictures were not inviolable.

A third example, revealing yet another kind of subordination, takes us to the University of Paris where two fourteenth-century scholars, Jean Buridan and Nicole Oresme, considered the possibility that the earth may rotate on its axis. Oresme's analysis shows that, within the scholastic tradition, there was room for new ideas, even if they were no more than conjectures. To demonstrate the possibility of the earth's rotation, Oresme had to counter the common objections. That we see planets and stars rising and setting proved nothing since they could be relative motions. That the surface of a rotating earth would experience a great wind rested on the assumption that the air could not rotate with the earth. How, on a rotating earth, could an arrow shot upright return to the spot from which it was fired? With remarkable prescience, Oresme recognized that, if shot from a rotating earth, the arrow would have a motion compounded of two simultaneous motions. Because it would continue to share in the earth's rotating motion, even after leaving the bow, it would fall to its origin even if the earth were moving. Oresme even drew an analogy with what would happen on a moving ship, as Galileo was later to do. Some arguments actually favored the rotation. A spinning earth eliminated the excessive speed of rotation currently ascribed to the outer spheres. Indeed, the outermost, ninth sphere might be dispensed with altogether since it was only required to produce the daily motion of the planetary spheres and fixed stars — an economy that would be in keeping with a God who did nothing in vain.

What was the outcome? Did Oresme *propose* a rotating earth? Not at all. He clung to the traditional view, stating that the problem was scientifically indeterminate. Why would he not commit himself? Because the whole critical exercise was subordinate to a theological objective. His purpose was to demonstrate not that the earth did spin but that it was impossible for human reason to prove that it did not. And if reason and experience could not settle so basic a question as this, how much less could they be relied upon to arbitrate on matters of faith?

The relations between Christianity and science have indeed been complex, with religious motives acting as both catalyst and inhibitor, sometimes simultaneously. In this case a trenchant critique of Aristotelian orthodoxy was both motivated and (to our eyes) deflected by religious concerns. This subordination of criticism to a Christian theory of knowledge did not, however, preclude a degree of differentiation. The arguments invoked by Oresme to spin the earth had an integrity of their own. Some were later exploited by Copernicus in his attempt to establish that the earth both turns on its axis and revolves around the sun. The popular notion of a complete fusion of Christianity and Aristotle, obstructive to all criticism, cannot stand.

A critical impulse could also derive from the clash between Aristotelian dogmatism and a Christian emphasis on the omnipotence of God. Aristotle might affirm that it was impossible for another cosmos like our own to exist. But surely, if God were omnipotent, He could make as many as He liked? In 1277, the bishop of Paris, Etienne Tempier, had condemned no fewer than 219 propositions that he considered prejudicial to divine omnipotence. One was that God could not make several worlds. The fourteenth-century Franciscan William of Ockham went further in declaring it probable that God could make a *better* world than this. In his reasoning he referred to Augustine's assertion that God could make a perfect man who would not wish to sin. That God could create other worlds *identical* to ours was, for Ockham, beyond question, for God had the power to produce an infinite number of identical individuals.

From the fourteenth century onward there was undoubtedly a tradition in which Aristotle was continually being challenged. In the works of the fifteenth-century German theologian, Nicholas of Cusa, the universe was even deprived of an absolute center. The possibility of earthlike planets was considered, and he envisaged illustrious extraterrestrials close to the sun. On the moon might be lunatics! To affirm the possibility of plural worlds was not, of course, to affirm the reality. Although it became heretical to deny that God *could* have created other worlds, it was as dangerous to say that He *had*. Nevertheless the subordination of science to theology had not resulted in a sterile fusion. Many of the arguments later deployed in smashing the geostatic cosmos had already been forged.

Magic, Science, and Religion

To approach the scientific revolution with preconceived notions of what constitutes science and what constitutes religion may help in streamlining historical reconstruction. But to do so always runs the risk of missing the point in particular intellectual debates. Nowhere is this more apparent than in the ambivalent role played by the magical philosophies of the Renaissance. The displacement of an Aristotelian world-picture was not simply the result of some modernizing movement of the seventeenth century. A critical impulse and a dream of power over nature sprang in part from natural magic − a somewhat disparate constellation of beliefs and practices (including forms of alchemy and astrology) rooted in the conviction that there were hidden, *natural* powers in the cosmos that could be tapped and channeled to human advantage. Partly inspired by a set of texts known as the Hermetic Corpus, the elite magicians of the Renaissance rejected what they saw as stultifying in Aristotelian science: a neglect of the particular for the general and an indifference toward practical application. For the Elizabethan magus John Dee (1527−1608), it was as imperative to study the *specific* characteristics of everything in nature as it was to understand the controlling influence of the stars. The invisible powers of nature were analogous to the visible powers of the magnet, which could act at a distance and penetrate matter with its rays. The objection that the scientific revolution saw the separation of science from *both* religion *and* magic hardly clears the air, because religious beliefs played a role in the assault on magical practices, with hard-headed Protestants putting magic and Roman Catholic rituals in much the same category.

It is tempting to draw sharp contrasts between magic and science, just as it is between science and religion. It is sometimes said that whereas occult sciences, such as alchemy and astrology, were static in their theoretical content, modern science has been progressive. Whereas practitioners of the occult would explain everything by magic, modern science has recognized the limitations of its knowledge. Whereas the esoteric practices of the alchemist were insufficiently specified to be repeatable, modern science has been characterized by the repeatability of control experiments. Whereas there were entrenched, almost sacred, propositions within the web of magical belief, modern science keeps such entrenched assumptions to a minimum.

Using such stereotypes, one can then argue that the occult sciences of the Renaissance could contribute nothing to the new sciences of the seventeenth century. The problem is, however, that real history rarely conforms to later stereotypes. It is more appropriate to accept the view of D. P. Walker that magic was always on the point of turning into art, science, practical psychology, or, above all, religion. It is the fact that it could turn into science, and promote a beguiling vision of mastery over nature, that has forced historians to reconsider the usual antithesis between occult and rational mentalities. In the works of Paracelsus, for example, a new status was given to chemistry and medical practice was reformed, from within a magical philosophy that was intensely spiritual in orientation.

Magic and medicine were often in the same melting pot. To make the most of beneficial planetary influences had been a principal object of Marsilio Ficino (1433–99), one of the first scholars to give the practice of magic the semblance of respectability. Effective treatment required that the body be in a receptive state to beneficial forms of spirit, which could remedy deficiencies resulting from temperament or life-style. For those of melancholy or saturnine disposition, the means had to be found for inducing receptivity toward Jovian or solar influx. Music, verse, diet, and scent could all be used. The powers residing in herbs, stones, and aromas were both natural and divine. As a gift of divine origin, there was nothing sacrilegious in their use.

Figure II. 1. Illustration from Thomas Norton's *Ordinall of alchimy* (1652). The picture illustrates the transmission of the secrets of alchemy from master to student by an oral convention. In his account Norton claims to have arrived at the Elixir of Gold but to have had it stolen by a servant. He then turned to the Elixir of Life, only to have the magic formula stolen again, this time by a woman. Hence his resignation:

> Remembring the cost, the tyme, and the paine,
> Which I shulde have to begin againe,
> With heavie hearte farewell adieu said I,
> I will noe more of Alkimy.

Reproduced by courtesy of the Science Photo Library, London.

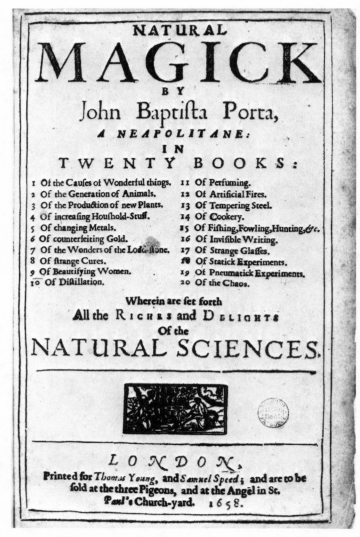

NATURAL

MAGICK

BY

John Baptiſta Porta,

A NEAPOLITANE:

IN

TWENTY BOOKS:

Wherein are ſet forth

All the Riches and Delights

Of the

NATURAL SCIENCES.

LONDON,

Printed for *Thomas Young,* and *Samuel Speed*; and are to be
ſold at the three Pigeons, and at the Angel in St.
Paul's Church-yard. 1658.

Figure II. 2. Title page from an English edition (1658) of J. B. Della Porta's *Natural magick,* first published in Latin in 1558. The list of contents for this popular compendium shows the wide range of practices discussed under the heading "natural magic": from domestic and craft skills to the conduct of new experimental investigations; from the display of nature's marvels to the imitation of natural processes. Reproduced by courtesy of The Bodleian Library, Oxford; Shelfmark Douce P Subt. 107.

That magic could be both proto-scientific and proto-religious is most transparent in the literature of the Paracelsians where a critique of traditional, Galenic medicine was developed within a magical philosophy that demanded an initiation into the secrets of chemistry. The task, assigned to humanity by God, was that of transforming, by chemical means, the raw products of nature into a state that made them serviceable. For Paracelsus, the religious dimension was paramount. The alchemist's work amounted to nothing less than the redemption of the physical world. As Christ had brought spiritual redemption to humanity, so humanity in turn must redeem the whole of nature, which, as a consequence of Adam's sin, was also fallen. Chemistry was given new status as the art of separating the pure from the impure. And the challenge was social as well as religious; for Paracelsus both exalted the laborer and cut across the educational divisions that conventionally separated physician from apothecary.

The theology of Paracelsus celebrated the mercy of a God who had granted the human mind sufficient illumination to cultivate nature and to extract those gifts necessary for subsistence. It was also a thoroughly Christocentric theology because it was through Christ that the light of grace had entered the world as the profoundest expression of divine mercy. Humankind, reborn into the life of Christ through baptism, was under obligation to continue Christ's redemptive and healing ministry. This association of the humanity of Christ with the purifying goals of the practical chemist underlines the artificiality of retrospective analyses that seek to place magic, science, and religion into separate compartments. The historical reality during the sixteenth and much of the seventeenth centuries was usually some kind of amalgam, such as that expressed by Martin Luther in his evaluation of alchemy:

The science of alchemy I like very well, and indeed, 'tis the philosophy of the ancients. I like it not only for the profits it brings in melting metals, in decocting, preparing, extracting, and distilling herbs . . .; I like it also for the sake of the allegory and secret signification, which is exceedingly fine, touching the resurrection of the dead at the last day.[2]

As the earthly alchemist purified through fire, leaving the dregs at the bottom of the furnace, so, at the Day of Judgment, the divine alchemist would separate all things through fire, the righteous from the ungodly.

Figure II. 3. Circular plate from Heinrich Khunrath's *Amphitheatrum sapientiae aeternae* (1598); "The oratorium." An intimacy between prayer and alchemical practice is suggested by this plate in which the "oratorium" (the place of prayer) stands opposite the "laboratorium" (the place of work). The musical instruments symbolize an underlying harmony behind nature's powers, to which the successful alchemist must himself be attuned. Reproduced by courtesy of The Bodleian Library, Oxford; Shelfmark R. 1. 9. Med.

When one reads a Paracelsian recipe for an amulet to ward off the plague, when one learns that the prime ingredient is a paste requiring eighteen dried toads, one could be forgiven for rejecting continuity between magic and science. Because there were discontinuities as well as continuities, it makes good sense to say that modern chemistry emerged through a dialectical process in which the enthusiasms of the Paracelsian chemists were tempered by their critics. Nor was it through magic alone that the impulse arose to gain power over nature. Recent research has shown that the chemical experiments of the magician were often in the contemplative mode: devised to bring out divine harmonies and correspondences between metals and planets. Conversely, to enter the chemical laboratory envisaged in 1606 by Andreas Libavius – a respectable Lutheran humanist, critic of the Paracelsians, and author of what has been described as the first chemical textbook – would be to enter an open house in which the products of the furnace were investigated with a sense of responsibility for the welfare of one's fellow men.

The magician, with his emphasis on private illumination, was not alone in seeking control of nature's powers. But this is not to dismiss his vision as an irrelevance in the forging of a nascent science. There was a degree of continuity in the sense that specific advances in understanding were achieved from within a magical framework. In rejecting the Galenic theory that diseases arose from an internal imbalance of the four humors, Paracelsus switched attention to agents invading the body from outside. These were envisaged as seedlike entities, introduced through air, food, and drink. Carriers of specific diseases, they invaded specific organs. Accordingly, Paracelsus developed the view that diseases are localized and require specific cures – not the elaborate mixtures commonly prescribed as cure-alls. The practice of blood-letting, associated with the Galenic theory that disease arose from an imbalance of the four humors, he duly rejected in favor of specific chemical remedies. Not that Paracelsus always found the right ones: One wonders how many unfortunates owed their demise to his prescription of mercury to clarify the spleen. Yet there clearly were affinities between the Paracelsian approach to disease and later models – not least in the insistence that a chemical cure should always be sought rather than accept the physicians' word that certain conditions would remain incurable. The analytical techniques, employed by the Paracelsian chemists to extract their active ingredients from otherwise dangerous chemicals, contributed to the excitement of a science that the young Robert Boyle would embrace with an almost religious fervor. In 1649 he confessed that he had been transported and bewitched by Vulcan.

There is a further reason why the image of a revolutionary break from both magic and religion will not do. Statements made in the seventeenth century about the desirability of separating science from religion have to be read against a background in which the excesses of an enchanted universe were straining credulity. When Francis Bacon, Boyle, and others warned against the mixing of science and religion, they were, in part, reacting against the worst excesses of Renaissance magic. Thus the Paracelsians, who boasted that their account of creation as a process of chemical separation was the *only* legitimate interpretation of Genesis, incurred Bacon's displeasure not only for their hijacking of the Bible, but also for the implication that knowledge of nature was in need of biblical support. Similarly, for Boyle, there was a sense in which matter had to be divested of spirits and other inherent powers in order to be seen in proper spiritual light. The critical point is that religious beliefs were often invoked to sharpen an attack on the more extravagant features of an enchanted universe.

At first sight, it may not be clear how this could be so. After all, it is possible to argue that the most influential magicians in Catholic countries were the clergy. Insofar as they condoned popular religious rituals, belief in the efficacy of relics, supplication to saints who would reinforce earthly petitions, and the use of images of the Virgin Mary to halt the rain or even to solve political problems, their sacred system could serve as a template for rival systems of magic. When Cornelius Agrippa (c. 1486–c. 1534) had discussed Christian prayers and ceremonies in relation to magic and pagan religions, he had regarded them all as examples of the same basic activity. How, then, would a learned Roman Catholic scholar discriminate between the religious practices of his Church, which were acceptable, and magical practices, which were not?

With difficulty – to judge from the efforts of the Catholic scholar Martin del Rio, who actually addressed the issue. He had to say that any effect produced by a talisman around the neck was to be ascribed to the devil, whereas effects due to Christian amulets came from God's goodness. Sacramental formulas in Christian worship were effective, not because any power resided in the words themselves, but because God had promised a certain effect only if such words were pronounced. In addition, del Rio rejected the theoretical basis on which various forms of magic were practiced, including the concept of planetary correspondences. It is difficult to escape the conclusion, however, that hard-headed Protestants found it easier to attack what they did not like in magic without compromising their religion.

One such was the sixteenth-century critic Johann Wier who,

in condemning every kind of magic, included in his denunciation those Catholic practices he considered superstitious: most forms of exorcism, the consecration of bells, and the use of relics or recitals from Scripture to cure disease. His Protestantism was an intensely private religion in which the effects of devotion were directed inwardly to produce change of heart or mental illumination. He still had to distinguish the miracles of Christ from the effects of magic, but this he did by stressing that the former were beneficial whereas the latter, especially in the case of witchcraft, could be detrimental.

Protestant critics, looking for a religion denuded of magic, would enlist the Bible on their side. Thus Thomas Erastus (1523–83) appealed to the first chapter of Genesis to demonstrate that God had created plants before planets. Consequently, the former could not owe their form to the latter, and so magical correspondences between them were eliminated. For Erastus, God, not the heavens, was the giver of form. It was He who had instructed the animals to multiply; He had given no instructions to the stars. The God of the Bible was not a God who had delegated control of the world to celestial intelligences or other "Platonic godlets" as Erastus called them. A biblical literalism could therefore assist in the disenchantment of nature. A particular brand of Protestant religion correlated with a science that was at once emerging from, and emancipating itself from, the organic and spirit-filled world of the magician. The refusal of Erastus to derive the forms of bodies from the stars, or from anything within nature, pointed toward the empiricism of Bacon, for whom the only guides to knowledge were experience and the Bible. Because the forms of bodies depended entirely on the will of God, it was impious, as well as impossible, to construct a science of nature based on preconceived opinion.

This association of empiricism with piety underlines the difficulty facing theses that affirm a separation of science from religion in the seventeenth century. It is, however, perfectly consistent with the view that the content and practice of both were differentiated and reassessed in new and portentous ways. Just as religious beliefs intruded in the critique of magic, so they intruded too in the presentation of new scientific concepts. In the next section we shall examine three examples in more detail.

The Theological Presentation of Scientific Innovation

The principal problem with the separation formula is that many significant innovations in seventeenth-century science were introduced with theological connections to the fore. Our three examples are J. B. Van Helmont's development of the Paracelsian theory of disease, Giordano Bruno's revolutionary claim for a plurality of worlds, and Descartes's presentation of a concept of linear inertia.

In Van Helmont's contribution to medical theory, it is possible to discern a positive debt to a devout belief in God. It is visible in his challenge to the academic learning of his day. He attacked the dominance of formal logic because it reflected a misplaced confidence in the powers of human reason. Nature had to be investigated, not subsumed under logical categories. His attitude was one of religious pragmatism. It was important to make available the hidden resources in nature that God had created for suffering humanity. There was even the suggestion that it was in the nature of God's goodness that there *would* be a cure for every disease, an optimism that was not always in the best interests of physicians to share.

Van Helmont envisaged the natural world as a system of seedlike entities, seminal spirits that penetrated matter giving it specific properties. In fact all matter had within it a divine spark of life. The word *gas,* which he coined, had its origins in this view of the world in which the animation of matter depended ultimately on God's creative role. A conception of nature in which matter was suffused with spirit allowed Van Helmont to articulate his contributions to physiology. His belief that specific principles were responsible for each phase of digestion lay behind his discovery of the action of acid in the stomach and alkali in the duodenum. In his analysis of disease, he presented a similar critique of the four-humor theory to that of Paracelsus, developing the view that diseases should be treated as specific disorders of specific organs requiring specific chemical remedies. A further contribution stemmed from his insistence on the individuality of organic function. A characteristic *archeus* – a kind of overseer – was postulated in each organ, responsible for organizing its chemical and metabolic activity. Failure of the archeus followed the intrusion of some alien ferment, leading to disease of the organ. With this conceptual framework, Van Helmont opened up neglected avenues of pathological research, and chemical cures were sought as part of a fully causal therapy.

The religious grounds of Van Helmont's philosophy became explicit in his attack on Aristotle. He could not accept that the goal or end of a process, the final cause, was the principal cause. With God there was no priority of causes. The ends for which the seeds, the natural causes of things, acted were not known to themselves but to God alone. "From a necessity of Christian philosophy," he wrote, "a final cause has no place in nature." Scriptural authority was a further resource in his assault on Aristotelian philosophy, for there was no reference in Genesis to the creation of fire, which he wished to see excised from the elements. What there was in Genesis was a clear statement concerning the Creator Spirit hovering over the waters, and this provided the model for Van Helmont's conception of matter, with water as its primary form.

The immediate reason for his assault on Aristotle was his hostility toward the Jesuits whose presence in Flanders, in the wake of the Spanish conquest, he vehemently resented. When antagonized by Jesuits proffering supernatural explanations for cures that he considered natural, he would retaliate with "let the divine inquire concerning God, but the naturalist concerning nature." With such remarks a differentiation between two forms of study was implied. But it would be misleading to speak of separation given the religious foundations of his natural philosophy.

Van Helmont's integration of science and religion was not exactly orthodox. His attack on a prominent Jesuit and his expansion of natural magic to embrace the allegedly miraculous powers of sacred relics led to denunciation by the Louvain medical faculty in 1623 and a subsequent appearance before the Spanish Inquisition. His heresies spelled prison. A similar pattern, though etched more deeply, may be seen in the case of Bruno. His infinite universe, with its infinite worlds, looks at first sight like the apotheosis of free thought – untrammeled by theological constraint. Yet, on closer inspection, the key to understanding his extrapolation of Copernican astronomy turns out to be an innovative theology that he took extremely seriously. To detach his innovations from the metaphysical theology in which they were embedded would be a violation of his integrity as a thinker. His theology was, however, deeply unorthodox and, as we saw in Chapter I, he paid the highest price for wishing to supplant contemporary Roman Catholicism with an alternative religion, which he believed was of older pedigree.

The unique features of Bruno's universe arose from an original blend of several philosophical traditions. He was attracted by the atomic theories of antiquity, which had themselves been associated with the possibility of plural worlds as different combinations of atoms passed in and out of being. He was attracted too by the reasoning of Nicholas of Cusa, who had imagined a universe with no center and in which space was homogeneous. The Copernican system, Bruno believed, fitted perfectly in such a universe and provided a model for other planetary systems extending to infinity. Space was unbounded; for, repeating the question of Lucretius, Bruno asked what would happen if one thrust one's hand through the supposed boundary. An infinite universe containing infinite worlds was philosophically the most coherent.

It was also theologically the most coherent. If divine omnipotence had been really displayed, the deity would have had to create those other worlds that Bruno's scholastic predecessors had acknowledged He *could* have (but actually had not) made. God's infinitude, so Bruno argued, could only be expressed by creating infinite worlds – worlds that were real, not hypothetical. It was not enough that God could have done what Aristotle had deemed impossible. The immensity and perfection of God required that it had been done. A physics of the infinite was the correlate of a theology of the infinite, however heterodox that theology was perceived to be. In effect, Bruno was arguing that divine attributes could be given physical meaning – as Newton was to do later when he reconstructed space in terms of God's omnipresence. Such transformations of metaphysical axioms into prescriptions for the natural world were extremely common in early modern science.

In Bruno's case, the transformation eventuated in an infinite universe. In the case of Descartes, it eventuated in laws of motion, one of which resembles the principle of linear inertia that was so basic to Newton's analysis of the relationship between force, mass, and acceleration. At first sight, Descartes may seem an unpromising example. Critical of the quest for final causes in nature, he was inclined to distinguish between revealed truths that were beyond the realm of intellectual inquiry and scientific truths that were not. He developed a mechanistic interpretation of nature so extensive in scope that, to all appearances, the physical world acquired a complete autonomy. His rationalist mentality is commonly seen as a secularizing force, despite the fact that one of his aims was to combat a pervasive skepticism.

Descartes's mechanical philosophy will be examined in detail in Chapter IV. For our present purpose, however, it is sufficient to note that he did not describe the physical world as autonomous. There may have been few spiritual insights in his system, but in his *Principles of philosophy* (1644) he did insist that bodies continue in existence only because God preserves them in being. And this had implications for their motion:

From the mere fact that God gave pieces of matter various movements at their first creation, and that He now preserves all this matter in being in the same way as He first created it, He must likewise always preserve in it the same quantity of motion.[3]

Because God is immutable, the bodies He created would remain in the same condition unless they were subject to external causes. His conclusion was that a moving object, so far as it can, goes on moving. That it goes on moving in a straight line, rather than in circles, followed from what Descartes described as the immutability and simplicity of the conserving operation. Motion was conserved in the precise form in which it occurred at the instant of its preservation. The instantaneousness of God's preservation underwrote the linearity of the inertia. What we may recognize as a scientific principle was enunciated via the theological concept of divine immutability. The point, of course, is not the validity but the form of the argument. A physical insight was presented in a context where science and theology were far from separate. Differentiation on certain levels did not prevent fusion on others.

Differentiation and the Difficulty of Separation

In this concluding section, we shall look at three quite different examples to see how difficult it was for scholars to achieve a total separation, even assuming they wanted it. Our first example, Robert Boyle, did not. But he was puzzled to find in the writings of Descartes that argument for the conservation of motion that we have just described. After all, Descartes had decreed that God's purposes in nature were not accessible to reason. Boyle felt there was an inconsistency. Unless one presumed to know God's purposes, how could one know that He might not achieve them by varying the quantity of motion? In common with Descartes, Boyle developed a mechanical philosophy in which the properties of bodies and the interactions between them were explained in terms of material particles and their motions. In common with Descartes, he visualized the universe in terms of clockwork rather than as a living organism. Like Descartes, he denied the autonomy of the clockwork: Divine preservation held it in being. In rejecting Descartes's theological argument for a physical principle, he did, however, extend the process of differentiation between the provinces of science and religion. Yet the differentiation, once again, was conducted by reference to theological as well as empirical principles.

Boyle's essays reveal an unprecedented series of distinctions, of the utmost importance in the promotion of natural philosophy. One finds a sharp differentiation between God and nature, between different types of knowledge and the degrees of probability to be accorded them. There was a clear distinction between the book of God's words and the book of His works. He distinguished between different kinds of final cause, some of which, like the purpose of the eye, *could* be identified. He also insisted that the belief that everything was made for human benefit should not intrude when assessing the merits of rival theories. In contrast to Van Helmont's involvement of God's power *within* matter through the agency of seeds, Boyle wished to divest nature of all inherent powers, making brute matter subservient to God's immediate will, controlled in its motions by laws externally imposed. In such a universe God's sovereignty could be celebrated.

But for that very reason, the differentiation achieved by Boyle cannot be equated with separation. The powers that others had invested in nature, Boyle placed in God. Forms of words that suggested, for example, that nature abhorred a vacuum were abhorrent to him, because they amounted to a deification of nature. In contrast to Van Helmont's deduction of a water–spirit ontology from Genesis, Boyle resisted a constitutive role for biblical texts. Chemistry was not to be conflated with theology, for God was the Author, not the

Spirit of creation. In contrast to Van Helmont's confusion of the two books, Boyle distinguished them but without setting them in opposition. The study of Scripture, he suggested, did nothing to hinder an inquisitive man's delight in the study of nature.

The study of nature had an integrity of its own, as one sought to uncover the laws that governed the clockwork. But there were still the strongest links with the defense of Christianity; for the excellence of the machinery could only confirm the excellence of its Maker. Organic models of nature had often been sustained by analogies between microcosm and macrocosm, between man as an organism and the rest of creation. In criticizing such analogies, Boyle again resorted to theological argument. He was adamant that man had been made in the image of God, not in the image of nature. Insofar as he separated science from religion, he – like Francis Bacon – was reacting against what he perceived to be excessive conflation in certain spiritualist philosophies. Among the Paracelsians there had been an emphasis on inner illumination, which could be perceived as a threat to established forms of religion. Such an emphasis on the inner light might also threaten the collaborative aspects of scientific research, which Boyle considered essential for the public good. His separation of science and religion was more accurately a differentiation in which theological arguments played a prominent role.

Our second example takes us back fifty years, to the second decade of the seventeenth century when Galileo was struggling with the question to which Boyle gave so easy an answer: whether the study of Scripture need hinder the study of nature. Galileo knew only too well that it could, if the province of biblical authority were not scrupulously defined. In his writings, a mutation occurred in the analogies customarily drawn between the two books. The search for signs of God in nature had often been based on the assumption that the two books had been written in essentially the same language. Galileo, however, achieved a telling differentiation when he argued that nature had a language all its own. The book of nature, he insisted, had been written in the language of mathematics. No amount of theologizing could be a substitute for mathematical analysis. In evaluating the Copernican system, for example, mathematical criteria should take precedence over interpretations of Scripture, which may have become normative but only through ignorance.

The differentiation went one stage further because Galileo found it necessary to distinguish two senses in which a biblical text could be construed. If it referred to the natural world, it could be construed as a statement in the language of commonsense observation, accommodated to the understanding of all men and independent of scientific theory. On the other hand, there would be a deeper meaning, which wise expositors had a duty to uncover for the benefit of those above the common herd. An advantage of the first point was that it allowed Galileo to say that the sacred scribes had intentionally refrained from imparting complex scientific knowledge, despite having it at their disposal. In this way, apparent contradictions between Copernican astronomy and biblical texts would be eliminated. If Scripture referred to the sun's motion, and even to its standing still during the miracle of Joshua's long day, this was merely the everyday language of common sense. It described the appearances but could not be prescriptive against true astronomical knowledge.

This theological argument for differentiation was to assume the greatest importance in spreading the Copernican theory. But it was not an argument for ultimate separation. In fact, Galileo's *Letter to the Grand Duchess Christina* (1615) shows how difficult it was for him to pursue such an argument, even had he wished. The irony is that he referred to the mutual relevance of science and religion in at least three different respects, each of which magnified, rather than resolved, his difficulties.

First, instead of being content to rest his case on the distinction between commonsense and scientific language, he insisted that certainties in science, having once been established, should be regarded as aids in biblical exegesis. And not merely aids, but the "most appropriate aids." The wise expositor, in short, had to be one such as Galileo, having a sound knowledge of the natural sciences – a somewhat startling implication for custodians of a traditional theological education.

The second difficulty arose from a proposition in St. Augustine in which Galileo saw some mileage. Demonstrated truths concerning physical matters could not contradict the Bible. Consequently, any teaching about nature that clearly went against Scripture could not be a demonstrated truth. In fact it could be declared false. Galileo was attracted by the implicit contrast between physical propositions that were demonstrated and those that were merely affirmed. As to the former, Galileo deduced, it was necessary for divines to show that they did not contradict the Scriptures. As for those "not rigorously demonstrated," anything contrary to the Bible involved by them was to be considered "undoubtedly false." The problem lay with this last inference; for it was not entirely consistent with Galileo's earlier argument for differentiation. Here, he *is* allowing the Bible a degree of jurisdiction over scientific statements, which, though merely affirmed in the first place, might after all turn out to be demonstrable. Galileo had his reasons; for he wished to say that it was up to his clerical opponents to prove that the Copernican system had *not* been demonstrated. But the irony is considerable. To those who believed that he had not rigorously demonstrated the earth's motion, it would look as if he had condemned himself.

As to demonstrated propositions, Galileo had written, it was the duty of wise divines to show that they did not contradict the Bible. The third respect in which Galileo's theologizing backfired arose from his taking that office upon himself. Not content to rest his case on the prudence with which the Holy Spirit had left astronomy out of the Scriptures, he proceeded to show that the miracle of Joshua's long day made more sense if the text was read from a Copernican standpoint. Joshua's command to the sun to stand still was best understood, according to Galileo, as a command to halt the sun's rotation on its axis — a rotation in which Galileo took special interest since it was required by his interpretation of the sunspots. Because the sun's rotation drove the planets around, the whole system would have come to a gentle halt for the duration of the miracle. By contrast, the sun could not have stopped, on the Ptolemaic interpretation, without violating the interconnected motions of the spheres.

Galileo's argument reflects neither an ultimate separation of science from religion nor the use of science to impugn the miraculous. The reference in Joshua 10:13 to the sun standing still "in the midst of the heavens" was surely more consonant, Galileo suggested, with the Copernican than with a geocentric system? His point was that, even if physical propositions were a matter of faith (which he wanted to deny), the Copernican system would have the edge. It was a clever but risky move. Moreover, to tie a biblical text, even conjecturally, to a particular scientific theory was the very tendency he had warned against. On one interpretation, Galileo was trying to protect his Church from condemning a theory that, by its evident truth, would prove an embarrassment if not embraced. If there is any truth in this, it only magnifies the irony, for it was the very zeal with which he executed his mission that would activate the machinery of condemnation.

Our third example concerns Newton's contemporary, Edmond Halley (c. 1656–1743), whose efforts to investigate the earth's history independently of Scriptural authority show how difficult it was to achieve a full separation even by the end of the seventeenth century. Even in a Protestant society, social and institutional pressures still tended toward integration. In fact it has often been said that Halley's reputation for atheism may have cost him the Savilian chair of astronomy in 1691–2. This was probably not the whole reason; for his levity and jocularity were well tuned for making enemies. The chair was, however, under the control of the archbishop of Canterbury, Tillotson, and the bishop of Worcester, Stillingfleet, both of whom were alive to the dangers of unbridled scientific speculation. So there were pressures that would have encouraged Halley to wear an orthodox face in public. These were intensified by the fact that he had been accused of believing that the world would continue to exist for eternity.

The striking point, however, is that instead of merely denying the charge, he had set out to vindicate himself by finding a *scientific* reason why the world could not be eternal. He chose to make a serious attempt at integrating scientific and religious perspectives, even though his clerical critics would almost certainly not have welcomed the intrusion of science into their domain. Despite pressures for differentiation, the result was the very opposite of separation. By adopting the view that the motion of the planets would be retarded by the ether through which they moved, Halley constructed a scenario in which the solar system would eventually collapse. The world would not be eternal.

The reason that has been given for saying that Halley's attempt was serious, rather than merely placatory, is that he confessed in public that his efforts had not, after all, produced decisive results. Despite his proposal that the earth was spiraling toward the sun, he could say that there was still no conclusive evidence derived from nature to prove that the earth had a beginning and would have an end. His argument may have turned out indecisively, but Halley evidently believed that scientific data were relevant to theological questions. In his pronouncements, science was no longer subordinate to theology in the way it had been for Roger Bacon, Thomas Aquinas, or Nicole Oresme. But it was not bereft of theological significance. His attempt to prove a point of theological orthodoxy by scientific argument may have heightened the suspicion in which he was held, but it hardly points to a successful separation of the two spheres.

I have argued in this chapter that conventional references to such a separation during the seventeenth century are defective in two respects. They imply an earlier fusion, when it is more accurate to speak of subordination. And they imply divorce when what was achieved during the seventeenth century was a differentiation often conducted on theological grounds. The question we have not yet addressed is the nature of the process whereby the natural sciences were released from their subordination. In the next chapter, we shall see whether insubordination in science was encouraged by insubordination in religion, whether the Protestant Reformation had contributed to the conditions in which a reformed philosophy of nature could flourish.

CHAPTER III

The Parallel between Scientific and Religious Reform

Introduction

Allied with Francis Bacon's proposal that a properly organized and properly conducted science could increase the power of the English state was his observation that a new vitality in the sciences had coincided with the Protestant Reformation:

When it pleased God to call the Church of Rome to account for their degenerate manners and ceremonies, and sundry doctrines obnoxious and framed to uphold the same abuses; at one and the same time it was ordained by the Divine Providence that there should attend withal a renovation and new spring of all other knowledges.[1]

This was, of course, a Protestant view of the matter. But it raises the question whether a celebration of the sciences might have been one means of differentiating Protestant cultural values from those of Roman Catholicism. And if – as many historians have argued – there were connections between Protestantism, capitalism, and the rise of science, we have to ask what form these connections took. If the sciences were less subordinate to theology at the time of Newton than at the time of Copernicus, had the Reformation in religion created favorable conditions for a reformation in science?

The possibility of a parallel between scientific and religious reform is the principal theme of this chapter. We shall also be concerned with the problems that arise in testing a historical hypothesis of this kind. One immediate difficulty is that the weight attached to it by different historians has varied enormously. At one extreme are those who would say that Protestantism was more propitious for science than Catholicism only in the sense that it was less obstructive. They would stress the doctrinal fluidity within some, if not all, Protestant movements and the absence of a mechanism of censorship as centralized or as effective as that supervised by the Holy Office in Rome. At the other extreme are those who have argued that specific Protestant doctrines gave a direct and positive stimulus to scientific research. The doctrine of the priesthood of all believers, for example, might have encouraged independence of thought on the interpretation of nature as in the interpretation of Scripture. Among many intermediate positions would be the view that puritan ideals resulted in a higher value being placed on certain forms of science, notably those promising practical benefits, as in agricultural and medical reform.

If Protestantism was more conducive than Catholicism to the expansion of science, one would expect this to manifest itself in a greater receptivity toward new and controversial ideas. This suggests that a good test would be to take a potent scientific innovation and compare its reception in Protestant and Catholic communities. It is suggestive, for example, that in Copernicus's native Poland, a sun-centered cosmology was taught in Protestant schools many years before it was taught to Catholics. Galileo was humiliated in Rome in a manner that has no exact parallel in Protestant countries. Accordingly, we shall begin with the Copernican innovation as a test case. This has the advantage that we can examine two major issues at the same time: the extent to which the Copernican revolution destroyed hallowed perceptions of humanity's place in the universe, and the extent to which Protestant audiences were more receptive. The results of the test are instructive, but as much for the complications they reveal as for any neat conclusion.

The Challenge of a Reformed Astronomy

There can be no doubt that the heliocentric astronomy of Copernicus did pose a challenge to a cosmic geography that Christian theologians had largely taken for granted. The voice of reaction was heard well into the seventeenth century. Here is an English schoolmaster and divine, Alexander Ross:

Is it possible that the world should last for about five thousand years together, and yet the inhabitants of it be so dull and stupid,

as to be unacquainted with its motion? Nay, shall we think that
these excellent men, whom the Holy Ghost made use of in the
penning of Scripture, who were extraordinarily inspired with
supernatural truths, should notwithstanding be so grossly ignorant
of so common a matter as this?[2]

The answer, of course, was that the motion of the earth
was not a common matter – that it required considerable
proficiency in mathematics to appreciate the arguments in its
favor. It also required a feat of imagination that was exceedingly difficult for those imbued with Aristotelian physics. On
the Copernican model, the earth was placed among the planets.
Yet this move was inconceivable on the basis of an Aristotelian cosmology with its fundamental division between the
sublunar and superlunary regions. Defenders of a traditional
cosmology simply could not allow that the earth was made
from the same materials as the planets, for the latter were
composed of a fifth element not found in the corruptible,
sublunar world. To adopt the Copernican system entailed a
rethinking on matters of biblical interpretation, but it also
involved a reconsideration of what was meant by terms like
corruptibility. Here is Ross again, in one of his outbursts
against the Copernican John Wilkins:

The heavens (you say) are subject to that general corruption in
which all creatures shall be involved in the last day. But you cannot
tell us what that corruption shall be, and so you speak at random;
you do not mean (I hope) that the heavens shall be involved in the
same corruption with snakes, rats, toads.[3]

Ross protested that the earth was "no wandring star except
in the wandring heads of Galileans." His reasoning shows that
the incommensurability between the two systems extended
to such theological issues as the location of heaven and hell.
An attraction of the older cosmology was that heaven and
hell were as far apart as possible – the former beyond the
outermost sphere, the latter in the bowels of the earth. To
set the earth revolving around the sun was to set all hell on
the move. The analogy between earth and planets also raised
delicate questions about their habitability, creating discomfi-
ture on such doctrinal matters as whether extraterrestrial beings
might not be in a similar spiritual state to humankind, and
whether God's Incarnation in Christ was truly unique.

The incommensurability of the two worlds has been well
expressed by C. S. Lewis with the observation that, in the

pre-Copernican universe, as man looked upward, he looked *inward* to a harmonious and animated world in which all celestial movement derived from the spheres, and the motion of the spheres from an inner drive on the part of the *primum mobile* to share in the perfection of God. By contrast, in the post-Copernican universe, man looked *outward* on the night sky and, if he was Pascal, expressed terror at the eternal silence of infinite space. To look into the universe of Dante was like being conducted through an immense cathedral. To look out upon the infinite worlds of Bruno was to be lost on a shoreless sea.

Yet it is possible to exaggerate the implications of the new cosmology. The notion that humanity was dethroned from a central place in creation requires careful qualification. Moreover, it was possible to reconstruct a universe in which Christian doctrines remained intact, even if the imagery through which they had been expressed was gradually discarded. During a long process of assimilation, the means were found for reinvesting the physical world with spiritual significance. One such arose from a technicality associated with the Copernican system. If the earth were in orbit, it would surely be possible to detect stellar parallax – a change in the perceived position of one star relative to another as a consequence of the earth's motion through space. Failure to observe this effect implied that the diameter of the earth's orbit must be negligible compared with stellar distances. Accordingly, there was a sense in which humanity remained close to the center, if not at the center itself. Such reasoning assumed the centrality of the sun. But this was an assumption that Kepler, for one, was happy to make. In Kepler's interpretation there was even a sense in which the earth remained central: Just as the sun had occupied the central orbit in the sequence radiating out from earth, so now the earth enjoyed that same centrality with respect to orbits around the sun.

Even if men and women were removed from the center of the cosmos, this was not necessarily to diminish their status. The center of the geocentric cosmos had not been salubrious. It was the point to which earthly matter fell, the focus of change and impurity, the physical correlate of humanity's fallen state. To be placed on a planet was to move upmarket. It was to be delivered from a dump that was, in reality, diabolocentric. Galileo was certainly conscious of this, rejoicing that there was an escape from the refuse. Kepler, too, spoke of an en-

Figure III. 1. An engraving, dating from 1493, showing the cosmography of a geocentric world system. Here one sees a typical synthesis of physical and theological motifs in a pre-Copernican cosmos. The sublunary spheres corresponding to the four elements are succeeded in the superlunary region by the concentric spheres of the sun and planets. Presiding over the cosmic hierarchy are God and a corresponding hierarchy of angels. Reproduced by courtesy of J-L Charmet/Science Photo Library, London.

net, in quo terram cum orbe lunari tanquam epicyclo contineri
diximus . Quinto loco Venus nono menfe reducitur. Sextum
deniq; locum Mercurius tenet, octuaginta dierum fpacio circũ
currens. In medio uero omnium refidet Sol. Quis enim in hoc

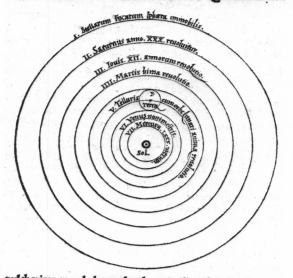

pulcherimo templo lampadem hanc in alio uel meliori loco po
neret, quàm unde totum fimul pofsit illuminare? Siquidem non
inepte quidam lucernam mundi, alij mentem, alij rectorem uo-
cant. Trimegiftus uifibilem Deum, Sophoclis Electra intuentẽ
omnia. Ita profecto tanquam in folio re gali Sol refidens circum
agentem gubernat Aftrorum familiam. Tellus quoq; minime
fraudatur lunari minifterio, fed ut Ariftoteles de animalibus
ait, maximã Luna cũ terra cognatione habet. Concipit interea à
Sole terra, & impregnatur annuo partu. Inuenimus igitur fub
hac

Figure III. 2. Diagram from Nicolas Copernicus's *De revolutionibus orbium
coelestium* (1543), folio 9 verso. There were highly technical reasons why
Copernicus believed that a heliostatic model for the universe would ease
the task of predicting planetary positions. He also stressed the elegance of
a system in which orbital periods increased in sequence according to the
distances of planets from the sun. The daily rotation of the earth on its axis
had the advantage that it eliminated the need for the huge outermost sphere
to revolve every twenty-four hours. In the text adjacent to the diagram,
Copernicus justified his innovation by invoking ancient authorities (including Hermes Trismegistus) and by stressing that the center was the most
fitting place for the lamp of the universe. Reproduced by courtesy of The
Bodleian Library, Oxford; Shelfmark 3 Delta 426.

hanced status for the earth: At last it enjoyed legal citizenship in the heavens. Not surprisingly, John Wilkins was to say that a prevalent objection to the Copernican system was not man's dethronement but an elevation above his true station. And it was an elevation actually experienced. Among the Copernicans there was exhilaration at the thought that man, in his astronomical understanding, had now surpassed the ancients.

Need human significance depend on cosmic location? Was not a sense of religious destiny derived from sources that were essentially insulated from the effects of interchanging sun and earth? Was mankind not made in God's image? Was there not the assurance that God had demonstrated a concern for humanity, irrespective of spatial coordinates? It is difficult to believe that the Copernican innovation did not encourage questions of this kind. But it hardly prevented conciliatory answers. In the long run, one may see the seeds of a secularized universe. But in the short run, as John Donne observed, men lived and believed just as they had before.

The universe had, after all, been of unimaginable size before. The smallness of the earth was arguably felt more vividly in the medieval than the modern system, for there was then an absolute standard of comparison – the immense size of the outermost sphere. A hundred years before Copernicus published, Nicholas of Cusa had argued that man's distinctiveness consisted not in his location but in his ability to cultivate a state of learned ignorance. True wisdom recognized the limitations of knowledge. The mental powers, which made that spiritual quest possible, were the fount of human uniqueness – not a cosmic position. In such a manner it was possible to disconnect spiritual significance from spatial coordinates. Having contemplated the relativity of all motions within the universe, Nicholas would have been astonished by any suggestion that a moving earth was destructive of a Christian world-view.

Even the problems posed by a plurality of worlds were negotiable. John Wilkins insisted on the distinction between a plurality of worlds within one universe and a plurality of distinct universes. Although the latter would be prejudicial to a Christian monotheism, the former would not. The silence of Scripture on other worlds simply became a license for their admission, in that they were not explicitly denied. As for the doctrinal impediments, Wilkins pointed out that intelligent inhabitants need not be like humans. Even if they were, they had not necessarily fallen from grace. And even if they had, who could say that Christ had not died for them also? It is a mistake to minimize the impact of the new cosmology, but resources were available to protect a spiritual destiny.

Our question was whether it was easier for Protestant than for Catholic scholars to embrace the new system. In seeking an answer, three points emerge. First, there are too many complications to allow the conclusion that individuals were more or less likely to be receptive according to whether they were Protestant or Catholic. Second, if they were sympathetic, they were more likely to enjoy freedom in publicizing their science if they were Protestants. Third, the fortunes of the new cosmology were more deeply affected by the antagonism *between* Catholic and Reformed Christianity than by the doctrinal peculiarities of either. In this there was an indirect effect of religion on science.

Receptivity toward a Reformed Astronomy: Testing a Hypothesis

To compare receptivity toward the Copernican innovation as a means of testing the hypothesis that Protestantism created a more favorable climate for science leads to few definite conclusions. But the reasons for this are instructive because they illustrate difficulties that arise in analyzing attitudes to almost any new scientific theory. The first concerns the pretensions of the theory itself. Is it envisaged as a calculating device, a mathematical model geared only to prediction? Or does it purport to describe the way the world is?

This distinction is basic to an understanding of what was revolutionary in the Copernican achievement. Whereas the role of the astronomer within the Ptolemaic tradition had been custodian of mathematical, predictive devices, Copernicus transformed that role by importing physical arguments into the astronomer's domain, thereby raising questions about the real motion of the earth. Because hypotheses, construed as provisional descriptions of reality, had occupied a place in physics, Copernicus, Kepler, and Galileo transgressed traditional subject boundaries by allowing an astronomical hypothesis this representational role. To say of a particular astronomer or churchman that he was prepared to accept the Copernican system as a *hypothesis* is therefore to say something deeply ambiguous. One could accept the mathematical models of Copernican astronomy without even considering whether the earth really moves. Or one could treat the hypothesis as a provisional truth-claim. The nice complication then arises that to entertain the Copernican system seriously as a potentially true physical description, and subsequently to *reject* it, could be a more radical position than to *accept* it in the former sense.

At the Lutheran University of Wittenberg, parts of the Copernican system were accepted as improved tools for predicting the angular position of planets. But with the exception of Georg Joachim Rheticus (1514–74), who had been seduced by the harmony that an exchange of earth and sun could bring to the sequence of planetary periods, the Lutheran circle inspired by Philip Melanchthon (1497–1560) played down the cosmological aspects of *De revolutionibus* no less than Catholic observers. Not until the 1570s was this emphasis called into question by a new generation of astronomers, including Tycho Brahe and Kepler's teacher Michael Maestlin (1550–1631). Through their teaching and their textbooks, the Wittenberg astronomers did at least promote Copernican mathematics, but it would be wrong to portray them as receptive to a new cosmology. Much the same distinction was to assume importance in the Galileo affair, for when Niccolo Riccardi wrote to the Florentine inquisitor about the license for Galileo's *Dialogue,* he indicated that the pope, Urban VIII, would only allow "mathematical considerations of the Copernican positions."

Between 1543 and 1600 only ten Copernicans have been identified in the strong sense of having advocated the earth's physical motion. Seven were Protestants, three Catholics. It is difficult to draw any hard conclusions from this, especially when one of the Catholics, Diego de Zuñiga, who taught philosophy and theology at the University of Salamanca, adopted the Copernican system in the context of expounding Job 9:6. The text referred to the power of God in shaking the earth "out of her place." In contrast to scholars, both Catholic and Protestant, who developed the concept of biblical accommodation, de Zuñiga saw a literal congruence with what Copernicus had proposed.

A further complication is that included in the Copernican theory were several proposals. One could therefore accept some and reject others. An attractive option was to accept the reality of the earth's axial rotation, thereby avoiding the cumbersome daily rotation of the outermost sphere, but to dismiss its orbital revolution, thereby saving its central position. Such a model is to be found in the work of Nicolai Reymers Baer, official mathematician to the Holy Roman Emperor Rudolf II. Baer, better known as Ursus, claimed originality in 1588 for a scheme in which planets other than the earth revolved about the sun, while the earth merely spun on its axis. He would not confer other motions on the earth, as Copernicus had, because this would violate a cherished Aristotelian principle that bodies could not have more than one natural motion.

The idea that planetary orbits were centered on the sun — which, with its attendant planets, revolved about a stationary earth — was also formulated in 1588 by the Danish astronomer Tycho Brahe, with the consequence that claims and counterclaims of plagiarism were elaborated by both parties. Ursus differentiated his scheme from that of Tycho by insisting on the daily rotation of the earth. Tycho accused Ursus of stealing his idea that the sun revolved about the earth, carrying the planets with it. For our purposes, the important point is that variants of these intermediate or compromise schemes were propounded by both Protestant and Catholic scholars. Tycho was a Protestant, beholden only to the king of Denmark, Frederick II, who had granted him free use of a small island to pursue his astronomical studies. But the retention of a central earth, coupled with the mathematical advantages to be gained by having other planets orbiting the sun, commended his system to Jesuit astronomers who, by 1620, were making it their own. In rejecting the full Copernican scheme, Tycho had been swayed by biblical considerations and by the failure to detect stellar parallax, which, if Copernicus were right, would entail an enormous gap between the outermost planet and the closest of the stars — a gap that he, and Catholic scholars too, found unacceptable on aesthetic grounds.

The viability of the Tychonic system makes it impossible to divide the intellectual world of the early seventeenth century into pro- and anti-Copernicans, still less to countenance the remark of one historian that the English Civil War was fought between Copernicans and Ptolemaics. The multiplicity of mathematical models did, however, have consequences for the perception of astronomy as a science. It could reinforce that conventional distinction between astronomy and natural philosophy, which allowed one to be skeptical concerning any claims for physical representation that the mathematical astronomer might make. This type of skepticism had been sustained by the knowledge that the mathematical devices used by Ptolemy could not be squared with the homocentric spheres of Aristotelian cosmology. And, as with Ursus, it was a convenient posture when disparaging other people's systems. By the end of the sixteenth century, however, this rigid distinction between astronomy and natural philosophy was beginning to collapse, until with Kepler the quest for physical causes had become part of the astronomer's task.

The shift that Kepler represents was not merely the collapse of a traditional boundary between two academic disciplines. The suggestion that an astronomical innovation might require a new physics was subversive of a hierarachy in which physics had enjoyed greater prestige than astronomy. This subversion was largely achieved by astronomers enjoying the freedom of court patronage, rather than by those in university posts where traditional boundaries acted as constraints. In Kepler's defense of Copernicus, mathematical, physical, and theological considerations all carried weight. The orbital motion of the earth provided a causal explanation for why the planets *appeared* to meander across the sky. If the astronomer could display the harmonious motions underlying the appearances, he surely deserved gratitude for revealing the divinely ordained harmony.

Kepler knew that the argument for skepticism was persuasive. If the same set of appearances could be saved by each mathematical model, then why should one be privileged as that which corresponded to physical reality? The history of astronomy also disclosed a succession of hypotheses, which made claims for physical truth look precarious. In reply, Kepler argued that although several different systems might appear to be equally powerful in predicting a specific range of phenomena, each would nevertheless be distinctive in generating further predictions peculiar to itself. These could then serve as the basis for further testing, allowing the best candidate for a true representation to emerge. He also argued that, despite the manner in which hypotheses were supplanted by others, astronomers could, and had, reached agreement on certain fundamentals. There had been progress.

Were such arguments for astronomical realism the exclusive property of Protestants? Again, there are too many complications to permit a straightforward answer. Kepler's theology was Protestant; but it was also idiosyncratic, in that the doctrine of the Trinity was given physical meaning in the structure of the cosmos. The central sun symbolized God the Father; the surface of the spherical universe, God the Son; the intermediate space, God the Holy Spirit. In this respect, a heliocentric cosmos was a direct expression of a Trinitarian theology. But the realism to which Kepler's arguments pointed was also embraced by Galileo, who – for all his anticlericalism – at least claimed to be a devout Catholic. To compound the difficulty, a skepticism concerning the propriety of scientific realism on astronomical issues continued to be fashionable among radical Protestants as well as conservative Catholics:

> Whether the sun predominant in heaven
> Rise on the earth or earth rise on the sun . . .
> Solicit not thy thoughts with matters hid:
> Leave them to God above, him serve and fear.[4]

So wrote John Milton in *Paradise lost*.

Applying our test has generated complications rather than answers. In another respect the test proves defective. What if there were Protestants campaigning vigorously for the empirical sciences who nevertheless rejected the Copernican theory? They would be disqualified by the test and yet might still testify to some correlation between Protestantism and science. Peter Ramus, the sixteenth-century humanist and champion of educational reform in the French universities, would be one example. Whereas contemporary French Catholic institutions devoted only one-sixth of their curriculum to what might be termed science, Ramus allocated a half, praising the value of the mechanical arts. Committed to an ideal of "astronomy without hypotheses," he nevertheless chided Copernicus for having employed the "most absurd fable in order to demonstrate the true facts of nature from false causes."

In his *Apologie or declaration of the power and providence of God* (1627), the Oxford scholar George Hakewill insisted that

there were no theological grounds for supposing that his contemporaries were any less capable of lasting achievement than
the ancients. There was no reason why God should not have
reserved opportunities for fresh discoveries to be enjoyed by
a later generation. Here was a Protestant vision amenable to
an emerging concept of scientific progress. But in that progress, the Copernican theory had no part; for Hakewill rejected a moving earth and showed no inclination to disband
the celestial spheres. Even more striking is the example provided by Francis Bacon, that most ardent apologist for useful
science – but not for a Copernican universe. Although he
wanted a physical astronomy that would deal with real motions, there were too many "great inconveniences" in a system that saddled the earth with three separate motions: an
orbital revolution, an axial rotation, and a change in the orientation of the axis itself. Bacon described the triple motion
as "incommodious" and protested at so radiant and exalted a
body as the sun having to suffer the indignity of immobility.

Any hope that responses to the Copernican innovation might
prove a sensitive indicator of Protestant and Catholic attitudes to the value of natural philosophy is clearly frustrated.
It does not follow, however, that the division between Catholic and Reformed societies and the antagonism between them
were irrelevant to the fortunes of the Copernican system. On
the contrary, there were numerous indirect effects of the
rupture within Christendom that are vital to an understanding of Galileo's predicament.

Reformation and Counter-Reformation: Indirect Effects of Religion on Science

The great painter Albrecht Dürer rejoiced that Luther had
delivered him from a "terrible distress," from bondage to rituals that the Catholic clergy prescribed as necessary for salvation. The liberating message of Luther's theology was that
salvation could be obtained only by faith, not by works –
faith in a God whose love for humanity had been demonstrated in the person of Christ. This Christocentric theology
not only kindled the renewal of a Christian spirituality. By
implying that salvation was the responsibility of the individual, by insisting that the only mediator between God and man
was Jesus Christ, Luther had posed a political as well as doctrinal threat to the power of the Catholic clergy. His vehe-

ment denial that God's favor could be earned through the sacraments, or bought by donations to an often grasping priesthood, had set up vibrations through Christendom, eventually winning him the protection of lay powers having a vested interest in a deflation of the papacy. By writing his tracts in German, Luther had appealed to an unlearned audience as well as to sympathetic scholars. A million copies are estimated to have been in circulation by 1524. An affirmative response meant the repudiation of one's religious heritage, which happened time and again as monks who had sworn obedience to the pope became his mortal enemies, as nuns who had taken vows of celibacy entered matrimony, and as priests denounced the Catholic Mass as an abomination.

Calvin's theology, no less than Luther's, illustrates that same capacity within Christianity for self-criticism and renewal. God's sovereignty was reaffirmed in a manner that made salvation by works untenable. Not that good works were unimportant. They were a Christian's duty in the disciplined life of his church. But so great was the gulf between the righteousness of God and the depravity of man that the hope of salvation, made possible by the obedient and sacrificial death of Christ, was ultimately dependent upon divine rather than human initiatives. Calvin's doctrine of election had the consequence that no earthly priesthood could arrogate the authority to determine whether an individual was of the elect. Christ alone knew his own. Any claim of the clergy to have control of one's spiritual destiny was effectively undermined. The Christian took the sacraments in the hope that, as he received Christ, so Christ would receive him. But those who claimed to know that they were chosen, Calvin viewed with suspicion.

In the doctrinal disputes, power struggles, and eventual warfare unleashed by the Reformation were forces that had a profound bearing on how a controversial scientific innovation, such as that of Copernicus, would be perceived. Its initial presentation indirectly reflected the growing polarities within Christendom. In an anonymous introduction, the editor of De revolutionibus, Andreas Osiander, had implied that the earth's motion was to be construed as nothing more than a convenient hypothesis. This intervention has often been seen as a nasty piece of clerical obstruction, the anonymity of Osiander's remarks compounding the crime by deceiving the reader into believing they were by Copernicus himself.

Recent scholarship suggests that the tactic of anonymity may have been employed for the best of reasons. By the early 1540s, Osiander had become notorious for his antipapal blasts, issuing from the church of St. Lorenz in Nuremburg. To have placed his name on the work of a loyal Catholic canon would have cast suspicion on Copernicus himself. The printing of *De revolutionibus* took place during a tense period when the freedom of Protestants and the independence of Nuremburg were being threatened by Catholic authorities. Every precaution had to be taken to ensure that the book would be read, not burned.

The dialectics of Reformation and Counter-Reformation had another indirect effect in that, during the second half of the sixteenth century, attitudes toward biblical interpretation hardened. A fundamental issue was whether religious authority was ultimately vested in an ecclesiastical succession or in the Bible alone. The consequence was that intransigent positions were soon staked out on the correct modes of biblical interpretation. Within Protestantism a shift occurred — away from the exegetical principles of the Reformers themselves toward a more legalistic attitude. The respect in which this shift affected attitudes toward the Copernican system may not be immediately obvious because Luther and Calvin had both made disparaging remarks about the new astronomy. Whether Luther really referred to Copernicus as a fool has been doubted, but in an off-the-cuff dismissal he remembered that Joshua had told the sun, not the earth, to stand still. Superficially at least, Calvin was even more disparaging. In a sermon on 1 Corinthians 10:19–24, he rebuked "those dreamers who have a spirit of bitterness and contradiction, who reprove everything and pervert the order of nature." There are some, he continued, "so deranged, not only in religion but who in all things reveal their monstrous nature, that they will say that the sun does not move, and that it is the earth which shifts and turns."

Calvin's target, however, was those who, in a spirit of contrariety, argued out of "pure malice." That he could even think of illustrating that perversity with purveyors of a contrary cosmology suggests that his opposition was born of the common conviction that a moving earth violated common sense. There is no reason to suppose that he was sufficiently interested in the Copernican system to set biblical authority against it. The important point is not whether Luther and Calvin happened to make peremptory remarks, exuding a lifetime's confidence in a pre-Copernican cosmology, but whether their exegetical principles implied an inevitable clash as the new system gained in plausibility. Calvin's did not. In his commentary on Psalm 136:7 he observed that the Holy Spirit "had no intention to teach astronomy" but had used popular language "that none might shelter himself under the pretext of obscurity." Nor was Luther tied to the verbal inspiration of Scripture in any way that demanded a strict literalism when interpreting verses that referred to the natural world. It has been well said of the Reformers themselves that the Bible was the sole norm and guide in matters of faith and conduct, not in everything under the sun.

During the later years of the sixteenth century, however, a Protestant scholasticism came into existence, perpetrating a less flexible approach to the meaning of God's "word." Through the systematization of Reformed doctrine, and through polemics with Rome, the Bible became a set of proof texts. Thus Philip Melanchthon, who took charge of Luther's educational reforms, was quite prepared to marshal the Scriptures against a plurality of worlds. To imagine other worlds was to imagine that Christ might have died and been resurrected more than once, or that their inhabitants might be restored to eternal life without a knowledge of the Son of God. But Genesis delivered the coup de grace, for it was stated that God had rested on the seventh day, not that he had immediately started work on other worlds.

It must not be supposed, therefore, that the desire of Protestants to dissociate themselves from Catholic Christianity automatically created a disposition in favor of free thought. And just as Protestant attitudes were shaped by the engagement with Rome, so the Catholic Church itself was deeply affected by the need to repel the Protestant heresies. The facility with which Protestants interpreted the Bible for themselves – opening, it seemed, a Pandora's box of subversive opinions – led to a declaration at the Council of Trent that was to create particular problems for Galileo, seventy years later. At its fourth session in April 1546, it was decreed that no one should dare interpret the Scriptures "contrary to the unanimous consensus of the Fathers." This ruling did not demand a strict literalism, but since the fathers and later commentators had interpreted the relevant passages of Scripture in geostatic terms, Cardinal Bellarmine would later argue that it was imprudent to abandon that consensus. Galileo would protest that the earth's motion was not a matter of faith. But Bellarmine had a reply at the ready: "For if it is not a matter of faith from the point of view of the subject matter, it is on the part of those who have spoken." Despite expressing a willingness to reconstrue Scripture if the earth's motion were proved, Bellarmine would not relent.

We have already seen how Galileo argued, on theological grounds, for the differentiation of scientific and theological propositions, and how he created difficulties for himself in the process. In order to complete his case he had to challenge this deference to a patristic consensus — which he did by observing that the church fathers had taken the motion of the sun for granted because the alternative had not, in that era, been laid before them. In which case they could not be said to have *decided* the issue for themselves, let alone for others. One may perhaps speculate on other respects in which Galileo's predicament was affected by the dynamics of Protestant –Catholic engagement. An argument for clemency toward him could be grounded in the belief that it would be a tactical error to declare Copernican doctrines heretical, because that might discourage Protestants from returning to the Roman fold. On the other hand, the fact that Protestant heretics were associated with the new astronomy could intensify the suspicion in which it was held. Galileo's association with Paolo Sarpi who, as leader of a Venetian revolt against the Roman Curia, was perceived as a crypto-Protestant could not have helped his case.

Part of the justification for censorship in the first place stemmed from competing conceptions of the priesthood. Against Protestant tendencies to emphasize the priesthood of all believers, Catholic scholars would reaffirm, as Bellarmine did, that many believers had neither the spiritual capacity nor the mental endowment to protect themselves from literature that might be damaging to faith and prospects for eternity. As for Galileo himself, the fateful urgency with which he mobilized the earth could reflect a genuine desire to uphold the reputation of Catholic scholarship. It would not do, he once wrote, for others to think that there were no Catholics capable of appreciating the Copernican arguments. He may even have considered that to establish the superiority of the Copernican over the Tychonic system would be to reaffirm the superiority of Catholic over Protestant science. He certainly never missed an opportunity to underline the Catholicism of Copernicus or to exaggerate his connections with Rome.

It is important not to underestimate these indirect effects of religion on science. The defensive measures taken by the Catholic church against what was perceived as a Protestant cancer altered the criteria of truth, allowing authority on scientific issues to be wrested from scholars and vested in a Roman bureaucracy. In condemning Galileo, Pope Urban VIII was condemning a former friend. That he was prepared to do so may reflect yet another aspect of the Catholic–Protestant contest. Urban's pro-French foreign policy during the Thirty Years War had left him in an exposed position when Richelieu joined forces with Protestant Sweden to thwart the restoration of Catholicism in Germany. Instead of condemning the alliance, Urban insisted that the war was one of state, not of religion, inviting the criticism that he was putting political expediency before religious necessity. It is not impossible that his severity on Galileo was one way of saving face – an attempt to placate prominent Jesuits still smarting from wounds that Galileo had inflicted on their colleagues in earlier debates about the nature of sunspots and comets. Relationships had been soured by Galileo's protracted debate with Horatio Grassi, the leading mathematician of the Jesuit College in Rome, who had followed Tycho Brahe in treating comets as real, self-luminous bodies located beyond the moon – in contrast to Galileo's theory that they were merely optical effects, produced by the reflection of sunlight in thin vapors. And there had been a long-standing feud with Christopher Scheiner, who had incensed Galileo by borrowing *his* telescopic discoveries for his own ends. In order to bolster his suggestion that sunspots were not (as Galileo affirmed) blemishes on the sun but small satellites in orbit around it, Scheiner had exploited Galileo's own discovery of the satellites of Jupiter. If a planet could have its satellites, why not the sun! Scheiner was still in the wings at the time of the trial.

The Question of Intellectual Freedom: Galileo and Wilkins

The indirect effects of religious controversy on attitudes toward science have been stressed because they indicate the contingency of those attitudes at particular times and places. One cannot generalize about a greater receptivity among Protestant clergy toward the new cosmology. As late as 1679, Lutheran orthodoxy in Sweden could still be obstructive. Biblical objections were wheeled out by the theological faculty at Uppsala against a young scholar, Nils Celsius, who had tried to minimize them. In Denmark, too, Lutheran clergy found it difficult to adapt. In 1661, when C. N. Lesle published on the mobility of the earth, he was considered by fellow clergy to be guilty of a dangerous departure. A resilient alliance between Protestant theology and Aristotelian philosophy, which had established itself in different parts of Europe, remained particularly strong in Scotland during the first half of the seventeenth century. In 1626, James Reid of Edinburgh University considered the possibility of the earth's axial rotation, suggesting that Scripture sometimes described phenomena as they appeared to the senses, rather than as they really were. In so doing he triggered a theological storm that led to his resignation.

In certain European countries, however, notably England and Holland, the freedom with which scholars could defend the Copernican system stands in striking contrast with the repression faced by Galileo. England in the 1640s saw the Copernican texts of Galileo, Kepler, and the Dutch astronomer Philip van Lansbergen (1561–1632) read with increasing approval. In Holland there were pockets of resistance. Gisbertus Voet (1588–1676), first rector of the University of Utrecht, was a notable critic. But Lansbergen's defense of Copernicus had already made considerable headway during the 1620s. That Lansbergen's opponent, the Catholic theologian Libertus Fromondus, spoke disparagingly of the "Calvinistic-Copernican" system suggests that a parallel was beginning to take shape between scientific and religious reform – at least in the minds of critics. Descartes was to enjoy a freedom to publish in Holland that was denied him in France. As for the Copernican system in America, it has been observed that the New England clergy were its principal propagators.

On the issue of scientific freedom a contrast can therefore be drawn between certain Protestant cultures and those under Roman Catholic jurisdiction – a contrast exemplified by the careers of two Copernicans: Galileo himself and the most fervent popularizer of the system in England, John Wilkins. Although they belonged to different generations, and though

Wilkins was deeply indebted to Galileo for his telescopic observations, the comparison is nevertheless illuminating. Galileo ended his days under house arrest; Wilkins ended his a bishop.

An initial reluctance on the part of Galileo to publicize the Copernican system should not automatically be ascribed to fear of clerical censure. In a letter to Kepler of 1597 he was more concerned by the prospect of being dubbed a fool by that "infinite number of fools" who found a moving earth derisory. If, during the first decade of the seventeenth century, he was not yet an outspoken Copernican, he was nevertheless a virulent critic of Aristotle, soon arousing the hostility of Ludovico delle Colombe who, in seeking to protect the immutability of the heavens, had to endure Galileo's exposure. Colombe was to lead the academic attack on Galileo between 1610 and 1614. This was the period following Galileo's move from Padua, where he had held the chair of mathematics since 1592, to Florence where he had been appointed philosopher and mathematician to Cosimo II, grand duke of Tuscany. Had he suspected, when he left the Venetian state, that his Copernicanism might one day bring him into confrontation with papal authority, he might have thought twice before leaving a region in which dissident scholars could flourish and where printers, for commercial reasons, were already disobeying papal rules. But at that stage he may have valued the intellectual stimulus that a closer intercourse with Jesuit astronomers seemed to promise. It was not long, however, before he learned of a meeting at the house of Archbishop Alessandro Marzimedici, at which strategies for dealing with Copernicans were reviewed. One present, possibly Colombe, had apparently asked that Galileo's opinions be condemned from the pulpit.

Whereas academic philosophers with a vested interest in preserving the Aristotelian world-picture were united against Galileo, there was no such unanimity among his clerical contacts, some of whom gave constructive advice. When asked about the admissibility of change in the heavens, Cardinal Carlo Conti replied that the Bible did not support Aristotle. He even referred Galileo to de Zuñiga's commentary on Job in which the Copernican system had been countenanced. Conti's opinion was that reconciliation could be achieved, but only so long as the Scriptures were understood to speak the language of common people. Unfortunately for Galileo, though there was no unanimity of reaction, there were clerics who would not leave the issue alone and who, by their denunciation, drew attention to its divisiveness. A young Dominican, Tommaso Caccini, embarrassed more discerning members of his order when, from his pulpit in December 1614, he lampooned Galileo, and all mathematicians, as magicians and enemies of the faith. The problem for Galileo was that, once suspicions were aroused, the machinery of censorship could be switched on at a moment's notice. He had to work in the knowledge that he could be reported to the Inquisition at any time, when his writings would become the subject of serious scrutiny.

He was in this predicament by February 1615 when his letter to Castelli (which in expanded form became the letter to the Grand Duchess Christina) was examined by the Inquisition in Rome. The prospect of censure intensified what, for Galileo, was fast becoming a dilemma. On the one hand he was receiving advice from Cardinal Barberini (later Pope Urban VIII) that he should steer clear of theological issues; on the other he was advised that Cardinal Bellarmine would welcome his comments on a passage in Psalm 18 that appeared inconsistent with Copernican doctrine. Trapped with one foot in and one foot out of the sacristy, it was during 1616 that his personal liberty was put at risk. When a papal commission decided that a Copernican cosmology was unacceptable, the pope's instruction to Bellarmine was that Galileo be required to abandon his Copernican beliefs − with the threat of jail if he refused. The minimum this required was that Galileo should be warned that a heliostatic universe had been found formally heretical and that he must no longer hold or defend it. If he agreed, the threat of proceedings against him would be unnecessary. The maximum requirement was that Galileo should be warned by the Commissary of the Inquisition that if he were to hold, defend, or teach the Copernican doctrine in any way, he would be imprisoned. There is much debate as to whether he was actually threatened with prison, as the document in the Vatican file reported. At his trial he would protest that, although he could remember being told not to hold or defend the Copernican system, he had no recollection of an injunction against teaching it. Whatever passed between him and Bellarmine in 1616, his freedom to champion a moving earth had clearly been curbed, even if at that stage

– as Bellarmine confirmed in a signed affidavit – there had been neither abjuration nor penance.

The decree of 1616 did attempt to discriminate between books that aimed to reconcile the earth's motion with Scripture and those that remained within the confines of astronomical hypothesis. Those of the first category, including a recent work by the Carmelite monk Foscarini, were to be destroyed. But those of the second, including *De revolutionibus* itself, were merely suspended, pending correction. As time passed, however, the Church censors began to act as if the distinction were of no importance. Furthermore, the very censorship under which Galileo was operating increased the nervousness of his opponents. There was mounting anxiety among the Jesuits lest Galileo had a decisive argument against the Tychonic system, which he was bound to conceal. The controversy with Grassi over the nature of comets created an atmosphere in which Galileo and the Jesuits finally became locked in an intellectual trial of strength. Any vestige of friendship Grassi might have retained was to disappear when he saw a letter from Florence claiming that the Jesuits would be powerless to answer Galileo's arguments. His reported reaction was that if his order could answer a hundred heretics a year, they would not be beaten by one Catholic.

Meanwhile that one Catholic entertained the hope that his freedom to defend the Copernican system might yet be restored. With the appointment of his friend Barberini to the papacy in August 1623, there were high hopes that an avowed patron of the arts and sciences might lift the restriction. But it is a measure of the difficulty Galileo was working under that, despite several audiences with the new pope, he was unable to ease the ban. Some clarification was, however, secured. Reassurance was given that he was free to *discuss* the rival systems as long as the arguments were confined to mathematical astronomy. The dialogue form of what became the *Two chief world systems* (1632) was a brilliant response to these restrictions, Galileo probably calculating that it would protect him from censure, even if his impartial discussion pointed to far from impartial conclusions. By incorporating his favorite argument from the tides – incontestably a physical argument – he nevertheless transgressed the second of the two conditions.

The issue of censorship arose again in connection with a

Figure III. 3. Frontispiece to Galileo's *Dialogo i due massimi sistemi del mundo* (1632). In Galileo's text, representatives of the Copernican and Ptolemaic systems (Salviati and Simplicio respectively) defined their positions in conversation with an ostensibly neutral third party (Sagredo). Salviati does his best to convert the reactionary Simplicio, who habitually resorts to arguments based on Aristotle's physics and commonsense observation. Galileo's controversial argument that the tides were the result of the combined motions of the earth, creating a jolting effect similar to that produced when a water-carrying barge hits the quayside, was conceivably vulnerable to the skeptical reply that tidal effects could be the result of other mechanisms that an omnipotent God might have devised. Perhaps to assist in conferring the semblance of impartiality on his treatment, Galileo placed this papal argument in the mouth of Simplicio – a tactic that tragically backfired. Reproduced by permission of the Syndics of Cambridge University Library.

license to print, which he was fortunate to obtain since he knew exactly what he had done. To one correspondent, Elia Diodati, he described his book as "a most ample confirmation of the Copernican system by showing the nullity of all that had been brought by Tycho and others to the contrary." That he did not escape a final censorship has been attributed to Urban's discovery of the 1616 injunction, as recorded in the file, and to his realization that Galileo had exceeded *his* perception of what had been allowed. Because it transpired that Galileo had not mentioned the 1616 prohibition when applying for his license, Urban felt betrayed on several counts. That the tides were a physical consequence of the combined motions of the earth, as Galileo maintained, was the very argument he had outlawed. And to add insult to injury, one of his own arguments celebrating the omnipotence of God had been placed in the mouth of Simplicio, that character in the dialogue who, as the spokesman for naive common sense, had been the subject of ridicule. Despite an extrajudicial bargain that, if Galileo confessed his error, he would be treated leniently, the final censorship was a sentence of life imprisonment. Although this was commuted to house arrest, Galileo remained embittered in the belief that the other side of the bargain had not been kept. What Federico Cesi once wrote of the members of his Lincean Academy applied to Galileo with a vengeance: "All we claim in common is freedom to philosophize in physical matters." There can be no denying this was an issue.

It is doubtful whether it was ever such an issue for Wilkins. Like Galileo, he was committed to the Copernican system as a cosmology and not merely as a mathematical hypothesis. As with Galileo, he wrote in the vernacular to reach a larger audience. For rejoinders to the commonsense objections against a moving earth, Wilkins was indebted to Galileo, as he was for striking analogies between the earth and moon. On the issue of intellectual freedom, however, contrasts begin to appear. Wilkins took the silence of the Bible, on a question such as the plurality of worlds, not as a ban but as an invitation to entertain the notion. Such license had effectively been denied Galileo who, by his friend Ciampoli, had been advised to suppress remarks that might imply extraterrestrial life: Would they not invite awkward questions such as how descendants of Adam and Eve could have reached the moon?

Saturnus.

Iupiter.

Mars.

Ceres et Proserpina.

Venus.

Mercurius.

Omnibus Do lucem, calorem, motum.

Sua fovent Mundo se illuminant

Vniversi ornant.

Quid si sic?

Vtinam et alæ.

Hic ejus oculi.

A Discourse concerning A NEW world & Another Planet In 2 Bookes.

Printed for Iohn Maynard, & are to be sold at the George, in Fleetstreet neare St Dunstans Church. 1640.

N. Copernicus.

W. Marshall, sculpsit.

Galilæus.

Keplar.

For the publication of his pro-Copernican *Discourse concerning a new planet* (1640), Wilkins had to obtain a license. But in Protestant England, even in Oxford under the chancellorship of Archbishop Laud, he encountered little resistance. His previous announcement of a *Discovery of a world in the moon* (1638) had already enjoyed two printings in its first year and a third in 1640. His proclamation that the moon might be habitable was bound to attract criticism from Aristotelian diehards such as Alexander Ross. But such retaliation seems to have been powerless to harm him. There *were* threats to his academic career, but they came from a different direction.

Having been appointed warden of Wadham College Oxford in 1648, he was criticized by less moderate puritans who objected to the manner in which he turned the college into a haven for those with Anglican and Royalist sympathies. His biographer has stressed the consistency with which he avoided extreme political alignments and maintained a steadfast dislike of religious persecution. Following the restoration of the monarchy in 1660, the same spirit of moderation lay behind his efforts to bring religious dissenters back to the Anglican fold and to oppose legislation that disadvantaged them. If the failed attempt to oust him from Wadham in 1654 was the work of stricter puritans who suspected him of reducing Christianity to morality, so in the 1660s his "club for comprehension" evoked the censure of high churchmen. As a religious moderate, he was attacked by extremists of different hues, but not, it would seem, because of his science. If he was persecuted, it was as a result of his avoiding persecution. He was "prepared for the embracing of any religion," one critic wrote, "rather than incur the hazard of persecution."

Wilkins's survival as warden of Wadham, his move to the mastership of Trinity College Cambridge in 1659, his be-

---

Figure III. 4. (*opposite*) Frontispiece from the combined edition (1640) of John Wilkins's *Discovery of a world in the moone* (1638) and *Discourse concerning a new planet* (1640). Copernicus, on the left, raises the possibility of the earth's motion, Galileo produces his telescope, and Kepler expresses a desire for wings that he might visit the new world – as indeed he had dreamed of doing in his unpublished Copernican text *Somnium*. The sun is portrayed as source of light, heat, and motion. Reproduced by permission of the Syndics of Cambridge University Library.

coming bishop of Chester in 1668, and his appointment as Lent preacher to the king suggest that there was nothing particularly hazardous in being England's most conspicuous Copernican. His differentiation between "divine" and "natural" knowledge found receptive ears. The former related to the quest for spiritual happiness, the latter to the sciences which, in their practical application, helped to constitute a third category of "artificial" knowledge. Once again, however, it was differentiation without separation. Natural knowledge was concerned with the frame of the universe, but, as such, it disclosed the course of providence in the physical world. Indeed, one of Wilkins's arguments in favor of the Copernican system had been that the alternative – imputing a daily rotation to the heavens – would be to argue "improvidence in nature." The desire to see beauty, convenience, and harmonic proportion in the universe had clear religious overtones. In extending the principle of biblical accommodation and the scope of natural theology, Wilkins enjoyed a degree of liberty that Galileo had been denied.

It is important not to exaggerate the oppressive effects of Index and Inquisition. The Counter-Reformation did not prevent Italian scholars from making original contributions in classical scholarship, history, law, literary criticism, logic, mathematics, medicine, philology, and rhetoric. Nor were they isolated by the Index from European scholarship. Prohibited books entered private libraries where they would be consulted by those prepared to break the rules in the interests of learning. One such collection was in the hands of Galileo's Paduan friend, G. V. Pinelli. One can lose a sense of perspective if the condemnation of Galileo is taken to epitomize the attitude of Catholic authorities toward the natural sciences. Relatively few scientific works were placed on the Index. The attempt to put a stop to the moving earth stands out because it proved so tragic an aberration – a personal tragedy for Galileo and, in the long run, a tragedy for the Church, which overreached itself in securing a territory that would prove impossible to hold.

One can lose a sense of perspective, too, if one overlooks the positive side of Catholic educational programs. The Jesuit order, which had been established as the intellectual vanguard of the Church, made many enemies; but its educational institutions were greatly admired. A distinctive prominence was given to mathematics – not least because its value in astronomy, navigation, architecture, and surveying gave it particular relevance to those training for missionary work overseas, or to those preparing for high military or government service. The official condemnation of the Copernican system did put constraints on what could be taught, though the option of presenting it as merely a mathematical hypothesis was still open, and increasingly taken during the course of the seventeenth century. The important point, however, is that Jesuit scholars were active in other branches of science and particularly so in the emerging fields of electricity and magnetism. It would be a caricature to portray them as hidebound Aristotelians. In his commentary of 1646 on Aristotle's *Meteorologica*, Niccolo Cabeo reproached those who never questioned Aristotle's teaching: They were grammarians, not philosophers.

The scientific texts produced by Jesuit scholars were sometimes short on theory. This may reflect the belief that catalogs of data were more appropriate for teaching purposes or that there *were* pressures to shy away from controversial ideas. Even those who saw the need for theoretical models sometimes exulted in a pluralism that had the effect of giving priority to none. Francesco Lana, a pupil of the renowned Athanasius Kircher, confessed that he would use many models to express the truth to which the science of natural things could attain. What one model lacked, another would supply. If such eclecticism counts as a limitation, it hardly detracts from the achievement of Jesuits as patrons and teachers of science. Of the 195 members of the Paris Academy of Sciences to be honored with an official eulogy before the Revolution of 1789, at least twenty percent had received a Jesuit education. Two generations before the Paris Academy was founded, a Protestant crusader for educational reform had already conceded that the state of learning had been "much quickened and strengthened" by Jesuit schools. "If only they were ours," Francis Bacon had lamented.

Protestantism and Practical Science

We have been exploring the idea that Protestantism helped to create a climate in which the scientific enterprise could gain momentum. With reference to the fate of the Copernican system, we have encountered many complications, even in a test case that, because of Galileo's trial, might seem to

have been loaded from the start. We cannot, however, leave
the discussion with the ostensibly greater freedom of Protestants to subscribe to a new cosmology, because there are
stronger claims in the literature for a parallel between scientific and religious reform. Prominent among them has been
the thesis of the American sociologist Robert Merton to the
effect that the values associated with Protestant asceticism
gave a positive stimulus to the practical sciences. In conclusion we shall briefly consider Merton's thesis and the difficulties that arise in testing it.

Connections between ascetic Protestantism and the growth
of capitalism have been extensively discussed ever since Max
Weber drew attention to three ethical maxims commonly derived from Calvin's theology. These were that one should be
diligent in pursuing one's religious calling; that one should
renounce material satisfaction; and that one's time should always be used constructively. To abide by these norms was
not a way of earning salvation, for in Reformed theology there
was no way that salvation could be earned. But to follow these
maxims was the duty of the Christian whose ultimate destiny
was in God's hands. It is not difficult to see how capitalist
entrepreneurship might prosper under such ethical norms:
not by making monetary gain an end in itself but as an almost
incidental product of thrift.

Merton's thesis was that a godly involvement in the affairs
of the world would also encourage the growth of science.
There could be real connections between the spiritual injunction to glorify God and a quest for knowledge that would
not only demonstrate the Creator's power but also alleviate
suffering. To study the book of nature was a permissible use
of the gift of reason that differentiated humanity from the
beasts. It was also effective as a diversion from sensuality –
from bags, bottles, and mistresses, as Robert Boyle would
put it. With examples drawn from seventeenth-century England, Merton argued that the social utility of both science and
technology was increasingly recognized where puritan values
held sway. His thesis does not exclude the importance of wide
socioeconomic forces. Economic expansion in seventeenth-
century Europe, and the growth of the mining industry in
particular, are given special prominence in an analysis that
made generous allowance for the role of technical problems
in defining areas of scientific research. Nor did Merton imply
that the actual discoveries of a Boyle or a Newton could be

directly ascribed to the sanction of science by religion. The argument is not that Protestant religion constituted a primary, independent variable on which science depended, but that, through reciprocal interaction, a higher value was gradually placed on the practical sciences. Put another way, puritan values helped to create an audience receptive to programs for the improvement of man's estate.

A guide to the vast literature on Merton's thesis will be found in the bibliographic essay, but it is important to note that variants of it have survived intense criticism. There is evidence that Protestant millenarianism, enjoying a resurgence during the civil war period in England, could provide a framework within which Bacon's program for the empirical sciences gained an extra impetus – despite the fact that Bacon himself had been hostile to any religious movement that might threaten the fragile monarchy to which he had constantly vowed allegiance. A letter from John Beale to Samuel Hartlib in the late 1650s shows how the integration could be effected through the hope of a restored dominion over nature:

Here you must add the discovery of, or dominion over all the works of God; the conversion of stones into metals and back again; of poisons into powerful medicines, of bushes, thorns and thickets into wine and oil, and of all the elements to take such guise as man by divine wisdom commands.[5]

For some Protestant thinkers, experimental science promised a way of reversing the effects of the original curse, a way of making a better world that might in some small way mirror the perfection of God's heavenly kingdom, a way of restoring the world to a condition fit for Christ's earthly rule. Affirmations of a strong parallel between religious and scientific reform are not difficult to find. Thomas Culpeper remarked in 1655 that, as Reformed theology rejected a pope in religion, so a reformed science rejected a pope in philosophy. It was easy to claim, as did Thomas Sprat in his *History of the Royal Society* (1667), that the two reformations had this in common: Each prized the original copies of God's two books, nature and the Bible, bypassing the corrupting influence of scholars and priests.

It is difficult to discount the evidence that Protestant theology provided important resources for those who saw in experimental science the key to human progress. Merton himself could not discount statistical data, which indicated that, of the foreign associates of the French Academy of Sciences during the period 1666–1885, only eighteen had been Catholics, against eighty Protestants – despite Europe's far larger Catholic population. Detailed studies of English science during the period of the puritan domination have also revealed plans for a national agency to supervise technological innovation – plans that had taken shape in the mind of Samuel Hartlib. The English nation, wrote the parliamentarian soldier Walter Blith, "might be made the paradise of the World, if we can but bring ingenuity into fashion." An efflorescence of scientific publication in the period 1645–1660 – especially in agriculture, mathematics, and medicine – suggests that ingenuity did, for a while, come into fashion. Because good works were required of a puritan as a duty of thanksgiving, the appeal of Bacon's arguments to the intellectual leaders of the English Revolution can be readily understood. Bacon had insisted that the "rule of religion that a man should justify his faith by works applies also in natural philosophy; knowledge should be proved by its works."

The case for strong parallels between religious and scientific reform can therefore be made to appear quite plausible. But, in seeking to test the correlation, problems soon arise. The most obvious test would be to identify the religious allegiance of Europe's most illustrious natural philosophers of the late sixteenth and seventeenth centuries. A high preponderance of Protestant ascetics might then be suggestive. This is not, however, what one finds. Indeed, catalogs of Catholic contributions have been compiled to challenge the view that a Protestant spirituality provided distinctive motivation for scientific work. The value of such catalogs is that they prevent too parochial a view of the scientific effort. But they do not decide the issue because many of the Catholic scholars listed may have experienced the frustration of censorship to a degree not experienced by their Protestant contemporaries. In one such list the Jesuit astronomer Christopher Scheiner is included for his discovery of the sunspots. But there is no mention of the fact that Scheiner was put under pressure by his superiors to curb his anti-Aristotelian conjectures.

A second difficulty with this most obvious method of testing is that, even if a preponderance of Protestants were to emerge, it would not follow that it was their religious convictions that supplied the motivation for their science. Merton himself was aware of the problem when he raised this question:

To what extent did the old Puritans turn their attention to science
. . . because this interest was generated by their ethos, and to what
extent was it rather the other way, with those having entered upon
a career in science . . . subsequently finding the values of Puritanism congenial?[6]

Merton's answer was that both processes were at work, but to an unknown extent. The problem in a nutshell is whether particular forms of scientific and religious commitment might not separately depend on ulterior forces of social and economic change. It is striking, for example, that detailed studies of the relations between religious dissent and the promotion of science in eighteenth-century Britain have located the correlation in expanding northern towns such as Manchester where the ulterior forces were population growth, increase in wealth, a certain social and geographical isolation, a zest for political reform, and a concern for moral values prompted by the seamier side of city life.

To establish a correlation is not necessarily to establish a connection. Perhaps a more sensitive test would be to examine the writings of puritans who did take an interest in science in seventeenth-century England to see what connections they did make in their schemes for reform. But there are problems even with this more sophisticated test, because it could be argued that to examine those who were as a matter of fact interested in science is to skew the sample. If the object is to determine whether a puritan spirituality was conducive to scientific activity, it would surely be necessary to take into account the far larger body of puritans who were not so enthralled by the book of nature. The problem then becomes that when the puritan literature of the early seventeenth century is examined, the seeds of what might prove a flowering of science are difficult to find. To puritan divines such as William Perkins and William Pemble, it was not self-evident that the works of science were good works. An awareness of God's contrivance in the intricate designs of nature might confound the atheist, but it did not point him toward the God who had entered into covenant with sinful humanity. There was nothing in natural knowledge that, in Pemble's words, could "set straight the wryed and distorted image of God in us"; nothing that might assist the soul in achieving its sanctification. Add to this the imperfection of scientific knowledge, and it would then take a low place in the stratification of priorities. For puritan divines, the salvation of the human soul was of overriding importance, which suggests that some readjustment of priorities might be necessary for those who valued the study of nature.

Perhaps a more decisive test would be to examine the composition of England's first enduring scientific institution, the Royal Society, founded shortly after the restoration of the monarchy in 1660. In fact such a test has been performed to challenge Merton's contention that there was a puritan majority among its early members. From a sample of 162 eventual fellows of the society who had been over sixteen in 1642, and therefore old enough to have taken sides in the civil war, and discounting foreigners not in England during the war years (and a further 22 for whom there is inadequate documentation), 38 fought for, or supported, Parliament, while 85 were royalist in 1642. The high proportion of royalists looks embarrassing not only for Merton's thesis but for variants of it which have claimed that it was the political radicalism (not the puritanism) of the parliamentary radicals that made them receptive to revolutionary science. As few as one in twenty of the sample could be described as a utilitarian scientist of puritan middle-class background. Only three ministers in the sample were puritan enough to be ejected from their livings after 1660. The author of the test, Lotte Mulligan, has therefore concluded:

The typical background of a science enthusiast in the 1660s was not middle-class, mercantile, puritan, politically radical, unacademic or utilitarian. Rather, our typical Fellow was a royalist, Anglican, university-educated gentleman.[7]

But how decisive are such statistical data? Because the early Royal Society contained many passive members, encouraged for their lucre or social luster, would it not be a more sensitive test if the sample were restricted to the active nucleus? Of the ten most active fellows (including Robert Boyle, Jonathan Goddard, Henry Oldenburg, William Petty, John Wilkins, Lord Brouncker, Walter Charleton, John Evelyn, Sir Robert Moray, and Christopher Wren), the first five had collaborated with the parliamentarian regime, suggesting a higher proportion of puritan involvement than that indicated by the larger sample. And to compound the difficulty of how best to interpret the statistics, the author of the test has conceded that physicians, instrument makers, naval experts, agricultural reformers, and general applied scientists are more often to be found among the parliamentarians in the sample.

The suggestion that the typical science enthusiast was Anglican and not politically radical does not of course affect the wider generalization that Protestant cultures were more amenable to freedom of thought. But it may imply that the kind of Protestant spirituality, for which the term *puritan* is commonly used, was not the only catalyst for the expansion of science. Perhaps a certain detachment from puritan enthusiasm, an insistence on moderation and toleration on religious issues, defined the mentality that most often coincided with an interest in science in seventeenth-century England. John Wilkins, whom we encountered earlier, would illustrate this alternative correlation. Willing to side with Parliament against Charles I, he had not wished to see the king executed. Wishing to rid the Anglican Church of the influence of Archbishop Laud, he would not go along with more zealous reformers who wished to break away. Deeply committed to the pursuit of mathematics and the physical sciences, he stood up for the universities when they were attacked by vituperative radicals such as Cromwell's army chaplain, John Webster.

The moderate Anglicanism for which Wilkins stood is sometimes described as "latitudinarian," to indicate a breadth of mind and an attitude of religious toleration, at variance with the demands of puritan enthusiasts, whose overbearing self-assurance on the finer points of doctrine could be distasteful. The idea of a correlation between a latitudinarian and a scientific mentality can be appealing. They could be bound together by the belief, found in Bacon, that religious controversies were an impediment to science. There could be a suspicion of dogma, whether religious or scientific. And science itself might be seen as an instrument of religious union in that all parties could agree on the existence of a Creator whose power was visible in nature. In Wilkins one finds an elaborate and adaptable theory of knowledge, which differentiated between fundamental truths, provisional hypotheses, and areas of uncertainty in both scientific and religious domains. But we are still left with the problem of testing. And in this case it is particularly acute because to set up a typology in which science correlates with religious moderation risks the objection that what one means by moderation is going to change according to political circumstances. With the proliferation of puritan sects during the 1640s and 1650s, there was so great a range of extreme demands, from the nationalization of land to the emancipation of women, that it would be surprising if the natural philosophers had not begun to appear as moderates. Following the restoration of the monarchy in 1660, it would have been surprising if apologists for the newly founded Royal Society had emphasized a puritan lineage. In his *History of the Royal Society* (1667) Thomas Sprat reclaimed the name of Francis Bacon from the hands of puritan visionaries and reaffirmed the role of an inductive method, which, by promoting consensus rather than disputation, was at one with the quest for political stability.

The main thread running through this chapter has been the difficulty of testing hypotheses that link the reform of science to the reform of religion. That there are the difficulties we have uncovered does not mean that the hypotheses linking Protestantism to the expansion of science are necessarily false. But it does suggest the need for caution when faced with claims that a particular form of Christian piety was uniquely propitious. Certain developments in seventeenth-century science did prove more difficult for Catholic authorities to assimilate. A Copernican cosmology, as defended by Galileo, is the outstanding example. In the following chapter we shall encounter another: the recovery of an atomic theory of matter, which played a significant role in the mechanization of nature. But such is the fascination of the story that we shall find Catholic scholars in the vanguard of those who sought that very mechanization.

CHAPTER IV

Divine Activity in a Mechanical Universe

Introduction: A Historical Paradox

In 1704 a contributor to a French learned journal observed that a new style of scientific explanation had become all the rage. One heard of nothing else but *mechanistic* physics. Nor was the fashion confined to savants. Ladies familiar with the philosophy of Descartes would blithely reduce animals to machinery. "Please do not bring a dog for Pauline," a certain Mme. de Grignan begged in 1690: "We want only rational creatures here, and belonging to the sect we belong to we refuse to burden ourselves with these machines." The universe has become so mechanical, Fontenelle announced in 1686, that one might almost be ashamed of it.

This mechanization of the natural world became such a feature of late seventeenth-century science that historians have sometimes spoken of the death of nature, as organic analogies were displaced by images of clockwork. The impact of mechanical analogies varied from science to science, often proving premature in the study of living systems. In the long run, however, there was scarcely any branch of science that was not affected. Despite the revolution in physics, which in our own century has made the image of a rigidly deterministic universe less secure, the assimilation of natural processes to machinery continues to be a conspicuous feature of scientific investigation. The legacy of the seventeenth century mechanical philosophy is apparent in such terms as *genetic engineering* and in the description of computers capable of simulating aspects of human intelligence.

The philosophical implications of the seventeenth century

117

transformation were profound. A new conception emerged of what was real in the world. Particles of matter in motion defined the new reality. The world of appearances, of colors, odors, tastes was reduced to secondary status – as merely the effect of the interaction of particles on the human sensory apparatus. Moreover, if the real world was that which could be described in mechanical terms, new questions arose concerning the sensitive question of God's relationship to nature. What role could be left for God to play in a universe that ran like clockwork? Would one have to side with those who became known as "deists," who restricted that role to the initial creation of a law-bound system, and who attacked Christian conceptions of a subsequent revelation? Would God's special providence, His watchful concern for the lives of individuals, not be jeopardized if all events were ultimately reducible to mechanical laws?

Such questions will be the main concern of this chapter. They are not, however, as simple as, from a secular vantage point in the twentieth century, they might seem. Certainly to the free thinkers of the Enlightenment a clockwork universe had the great attraction that it could be presented as a universe that ran by itself. But the paradox is that among those seventeenth-century scholars who did most to usher in the mechanical metaphors were those who felt that, in so doing, they were enriching rather than emasculating conceptions of divine activity. Four examples will serve to illustrate the point. However paradoxical it may appear, the French Catholic Marin Mersenne promoted a version of the mechanical philosophy as a way of defending, not attacking, the miraculous. However paradoxical it may appear, Descartes mechanized the entire animal creation as a way of highlighting the spiritual uniqueness of humanity. However paradoxical it may seem, Robert Boyle compared the physical world with clockwork in order to emphasize, not detract from, the sovereignty of God. And, however strange, Isaac Newton saw in the very laws he discovered a proof, not of an absentee clockmaker, but of God's continued presence in the world.

These paradoxes are easily resolved, but a great irony remains. The philosophy of nature that, during the seventeenth century, was upheld as the most protective of a sense of the sacred in nature was the very one that, in later social contexts, was most easily reinterpreted to support a subversive and secular creed. This is the irony to which reference was made in the Introduction. It must now be explored in greater depth.

The Displacement of Organic by Mechanical Metaphors

The basic postulate of the mechanical philosophies was that nature operates according to mechanical principles, the regularity of which can be expressed in the form of natural laws, ideally formulated in mathematical terms. The contrast with previous approaches to the study of nature can be exaggerated. Indeed, one of the lessons of recent scholarship has been that to identify the mechanical philosophies with the triumph of rational over occult ways of thinking will not work. It is, nevertheless, useful to consider one or two examples of earlier, organic analogies in order to illustrate the transition.

Aristotle had had this to say in the context of explaining planetary motion: "We are inclined to think of the stars as mere bodies or units, occurring in a certain order but completely lifeless; whereas we ought to think of them as partaking of life and initiative." The peculiarities of planetary motion could then be understood because, as with a living organism, planets could exhibit different states of health, for which different modes of exercise were appropriate. Even sixteenth- and seventeenth-century critics of Aristotle had continued to imbue physical objects with a life of their own. Giordano Bruno, for example, had been happy to invest the stars with souls. William Gilbert (1540–1603), for all his apparent modernity in the manner in which he experimented on magnets, reproached Aristotle for not going far enough. The Greek philosopher had denied the earth a soul – a deficiency that Gilbert promptly remedied: "We consider that the whole universe is animated, and that all the globes, all the stars, and also the noble earth have been governed since the beginning by their own appointed souls and have the motives of self-conservation." Governed by souls: not yet by physical laws.

Where such organic analogies prevailed, it was common to suppose that there were sympathies, special affinities, between physical objects. These allowed one body to affect another, even though they were not in contact. The most celebrated examples come from popular belief in the effect of the stars on human destiny, or from the belief of the alchemists that certain planetary conjunctions were propitious for the success of particular experiments. But a striking example, still much discussed during the first half of the seventeenth century, concerned the efficacy of the weapon salve — a healing balm placed not on the wound but on the weapon by which it had been inflicted. The healing, according to the Englishman Robert Fludd (1574–1637), was effected by a sympathetic power transmitted from the blood on the weapon to the blood of the afflicted. Magnetic attraction was the paradigm for this kind of action at a distance, which contrasted sharply with the contact action on which Descartes would insist as the only means by which causes could produce effects. Scientific explanation for Fludd required a vocabulary of spirits, sympathies, messengers between heaven and earth, and invisible lines. By the end of the century a new vocabulary had been forged, replete with the cogs and wheels of human gadgets.

A contrast between the microcosm–macrocosm analogies of Renaissance philosophy and the new mechanics of the seventeenth century is visible in remarks made by Leonardo da Vinci and Kepler. For all his technical genius and flirtation with flying machines, Leonardo had still envisaged nature as a living organism:

We can say that the earth has a vegetative soul, and that its flesh is the land, its bones are the structure of the rocks . . . its blood is the pools of water . . . its breathing and its pulse are the ebb and flow of the sea.[1]

For all his number mysticism and flirtation with animism, Kepler, by 1605, was announcing a mechanistic program:

I am much occupied with the investigation of the physical causes. My aim in this is to show that the Celestial machine is to be likened not to a divine organism but rather to clockwork. . . . Moreover I show how this physical conception is to be presented through calculation and geometry.[2]

The power of geometry in the analysis of natural phenomena was shown by Kepler's reduction of planetary motions to the shape of an ellipse; by Galileo's study of the relations between speed, time, and distance when a body falls with uniformly accelerated motion; and by his analysis of the parabolic path of projectiles. It was to be supremely demonstrated by Newton when he showed that an inverse-square law of gravitation would explain Kepler's ellipses. As an analytical tool, however, geometry could only be applied to idealized representations of natural phenomena in which the multiple complexities of the real world were temporarily discounted. The geometric model, once constructed, could then be used as a control to see how closely events in the physical world conformed to it. This ability to create two worlds, to relate the real world to an idealized mathematical model, was one of the techniques that made modern science possible. The process of abstraction that was required came more easily in a mechanical than an organismic universe. And the more successful the method, the more it encouraged the view that the most fundamental elements of creation were precisely those amenable to mathematical analysis: the shape, arrangement, and motion of particles. Thus we find Galileo rejecting the notion that heat is a real property or quality residing in a body that feels hot. Rather, heat is a sensation produced in us by "a multitude of minute particles having certain shapes and moving with certain velocities."

This predilection for particles, so alien to Chinese philosophies of nature, became a distinctive feature of Western science during the seventeenth century. It can be seen in the efforts of the French philosopher Pierre Gassendi (1592–1655) to grapple with the phenomenon of gravitation, without recourse to action at a distance. Gassendi's proposal was that the earth emitted a continuous stream of particles from its surface. Like tentacles, these particles could attach themselves to the pores of any body suspended above the earth and so pull it down. The further the body from the earth, the weaker the gravitational pull, because fewer of the emitted particles would reach the object. Wishing to square his explanation with Galileo's law of falling bodies, he invoked an additional hypothesis – a downward thrust, effected by the air, complementing the effect of the gravitational radiation.

The particles deployed by seventeenth-century mechanical philosophers varied from one to another. Aristotelian philosophy continued to be one resource, for Aristotle's theory of matter required the postulation of minute particles, the *minima,* as a substratum. The English physiologist Walter Charleton (1620–1707), having become disenchanted with magical philosophies of nature, reverted to Aristotle before finally adopting an atomic philosophy similar to that propounded by Gassendi in France. Those who favored an atomic theory were, however, committing themselves to the view that matter is not infinitely divisible. For Gassendi's opponent, Descartes, this was to put constraints on the power of God. Accordingly, in the mechanized universe of Descartes, the particles were not atoms moving in a void, but were of such different sizes and subtlety that they filled the entire universe. In such a plenum, patterns of roughly circular motion could be established if each particle continually moved into the position that its predecessor was vacating – hence Descartes's mechanical model for the orbits of planets. Each was carried round in a whirlpool of subtle matter, like a cork floating on water. Despite its mathematical inexactitude, it was a model that was easily visualized, and one that had the virtue of explaining why the planets circle the sun all in the same direction and in roughly the same plane. The model could even be extended to explain the tides, for the earth had its own vortex of swirling matter. This subtle matter, in the vicinity of the moon, would be deflected as it encountered the moon as an obstruction – thereby setting up a downward pressure, which, when transmitted through the ether, compressed the ocean.

With so many different particles to choose from, some exponents of a mechanical philosophy preferred not to take sides, opting instead for the neutral term *corpuscle*. Robert Boyle favored this eclectic approach as he constructed an elaborate, hierarchical theory of matter in order to explain chemical phenomena. His corpuscles could be arranged in different ways, organized into different shapes, and were capable of different patterns of motion. Acids were acids not because they contained some essence, quality, or form of acidity, but because their particles had sharp points and were thus able to attack the surface of metals.

In the work of Newton, the particles of the mechanical philosophy achieved their greatest refinement. His theory of matter was explicitly atomic, but each atom was equipped with short range attractive and repulsive forces. These forces, in principle at least, were capable of quantification:

Since metals dissolved in acids attract but a small quantity of the acid, their attractive force can reach but to a small distance from them. And as in algebra, where affirmative quantities vanish and cease there negative ones begin, so in mechanics, where attraction ceases there a repulsive virtue ought to succeed.[3]

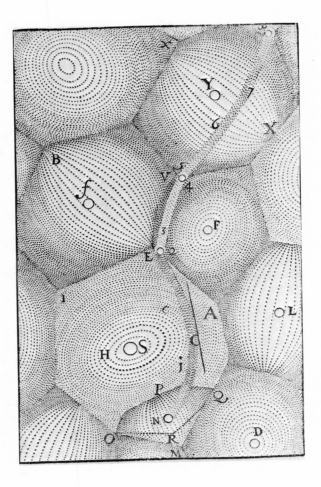

Figure IV. 1. Illustration from page 92 of Descartes's *Principia philosophiae* (1644). The vortices of subtle matter, swirling in contiguous solar systems, epitomize Descartes's mechanistic universe. The numbers in the diagram identify the path of a drifting comet, formerly a sun but now encrusted and defunct. Bereft of the power to drive its own vortex, it is simply carried along by the impact of particles with which it is at every instant in contact. Reproduced by permission of the Syndics of Cambridge University Library.

Newton's theory of matter was not a straightforward extension of earlier mechanical philosophies. His forces of attraction and repulsion were sometimes thought − especially by Continental critics − to mark a retrograde step, a reversion to the sympathies and antipathies that Descartes had excised. But Newton himself felt there was more than sufficient empirical justification for them. Short range repulsive forces would explain the reflection of a light ray as its constituent particles approached the reflecting surface. They would also explain processes of evaporation without having to picture air particles as springs or hoops.

Despite the lack of consensus among natural philosophers as to what was meant by the mechanical philosophy, they spoke of it as a new theory of matter, a new theory of causality, and a new theory of method. It could be presented as a new theory of matter in that the ultimate components of things were particles, stripped of qualities such as color, taste, or smell and divested of forms, seeds, spirits, and, in the philosophy of Descartes, of all inherent powers of activity. Among Descartes's critics in England the ultimate particles were sometimes allowed inherent powers of motion, as when Walter Charleton in 1654 referred to the "coessential motive faculty" of his atoms. But, in either case, a new theory of causality was stressed, whereby contact action, the mutual impact of particles, became the paradigm case of causal agency. A new theory of method was advertised in that the construction of mechanical analogues was extended even into the organic domain. Although the shift was gradual, and even in the case of Newton incomplete, a new image for the universe as a whole was taking hold. In this clockwork universe older beliefs were *not* automatically excluded. It was still possible to believe in the transmutation of base metals into gold, still possible to believe in the efficacy of the weapon salve, still possible to believe in witchcraft. There were, however, changing sensibilities as to what an acceptable mechanism might be in the rationalization of such effects.

Theological Justification for Mechanical Philosophies

However surprising it may seem, seventeenth-century natural philosophers had little difficulty in finding a theological rationale for their mechanical universe. In Chapter II we saw how scientific innovations were often presented in theological terms. In the case of the mechanical philosophies it was necessary to show that divine activity was not excluded by a clockwork universe. The task was all the more pressing because, in the ancient world, the atomic theories of Leucippus, Democritus, and Epicurus had been associated with the belief that nature has no need of gods, that worlds had come into being by the chance collision of atoms. In their defense of both a mechanical philosophy and divine providence, however, scholars constructed arguments that did more than merely reconcile the two. Advantages were seen in a mechanical philosophy for the defense of the Christian religion, despite the fact that during the course of the seventeenth century it was also seized by those who, like Thomas Hobbes (1588–1679), questioned the immateriality of the soul. Without committing himself to an atomic philosophy, Francis Bacon had insisted that it was *more* conducive to a doctrine of providence than the Aristotelian alternative:

Nay, even that school which is most accused of atheism doth most demonstrate religion; that is, the school of Leucippus, and Democritus, and Epicurus – for it is a thousand times more credible that four mutable elements and one immutable fifth essence, duly and eternally placed, need no God, than that an army of infinite small portions, or seeds unplaced, should have produced this order and beauty without a divine marshal.[4]

For Bacon, as for Boyle and Newton after him, it was simply inconceivable that, from a chance distribution and collision of atoms, a world of such order could have been produced – an order that the progress of science was confirming rather than destroying. It is appropriate, therefore, to examine the theological advantages that were perceived in a mechanized universe.

Mersenne and the Defense of the Miraculous

Mersenne has an important place in the history of seventeenth-century science because, through an extensive correspondence, he probably did more than anyone to coordinate the efforts of the early mechanists. He was particularly conspicuous in popularizing Galileo's physics. A French Catholic, and member of the Order of Minims, he was aware that his faith was under attack from two directions. First, Protestants and deists were drawing a distinction between religious truth and religious tradition that he, as a Catholic, felt bound to resist. Second, philosophies of nature had been emerging from Renaissance Italy that, explaining nature's marvels in purely naturalistic terms, threatened to explain away the truly miraculous. In seeking to combat both these threats, Mersenne found a mechanical philosophy useful.

In the preceding chapter we showed how religion could have an indirect effect on science. Continuing debates between Catholic and Protestant Christians helped to turn controversial scientific innovations into sensitive issues. But such debates also created issues of their own. Mersenne was wounded by Protestant allegations that the Catholic priesthood manufactured miracles in order to convert the masses. He therefore believed that it was in the best interest of his Church to tidy up the borderland between natural marvels and genuine miracles. A mechanical philosophy provided the means of doing so because it could assist in defining the boundaries of natural order. Miracles, after all, presupposed a natural order against which they could be judged to be miracles, against which they could be differentiated from mere marvels.

Mersenne's response to the second of the two threats helps to clarify the point. The challenge of a purely naturalistic philosophy had been issued by Pietro Pomponazzi (1462–1525), who had preserved the Aristotelian deity as a prime mover but at the same time rejected both providence and miracles. It was one thing to attack popular credulity as Pomponazzi had done, but to reject miracles altogether was, for Mersenne, going too far. For every prodigious effect in nature, Pomponazzi and his devotees had insisted there must be some explanation in terms of natural forces. But the forces he employed were precisely those that would offend the mechanists – forces whose causal agency was vested in stars, names, signs, images, thoughts. He frequently invoked intelligences associated with the stars. Nature knew no miracles but, in Pomponazzi's system, it had become a chaos of natural wonders. A mechanistic physics arguably appealed to Mersenne because it injected a ray of light, a sense of order and restraint into a naturalistic fantasy. The desire to discriminate between natural marvels and true miracles was religiously motivated and eventuated in a mechanical and less magical view of nature. If the ringing of church bells could disperse a storm, for the credulous masses it was a miracle; for Pomponazzi it was through the will of an astral intelligence; but for Mersenne it was through the will of God, which had expressed itself in the laws of fluid mechanics.

The coherence of Mersenne's position depended on an important, and long-established, distinction between God's power as displayed in the order of nature and His absolute power by which He could act in any manner He chose. A mechanical philosophy appealed to Mersenne, as it did later to Boyle, because it seemed the perfect way of expressing the manner in which God normally chose to act in the world. Physical laws were an expression of the divine will, the ultimate source of order. But they were in no way binding on God, who was always free to act in other ways if He chose. The possibility of miracles was not excluded. It was rather that the means of recognizing a true miracle was clarified. An event could be deemed a miracle if it was not explicable in terms of physical laws.

Descartes and the Defense of Human Dignity

In Chapter II we saw how Descartes promoted his mechanical philosophy in theological terms. The quantity of motion that God had put into the world at creation was conserved at every moment by His sustaining action. Other theological issues arose, however, in the context of applying mechanical analogues to the analysis of living things. Whereas organismic analogies had been habitually applied to physical processes, Descartes turned the procedure upside down, interpreting animals as pure machinery. This was a highly controversial move, but Descartes protested that his doctrine of the beast-machine was not so much cruel to animals as it was favorable to humans. It was favorable in that it emphasized a superiority over the whole of creation that had a parallel – the exactness of which continues to be a source of debate – with the reference in Genesis to a dominion over nature that was a human privilege and responsibility.

Descartes underwrote the privilege by insisting that humans are machines with a difference: They alone have the gift of an immortal soul. Descartes could see advantages in a mechanical philosophy because it helped to sharpen up the difference between the world of matter and the world of spirit. In the rationality of the human mind was evidence of a spirit world, but it was confined to humanity. The uniqueness of humankind, as alone made in the image of God, was not merely protected but positively enhanced.

Historians have hit on some grand explanations for why the mechanization of nature occurred when it did. It has been observed that it followed the establishment of absolute monarchies in Europe, which may have facilitated the projection of a divine sovereign, legislating for the universe as the monarch legislated for society. On the basis of such analogies, it has even been suggested that a mechanical philosophy could serve to undermine papal opposition to the divine right of kings. It is also argued that the experimentalists of the English Royal Society adopted mechanical metaphors as a way of legitimating their use of instruments, traditionally the preserve of the artisan. In the case of Descartes, one suspects that the universe had begun to seem self-evidently mechanical as he reflected on sophisticated machinery (clocks, looms, pumps, and fountains) of sufficient ingenuity to match some of nature's more recondite operations.

A smattering of hydraulics was already in the curriculum of the Jesuit college at which Descartes had received his education. As a youth he had planned mechanical models to simulate animal activity: a flying pigeon, and a pheasant hotly pursued by a spaniel! But what justification was there for regarding real animals as machines, for depriving them of their animal souls? From the very perfection of animal actions, Descartes observed, we suspect that they do not have free will. Compared with men and women who are often indecisive in their actions, animals behave in an apparently determinate manner, as if programmed to respond to external stimuli.

The idea that animal activity is determined by mechanical reflexes was sustained by a comparison with much that is involuntary in human behavior. A child placing its hand close to a fire would withdraw it without thinking. In fact the great majority of bodily movements appeared not to depend on the mind at all. Descartes listed the beating of the heart, the digestion of food, and respiration while asleep. Even walking and singing could be performed without the mind attending to them.

In the animal kingdom, the sheer regularity of instinctive behavior suggested a programmed mechanism. To speak of the birds and the bees was to speak of clockwork: "Beyond question when the swallows come in the Spring they act in this regard like clocks. All that is done by the honeybees is of the same nature." Why, Descartes asked, was it necessary to suppose that the essence of life consists in some special soul? Why could it not consist in the warmth of the heart? But the fundamental argument was derived from the fact, as Descartes perceived it, that pheasants and spaniels did not talk. Dogs were not devoid of vocal apparatus, but no dog had yet spoken French. Not a single brute speaks, Descartes declared; and this absense of communication implied an absence of rationality, an absence of a rational and immortal soul. Given that one would not even be tempted to attribute mind to a sponge or jellyfish, why confer it on more complex creatures? Man's self-consciousness in the perception of his self-consciousness made him the one exception.

What theological advantages were there in this rigid demarcation? To deprive the brutes of free will, Descartes observed, was no more than scholastic philosophers had already allowed. To deprive them of a sensitive soul might be to deprive them of feelings; but would it not be unjust if they did feel pain? It was humanity, not they, that had sinned and paid the price. To deprive the animals of an immortal soul was favorable to man in one final respect: Heaven would surely be more blissful in the absence of bites and stings. Applying his mechanical principles to the study of reproduction, he even absolved God of immediate responsibility for monstrous births. Physical deformity could be ascribed to some failure in the mechanism of generation.

It is tempting to ask why Descartes did not go the whole hog and turn men and women into robots. But that was a possibility he could never seriously entertain. He claimed that he had a clear and distinct idea of himself as a conscious being, from which he deduced that there was a sense in which he could be said to exist independently of his body. His mind was radically different from matter in that it exhibited the powers of imagination, understanding, and volition, each of which was foreign to matter. Consequently, there was a special substance in man, a thinking substance, in addition to the machinery of which his body was made. With some justice, later critics would object to the smooth transition from thinking substance to spiritual substance to immortal soul. But the transition was justified in Descartes's own mind because, once a unique immaterial substance was conceded, it became impossible to conceive how it could be destroyed by any physical cause. The existence of a soul, interacting with the body, and yet independent of it, was corroborated by the fact that different parts of the body could be affected by external agents without necessarily affecting the mind at all.

Descartes's most powerful argument, still entertained in some quarters, derived from the indivisibility of the human ego. One could conceive of the indefinite divisibility of matter, but not of the divisibility of the self-conscious personhood corresponding to the use of the word *I* in everyday speech. By way of clarification, Descartes insisted that he was not lodged in his body like a pilot in a ship. If he were, he would feel no pain when wounded. Despite the rigidity of his dualism between matter and spirit, he confessed to being "very intimately connected" to his body — so intermingled with it that he constituted "an entire whole with it." He pressed mechanical analogies so far, even in human physiology, that he has been called the father of neurophysiology. But his reasoning pointed toward, not away from, an eternal destiny for the human race.

Boyle and the Defense of God's Sovereignty

In the previous chapter it became clear that an unresolved dispute exists between those historians who see a correlation between puritanism and science in seventeenth-century England and those who insist that the correlation was rather between science and religious moderation. One way out of the impasse is to deny the value of attempts to assign natural philosophers to one or the other of such religious types and to recognize instead that the events of the civil war period could transform the perceptions of reformers who, from having been sympathetic to the early stages of revolution, eventually found themselves recoiling against the excessive demands of more radical campaigners. The demands of the extremists, which included disestablishment of the Church, abolition of tithes, the admission of lay preachers, the equality of women, extension of the franchise, property redistribution, and even sexual libertinism, were such that many who had once favored reform found themselves saying thus far and no further. To understand the relations between science and religion in mid-seventeenth-century England, it is well to recognize this continuing dialogue between conservative and radical reformers, the form of which could change as social and political circumstances changed.

One advantage of this approach is that it is more sensitive to the details of historical change than alternatives that aim to correlate an interest in science with some archetypal religious, or even hedonist, mentality. Another advantage is that it helps one to understand how certain strands within Descartes's mechanical philosophy could become attractive to the more conservative reformers who, like John Wilkins and Robert Boyle, were concerned about the threat to a stable society posed by puritan extremists. Because Descartes had poured a cold douche on those who claimed a hot line to God, his philosophy could be used against radicals who claimed the illumination of the Holy Spirit as their authority. Because he had also demonstrated the immortality of the soul, his dualism proved attractive in the assault on those who, like the Leveller Richard Overton, were arguing that the soul was mortal.

This does not mean that every aspect of Descartes's philosophy found favor in England. His reduction of animals to machines was generally considered eccentric. His reluctance to admit final causes into natural philosophy was often censured, as it was by Boyle, who felt that the French philosopher was unnecessarily depriving Christianity of one of the strongest arguments for God's existence. The vocabulary of English mechanical philosophers suggests that they found Descartes's emphasis on the complete passivity of matter too restrictive. But Descartes's vision of a universe running according to mechanical principles certainly took hold of the scientific imagination.

When Boyle spoke of the excellence of the mechanical philosophy, he produced many arguments in its favor. It would be wrong to imply that its advantages were purely theological or that Boyle adopted it merely for polemical purposes. Intelligibility and clarity were two intrinsic virtues he claimed for it. The case for reinterpreting traditional concepts, like that of *form*, in mechanical terms, had been developing over several decades. Some thirty years before Boyle pressed the case, the French chemist, Etienne de Clave, had already complained that the word *form* had come to mean at least ten different things. Versatility, in addition to clarity, was claimed for mechanical analogues as Boyle presented his corpuscular philosophy as a comprehensive alternative to that of Aristotle. Comparing his different corpuscles with the letters of the alphabet, Boyle was more than satisfied that there were letters enough to compose the book of God's works.

The case for a mechanical philosophy was not exclusively a theological one. Indeed, it was often justified in straightforward empirical terms. When the microscope revealed the compound eye of a fly, or the microstructure of small grains of sand, it gave a considerable boost to particulate theories of matter. The theological justification was nevertheless of fundamental importance, for Boyle found in his corpuscular philosophy the perfect antidote to all those heretical systems in which Nature (with a capital N) was seen as self-sufficient, as an active and productive source of all things. By stressing that his corpuscles behaved according to rules that God had freely chosen, he retained a central role for divine activity in the world. Nature was to be looked upon not as a distinct or separate agent but as a "system of rules, according to which those agents, and the bodies they work on, are by the great Author of things, determined to act and suffer."

We saw in Chapter II how Boyle used the mechanical philosophy to differentiate sharply between nature and God. But how exactly was God supposed to act in a world of whirling particles? Boyle gave an answer in an essay expressly directed against those who, in his own phrase, "would exclude the deity from intermeddling." Because the laws of nature, the "system of rules," were of God's devising, they were themselves an expression of divine activity. Because mechanistic explanations relied on the fact that matter exhibited certain properties, Boyle could argue that the choice of which properties to bestow on matter had been God's. Furthermore, for a viable universe, it was not enough that matter should have certain properties rather than others. The ultimate particles had to be put into certain situations and given certain motions. Divine activity was required to achieve that initial configuration. The phenomena that the divine will intended to appear would then follow in an orderly manner, but only through the assistance of His "ordinary preserving concourse."

Drawing on Descartes's definition of matter as extension, Boyle could make the further point that motion was in no way necessary to the essence of matter, for "matter is no less matter, when it rests, than when it is in motion." It could gain or lose motion; but no man, Boyle reported, had yet made

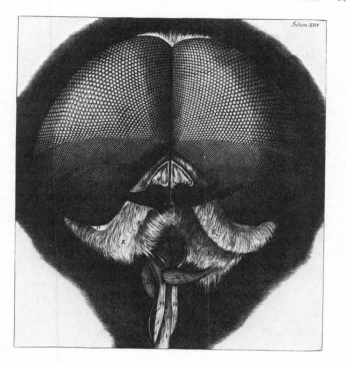

Figure IV. 2. Plate from opposite page 175 of Robert Hooke's *Micrographia* (1665); "Eye of a fly." The granular structure of a fly's eye was typical of microscopic phenomena that seemed to confirm an underlying reality of minute particles out of which the many edifices of nature had been constructed. Knowledge of mechanical structure, of nature's architecture, became a new form of active knowledge – knowledge by modeling and mechanical reconstruction rather than by reflection. In the preface to his *Micrographia,* Hooke declared that the "mechanical philosophy" alone was truly grounded in experiment and capable of giving access to physical reality. Reproduced by courtesy of The Bodleian Library, Oxford; Shelfmark Lister E 7.

out how matter could move itself. This had been an unwarrantable assumption, along with others such as the eternity of matter, on which the atomists of old had relied. The mechanical philosophy properly understood was supportive of divine activity. How, Boyle asked, could "so curious an engine" as this world possibly be produced by a "casual concurrence of the parts it consists of"? The design argument was reinforced, not overridden, by a philosophy of mechanism.

So far so good. But the argument from design was as compatible with the absentee god of the deists as with Christian theism. Among all these arguments, has Boyle yet produced one in favor of divine "intermeddling"? No; but he had one up his sleeve. Descartes once again came to his aid in establishing a powerful analogy, which Newton too would embrace. If there was a spiritual substance in man, then every voluntary human action was proof that an incorporeal and intelligent being could work upon matter. By analogy, why should God not do the same? There was no reason to discount the possibility that "there may be a spiritual deity, and that he may intermeddle with, and have an influence upon the operations of things corporeal."

This analogy with human voluntary activity may help to explain why Boyle's theology (and that of Newton) is sometimes called "voluntarist." It emphasized the freedom of God's will to create whatever world He wished, and His freedom to manipulate and dispose things as He saw fit. He was not bound by any kind of logical necessity, nor by the laws of nature, for they were simply expressions of the way He normally chose to act. The analogy may also suggest that what Boyle meant by "intermeddling" was not necessarily miraculous intervention. It did not require a miracle for a man to shift his leg or pick up his pen. Why should it require a miracle for God to shift matter?

Not that Boyle wished to exclude the miraculous. Quite the contrary: for he believed that the New Testament miracles were the surest proof of Christian doctrine:

For the miracles of Christ (especially his Resurrection) and those of his disciples, by being works altogether supernatural, overthrow atheism; and being owned to be done in God's name, and to authorize a doctrine ascribed to his inspiration; his goodness, and his wisdom, permit us not to believe that he would suffer such numerous, great miracles, to be set as his seals to a lie.[5]

In Boyle's clockwork universe there was therefore room for God's general providence, His particular providence when He chose to intermeddle, and His miraculous intervention when extraordinary effects were wrought by supernatural means.

Such distinctions were not purely academic. They helped to define the scope of mechanical philosophies. One could not define the province of mechanistic explanation without, tacitly at least, defining the province of "spirits." And this of course had been the preserve of priests more than experimental philosophers. Boyle himself would consult friends in the priesthood if his conscience told him that the investigation of phenomena commonly associated with the dark world of spirits might be illicit. He was inclined to divide the cosmos into "supernatural, natural in a stricter sense, that is, mechanical, and natural in a larger sense, that which I call supramechanical." The question of which processes should be placed in the supramechanical category produced a range of competing answers, with experimental programs in both pneumatics and chemistry designed to capture and reproduce the agency of subtle spirits. Boyle certainly made room for what he described as a "very agile and invisible" sort of fluid, for which the term *spirit* was commonly used. For this reason, it may be more accurate to see the English mechanical philosophers as capturing the domain of spirits with their experiments rather than discarding it. Boyle's contemporary John Mayow (1641–79) affirmed that there was an ingredient in the air, vital to human life. At times he would call it a nitro-aerial spirit. At times he would call it a nitro-aerial particle. But whatever its label, its very existence was proof of providence. It restored the atmosphere spoiled by respiration. Without it, "there would be no society at all . . . for we should be obliged to spend our lives single and separate, namely where a ration of nitro-aerial spirit sufficient for sustaining life might be obtained for each." There would be perpetual strife between mortals as they sought to partition their tracts of air. One could not implement a mechanistic program without confronting the continuum of meanings that the word spirit connoted. Boyle did not even demur from the view that experimental evidence in favor of intelligent and invisible beings might be useful in remonstrating with atheists, whose way of life implied a negligence of higher spiritual demands.

Newton and the Defense of God's Omnipresence

To say that mechanical philosophies had their origins in a voluntarist theology would be going too far, but the latter undoubtedly provided a convenient metaphysics for their promotion. In Newton, as in Boyle, one finds the same allowance made for a God whose sovereign will had not only dictated the properties bestowed on matter but was also capable of immediate action, whether directly or through the agency of natural causes. The irony is that it was only necessary to restrict the sources of activity to powers within matter itself to tear the veil and reveal a material universe in which divine activity was reduced to an initial act of creation and legislation. The irony runs deeper because Newton's gravitational force, and the short range forces of attraction and repulsion that he associated with atoms, were seen by later interpreters as just such inherent sources of activity. But not by Newton himself.

An extract from an early essay suggests that one of Newton's objectives had been to develop that analogy between human and divine volition that had been of central importance to Boyle:

I have deduced a description of this corporeal nature from our faculty of moving our bodies, so that all the difficulties of the conception may at length be reduced to that; and further so that God may appear (to our innermost consciousness) to have created the world solely by the act of will just as we move our bodies by an act of will alone; and, besides, so that I might show that the analogy between the Divine faculties and our own is greater than has formerly been perceived.[6]

In pursuing that theological goal, Newton was not merely philosophizing. His concern with the question of divine activity in the world was deeply informed by his study of the Bible. At a critical stage in his Cambridge career, he had had to come to terms with the moral demands that the divine ruler made of his servants. To retain his fellowship at Trinity College he expected that he would have to follow the usual pattern and take Holy Orders. But that would have meant swearing an allegiance to the thirty-nine articles of the Anglican Church, which in conscience he could not do. He had already come to reject the doctrine that Christ is of one substance and coeternal with the Father. In the event, he was let off the hook with a special dispensation (which may well have seemed like a special providence) allowing him to continue his work without being ordained. But he had lived through a pending crisis in which he must have weighed the cost to his soul were he to perjure himself.

Newton's philosophy of nature is sometimes discussed as if his conception of God were a mere appendage – a hypothesis to explain what his science could not, a god-of-the-gaps. But this is a superficial view, for Newton was deeply concerned with the action of God in human history. Fascinated by the history of ancient kingdoms, he would trace their degradation as they lapsed from monotheism into idolatry. His historical researches were designed to show that biblical prophecies had been fulfilled. They also reflected his conviction that he belonged to a faithful remnant whose Christianity, unlike that of the established churches, had not been defiled through centuries of political machination and doctrinal corruption. His aversion to the Roman Catholic Church was apparent in his treatment of Revelation 12:6, where 1,260 years of the reign of Antichrist were identified with the years of papal domination. Newton's adage that "the end of prophecy is not to make us prophets" was directed against enthusiasts who were forever predicting dates for apocalyptic events. But prophecies already and unmistakably fulfilled were evidence of the divine arm in human affairs. A belief in God's omnipresence even penetrated his analysis of space.

In Newton's mind, space was associated with the intimate presence of God, who knew and perceived all things, who knew when his servants disobeyed Him, who knew that Newton himself had once eaten an apple in Church, had once made a mousetrap on the Sabbath, had once lied about a louse to his roommate in Cambridge. His overbearing sense of a divine presence has been analyzed in psychological terms as a consequence of his having been a posthumous child. Never having known an earthly father, he found a heavenly substitute on whom every absolute was conferred. But whatever its origins, Newton's conviction that "God was everywhere from eternity" had implications for how space and time were to be conceived. They, too, became absolute rather than relative constructs. For Descartes there had been no space without matter; for Newton there was no space without God. It was not that space was an attribute of God, or the Body of God, or the sensory apparatus of God. But there was a sense in which space was constituted by God's omnipresence.

To help in defining God's relation to space, Newton employed the Hebrew word *maqom*. The Hebrews, he wrote, "called God *maqom* place, the place in which we live and move and have our being and yet did not mean that space is God

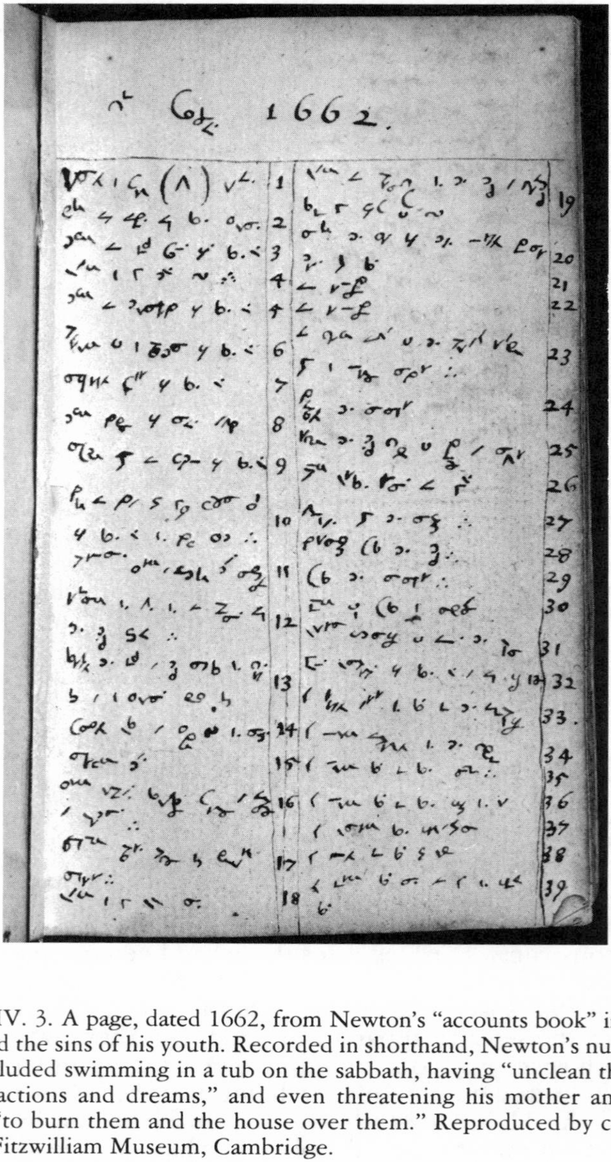

Figure IV. 3. A page, dated 1662, from Newton's "accounts book" in which he listed the sins of his youth. Recorded in shorthand, Newton's numerous sins included swimming in a tub on the sabbath, having "unclean thoughts words actions and dreams," and even threatening his mother and stepfather "to burn them and the house over them." Reproduced by courtesy of the Fitzwilliam Museum, Cambridge.

in a literal sense." In an early essay, *De gravitatione,* he argued
that space is eternal in duration and immutable in nature precisely because it is "an emanative effect of eternal and immutable being." The most perfect idea of God, he declared,
is one in which He necessarily exists everywhere, a substance
in which all other substances are contained – a substance
"which by His own presence discerns and rules all things."
When he wrote that "in Him are all things contained and
moved," the supporting references included eight from
Scripture, most notably St. Paul's speech in Acts 17:28. The
analogy that Newton reiterated, between the power of God
and our ability to move our bodies, was embedded in a long
tradition of Jewish and Christian speculation concerning God's
relation to space.

This might be considered of minor importance were it not
for the fact that Newton's association of space with the omnipresence of God provided one justification for regarding
his inverse-square law of gravitation as a universal law. In
Query 31 of his *Opticks* Newton explained that, because God
is present in all places, He is *more* able by His will to move
bodies (and thereby to form and reform parts of the universe) than we are to move our limbs. The link with the universality of the laws of motion was, however, made explicit
in an earlier, unpublished draft of that Query:

If there be an universal life and all space be the sensorium of a
thinking being who by immediate presence perceives all things in
it . . . the laws of motion arising from life or will may be of universal
extent.[7]

This may help to explain how Newton became so committed
to the universality of his law of gravitation, despite the existence of some evidence that could have told against it. The
stability of the fixed stars, for example, was all the more remarkable in a universe governed by gravity: One might expect that the most peripheral would experience a resultant
"pull" inward. If the universe were finite, a devastating im-
plosion would eventually occur.

Mechanical Philosophies and Religious Apprehension

The association of a mechanical philosophy with a voluntarist
theology served its exponents well – not least because it helped

to justify experimental methods. If the workings of nature
reflected the free agency of a divine will, then the only way
to uncover them was by empirical investigation. No armchair
science, premised on how God *must* have organized things,
was permissible. This meant that theological moves could be
made in criticizing any science that seemed too conjectural.
And this might include the very mechanical models favored
by one's predecessors! One of the claims made on Newton's
behalf was that he had discovered the laws of motion that
God had actually implemented, by contrast with Descartes
whose mechanical models, by his own admission, could never
provide more than a plausible account of how the clockwork
worked.

The promotion of a mechanical philosophy on religious
grounds was not, therefore, without its problems. And there
were to be more. The clockwork analogies of Boyle or Descartes, though lodged in theologies of nature that remained
Christian in inspiration, were to appear perfectly at home when
lodged in deistic philosophies – in the anti-Christian literature of the Enlightenment. For Voltaire, in the eighteenth
century, the clockmaker God was to be an attractive alternative to the gods of established religions. It would be surprising if such tendencies had not been discerned in the seventeenth century and religious apprehension expressed. But what
form did such apprehension take?

Although Mersenne had seen value in a mechanical philosophy that would help preserve the miraculous, it soon became apparent that if the mechanistic explanations he favored were imported into the naturalistic philosophies he was
attacking, they could so encroach on holy ground that they
might exclude the miraculous. This soon became a problem
with Descartes's account of how the solar system originated.
His detailed account of how the material vortices had assumed their present form was construed in some quarters as
disrespectful to the Bible. In England, in 1662, a future bishop
of Worcester, Edward Stillingfleet, complained that atheists
were busily exploiting "the account which may be given of
the origin of things from principles of philosophy without the
Scriptures." Five years later, a Cambridge tutor was complaining that his students were deriving from Descartes "notions of ill consequence to religion."

The nature of a mechanical philosophy might be theistic in
tone, but its nurture was liable to be otherwise. Given a cer-

tain arrangement of atoms, argued the French eccentric Cyrano de Bergerac, some object or other was bound to be formed. It was not in the least marvelous, he suggested, that among them would be trees, frogs, monkeys, and men. So, too, with Thomas Hobbes, a mechanical philosophy was associated with belief in a material soul, with a critical attitude toward the Scriptures, with an explanation for the origins of religious belief in an ignorance of natural causes, and with his discomfiting emphasis on self-interest as that which had to be accommodated in a social contract. Although Hobbes was not an atheist, he was commonly despised as one. He stood for a set of meanings that the term *mechanical philosophy* might denote, and from which Boyle and other Christian virtuosi had to distance themselves.

The beast-machine doctrine of Descartes was another source of apprehension. To regard God's creatures as no different in kind (however much Descartes might protest that they differed in degree) from human artifacts caused offense to Gisbert Voet, rector of the University of Utrecht. Another critic saw how the beast-machine could become the thin end of a wedge: "If one suppresses the vegetative and sensitive soul in brutes, one opens the door to the atheists, who will attribute the operations of the rational soul to a cause of the same kind and will give us a material soul to replace our spiritual soul." And he was right. Material souls, made of fiery particles, had made many an entry by the close of the seventeenth century. For thoughts to be produced, it was argued, all that was necessary was for fine particles to flow through the filigree ducts of the brain. This mechanization of the mind was to be a familiar theme in the clandestine literature of the Enlightenment. The ultimate fear was that it might strip humanity of free will. Further anxieties were raised by the beast-machine. It was often said that the test of a good Cartesian was whether he would kick his dog. Descartes's doctrine provided ample justification for a widespread cruelty to animals. For the English naturalist John Ray it was an indictment of Descartes's system that it was so viciously anthropocentric.

The question of divine activity in a mechanized universe became even more urgent in the context of Christian worship, and especially for the Roman Catholic Church, which took a distinctive view of the presence of Christ at the celebration of the Eucharist. With an Aristotelian theory of matter and form, it was possible to understand how the bread and wine could retain their sensible properties while their substance was miraculously turned into the body and blood of Christ. Drawing on the categories of St. Thomas Aquinas, it was said that the accidents of color, odor, and taste could continue to exist without their original substance. But if, as the mechanical philosophers argued, the sensible properties were dependent on an ulterior configuration of particles, then any alteration to that internal structure would have discernible effects. The bread and wine would no longer appear as bread and wine if a real change had occurred. From a recently discovered document in the Vatican archives, it is clear that the atomism of Galileo was suspect for precisely this reason. If, following the consecration, the particles of the bread no longer remained, then, as Galileo's critic observed, no accident of the bread could remain. Substance and accident were no longer separable as the miracle of transubstantiation required.

Descartes barely admitted the difficulty. As long as the surface texture of the bread and wine remained the same, they would continue to generate the same perceptions in us. As long as the real presence of Christ was confined within those surface boundaries, the transubstantiation could occur without perceivable effects. But Descartes's critics were not so easily appeased. After all, the Council of Trent had stipulated that *all* the substance of the bread was transmuted into the body of Christ. The theologian Antoine Arnauld (1612–94) warned Descartes that the derivation of qualities merely from extension and motion would prove impossible to reconcile with Catholic doctrine. The efforts of Claude Clerselier to promote the Cartesian reinterpretation were duly rewarded with censorship from Louis XIV and the archbishop of Paris. By September 1671 Descartes's doctrines were condemned by royal authority and by the University of Paris. Persecution was the word used by the Cartesians to describe their harassment. There was no comparable problem for Protestants, who often ridiculed the doctrine of transubstantiation as the epitome of Catholic mystification.

There was, however, another problem, which Protestants were not spared. This arose in the context of explaining what was meant by laws of nature. Despite Boyle's assurance that they were not binding upon God, that they merely reflected how He normally chose to act, there were occasions when Boyle implied otherwise. The pressure to do so came not so much from his scientific endeavors, as from a desire to say something constructive about the existence of suffering in the world. It was perhaps unreasonable, Boyle suggested, to expect God to intervene in every case when an individual was at risk. Was it not conceivable that God should subordinate the welfare of individuals to His "care of maintaining the universal system . . . and especially those catholic rules of motion and other grand laws, which He at first established among the portions of the mundane matter"? This may sound callous, but could one reasonably suppose that the deity would suspend a law of gravity just because someone fell over a cliff?

Whatever Boyle's intentions, it was possible to read into his remarks a diminution in the scope of God's particular providence. Again, for all his insistence that nature is not autonomous, that the clockwork has an external source of power, he was driven on occasions to use a terminology that gave the impression that the Almighty had created the world and then retired. He wanted to show that a world seemingly running by itself would be just as indicative of design, if not more so, than one requiring constant attention. But his references to a "self-moving engine" could be read as an invitation to deism. In fact, that is precisely how they were read, even by as sympathetic a scholar as John Ray, before he realized his mistake and duly apologized. That initial misreading is highly instructive because it shows how a mechanical philosophy designed to uphold the sovereignty of God could so easily turn Him, and literally so, into a *deus ex machina*.

The analogy between divine activity and human voluntary action was designed to prevent that from happening. But, as employed by Boyle and Newton, it too ran into difficulties. It is perhaps a measure of the problem with which they were wrestling that they were driven to this essentially organismic analogy to explain how God could be active in a mechanical universe. The real source of apprehension, however, was located in the inferences that might be drawn if the analogy were taken too literally. Newton did sometimes seem to be saying that space *is* the sensorium of God, inviting the objection that he was regarding the physical universe as the body of God. In other words, the analogy designed to protect God's freedom to act how He wished could be a slippery slope that led to pantheism. Newton's critic Leibniz was uneasy for that very reason. Newton might say that God, by His presence, perceives all things; but, for Leibniz, it was more appropriate to say that God knows all things because He produces them.

Such metaphysical squabbling shows how difficult it was to construct a model for divine activity in the world. In one respect the mechanical philosophies merely reinforced an ancient paradox: how God's foreknowledge of events could be squared with human freedom. For a Christian theist like Boyle, such problems belonged to a domain of "things above reason." They constituted paradoxes that could not be resolved by single analogies. Indeed he said as much when discussing the *nature* of God. Because it comprehended all perfections in their utmost degree, it was "not like to be comprehensible by our minds." For Boyle, the failure of analogies to explicate divine activity in the world was only to be expected. But for later generations less tolerant of paradox, less tolerant of things above reason, less tolerant of a realm of grace in addition to the realm of nature, a clockwork universe demanded nothing more than an original clockmaker.

Divine Activity in Newton's Universe: A Dilemma, an Ambiguity, and an Irony

Apprehension lest a fully mechanized universe might cripple divine activity was felt by no one more keenly than Newton. His disenchantment with the cosmology of Descartes was partly due to the boldness with which the French philosopher had presumed to show how an organized solar system could develop from a disorganized distribution of matter. Newton insisted that organization could not result from disorganization without the mediation of an intelligent power. As if to defuse the deistic tendencies of Cartesian philosophy, Newton scrutinized the universe for evidence of divine involvement. As he did so, however, he placed himself on the horns of a dilemma.

Because his voluntarist theology allowed events in nature to be explained *both* as the result of mechanism *and* of the divine will, there was a difficulty in determining what kind of event would most demonstrate divine involvement. The most spectacular evidence would surely come from extraordinary as distinct from ordinary phenomena. And yet, on a voluntarist theology, extraordinary events could still be envisaged as resulting from a divinely instituted mechanism. In which case, if a mechanism were specified for the extraordinary event, would not a skeptic say that references to divine activity are superfluous? Newton experienced this dilemma and there were respects in which his strategy proved self-defeating. By speculating, for example, on the mechanism by which God reformed the solar system, he drew attention to the role of providence in nature – but at a price. Those who did not share his religious sensibilities would look at the mechanism and see no further.

We shall consider this example in a moment, but it is necessary first to identify some of the ambiguities that arose in the interpretation of Newton's science. Because his own remarks left options open rather than closed, his references to divine activity in the world proved as rich a resource for Anglican clerics as for the deists they were attacking. Take the gravitational force itself. Though Newton had calculated how it varied with distance, he had little idea of *how* it acted. No fewer than four possibilities were explored, none of which he definitively abandoned. At times he toyed with the mechanism of an ether consisting of the most tenuous matter, the particles of which were associated with short-range repulsive forces. At other times he revealed his Platonist heritage, seeing in the diffusion of light a model both for divine activity in the world and for gravitational phenomena. On other occasions he used a vocabulary of "active principles" drawn from an alchemical tradition. These were sources of activity in nature, associated with matter but not inherent in it. Finally, the invisibility and immateriality of the gravitational force allowed him to write as if it were a direct expression of God's continual activity.

Newton's dilemma arguably arose from his resolute determination to avoid a fifth possibility, a possibility vetoed by his voluntarism but having the merit of clarity – namely that the source of activity *was* inherent in matter. But because he resisted that option, and because he left open the possibility of direct divine activity, his philosophy provided succor to theists as well as deists. In a letter to the Reverend Richard Bentley, who was pumping him for advice on how to turn the latest science to theological advantage, Newton admitted the ambiguity:

Gravity must be caused by an agent acting constantly according to certain laws, but whether this agent be material or immaterial I have left to the consideration of my readers.[8]

That different readers would draw different conclusions is clear from the remarks of two other contemporaries. William

Whiston (1667–1752), who succeeded Newton in the Lucasian Chair of Mathematics at Cambridge, identified the gravitational force with the interposition of God's "general, immechanical, immediate power." By contrast, the deist Anthony Collins (1676–1729) invoked Newton's authority to declare that gravity proved the activity of matter.

In his quest for evidence of divine involvement, Newton created a God in his own image, a divine intelligence "very well skilled in mechanics and geometry." He could do so because a gravitational force directed to the center of bodies was arguably insufficient to explain the origin of planetary orbits. A body falling toward the sun would have to acquire a transverse component of velocity if it were to go into orbit rather than fall into the sun or drift past it. Because the inverse-square law of gravitation was compatible with parabolic and hyperbolic paths, the planet's assumption of a closed, elliptical orbit depended on its receiving exactly the right "flick" at exactly the right time. It was an exquisite calculation on the part of the deity who had to consider the "several distances of the primary planets from the sun and secondary ones from Saturn, Jupiter and the Earth, and the velocities with which these planets could revolve at those distances about those quantities of matter in central bodies."

Newton had no difficulty, then, in arguing for the existence of an intelligent Being involved in the original creation. But a degree of ambivalence was still possible because few deists would have objected to a role for God at creation. To accept that initial role need not imply a continuing providence thereafter. And there were times when Newton seemed to hint as much. In Query 31 of the *Opticks,* after dismissing the view that the world could have arisen out of chaos by the mere laws of nature, he added that "once formed, it may continue by those laws for many ages."

Clearly, further evidence was required if divine activity since creation was to be affirmed. Newton found it in the stability of the fixed stars. Even in an infinite universe one would expect stellar movement because it was inconceivable that the resultant gravitational force acting on every star was zero. Newton's answer to the question What hinders the fixed stars from falling upon one another? involved divine providence. But once again an ambiguity arose. Was it that God directly held them in place? Or had He the foresight to place them so far apart that any resultant force would be negligible?

Newton's contemporary David Gregory opted for the former, noting that "a continual miracle is needed to prevent the Sun and fixed stars from rushing together." But Newton's conjecture that the problem might have been minimized at creation could again attenuate the argument for God's constant involvement. Newton's position has been described as one in which God had taken every precaution to minimize the destabilizing forces and yet had willed a world in which His intervention would also be required. There was no contradiction in such a view, but it smacked of an ambivalence that was always liable to be resolved in favor of divine foresight rather than divine intervention. For, as Leibniz objected, if God had to remedy the defects of His creation, this was surely to demean His craftsmanship.

That objection was directed toward Newton's belief that the solar system would require an occasional "reformation." If the planets were moving through an ether, however tenuous, they would surely be retarded by friction. Motion, in Newton's memorable phrase, was more apt to be lost than got. The intrusion of foreign bodies into the system could induce irregularity, pulling on any planet that might be in their vicinity. The sun, too, might lose mass through vaporization. The long-term security of the system needed a safeguard in the shape of divine providence. Without it, planets would stray or implode into the sun. But what provision had providence made? The same dilemma remained. Was this reformation achieved by direct fiat? Or was there a divinely controlled mechanism? The former might be the more spectacular, if the effects were visible. But, in a letter to Thomas Burnet, Newton had stated that "where natural causes are at hand God uses them as instruments." And Newton believed there were such causes at hand in the shape of comets. The irony is that, once a natural cause was found, the exponents of a thoroughgoing naturalism could claim it as an argument for dispensing with providence.

There is a popular view that, by reducing the paths of comets to regular laws, Newton divested them of religious significance. Formerly signs of divine displeasure, portents of catastrophe, they were stripped of that superstitious gloss once their return could be predicted. In fact, Newton reinvested comets with religious significance when they were already losing it. In 1683, four years before Newton's great book, the *Principia,* was published, the Congregational minister

Nathaniel Mather wrote to his brother Increase in a manner that shows that divines were already questioning a former superstition. Of comets, Nathaniel wrote they "do no more portend than eclipses, and eclipses no more than the constant conjunctions of the sun and moon, that is, just nothing at all, save only as they may be natural causes of alteration of air or weather." He did not deny that they were symbols of God's power, as were thunder and lightning. But no way were they "teachers from God to ourselves."

Newton reinvested comets with religious significance precisely because they were instruments of providence. Matter from their tails could replenish matter lost by planets, as he suggested in the first edition of *Principia,* or by the sun, as he suggested in the second. The comet would lose its matter through gravitational attraction as it passed close to the sun. Accordingly, "fixed stars, that have been gradually wasted by the light and vapors emitted from them for a long time, may be recruited by comets that fall upon them." The masses of suns and planets could be kept in balance, the orbits of the latter preserved.

With hindsight, Newton's strategy is apt to appear self-defeating. If comets fulfilled this ulterior purpose in a synchronized manner, was there any point, from the standpoint of natural philosophy, in trying to distinguish between ordinary and extraordinary providence? Could not the latter always be subsumed under the former? Were the reformations not, as it were, preprogrammed? For those who thought that a natural mechanism was sufficient, Newton had supplied it. By stressing periodic reformation Newton also invited the inference that a series of earths may have arisen, each from the ruins of its predecessor. If the processes of decay and reformation were cyclic, *might* they have been in operation from eternity? Hence the paradox and the irony, for it had been the fear of such a notion that had prompted Newton to underline a continuing providence. By the end of the eighteen century there would be an additional irony when the French mathematicians Laplace and Lagrange showed that irregularities induced in planetary orbits could be self-correcting. That did not have to mean that the universe was any less the work of design, but it was an embarrassing exposé of what could go wrong when a religious apologia was rooted in fallible science.

To that later generation of French secularists, Newton's

religious interests were essentially pathological. Even today, one sometimes encounters surprise that the man who laid the foundations of classical mechanics could have been so absorbed by biblical prophecy and the religious dimensions of alchemy. In his historical studies he did have prejudices that, if not idiosyncratic, were certainly lagging behind contemporary scholarship. The notion, for example, that any pagan civilization might have preceded the Jewish was anathema to him. Greek, Latin, Egyptian, and Persian chroniclers had "made their first kings a little older than the truth." But Newton was not schizophrenic. The rationalism characteristic of his scientific work was not so much *deflected* as *reflected* in his biblical studies. The same mind that set up rules for interpreting nature did the same for the correct interpretation of Scripture. With supreme self-assurance, he hoped to eliminate disputation both in natural philosophy and in biblical exegesis by achieving definitive truth. The existence of competing, speculative hypotheses was a symptom as distressing in natural philosophy as the existence of idolatrous deviation from a true, original monotheism in religion.

Newton formalized no fewer than fifteen rules for biblical interpretation. Just as one paid attention to the analogy and uniformity of nature, so one respected the analogy of prophetic style. A prophetic symbol such as the "beast" of the Apocalypse always signified a kingdom or similar body politic. Just as one sought certainty in the mathematization of nature, so one should choose interpretations of Scripture that converged on a unique and literal meaning. Just as, in debate with Leibniz, Newton would invoke a principle of simplicity to eliminate subtle matter from the heavens, so he insisted that constructions placed on Scripture were to reduce matters to the greatest simplicity. In his ninth rule, he drew the parallel himself: "It is the perfection of all God's works that they are done with the greatest simplicity. . . . And therefore as they that would understand the frame of the world must endeavour to reduce their knowledge to all possible simplicity, so it must be in seeking to understand these visions."

If one is looking for Newton the rationalist, one can still find him. He is visible in the manner with which angels, spirits, and devils were treated in his biblical study. There was a certain reductionism in his reading whereby the "cherubim" and "seraphim" became hieroglyphs of ordinary social groups. Evil spirits became mental disorders, and devils became the

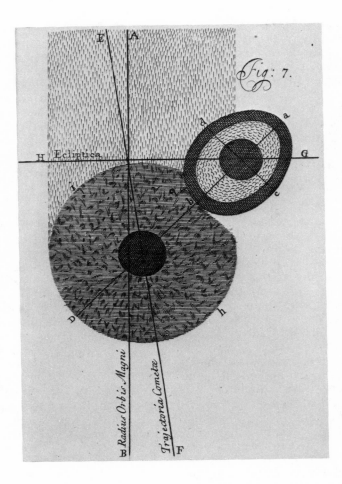

Figure IV. 4. Figure 7 from William Whiston's *New theory of the earth* (1696). Whiston proposed that, at the time of Noah, gravitational attraction between a passing comet and the earth had resulted in deformation of the latter, with a consequent cracking of its surface crust. Material from the comet's tail had then rained down. As in the scheme of Thomas Burnet, however, subterranean fluid also poured up through the cracks in keeping with the description that "the fountains of the deep burst forth." Reproduced by permission of the Syndics of Cambridge University Library.

imaginary ghosts of the departed. In his treatment of the six days of Genesis, a certain rationalism intruded. But — and this is crucially important — it was directed toward the protection of a literal, not a figurative, meaning. The first two days could be made as long as one wished — if the earth had not yet begun its rotation! The mechanization of nature may have given the sciences a higher profile in theological debate, but it would be quite wrong to imagine that God had yet been ousted from his universe. For Newton's successor, William Whiston, it was a matter of great excitement when he realized that the Genesis flood could be confirmed by retrospective calculations showing that a particular comet, actually seen by himself, would have been in the right place at the right time to have triggered it.

CHAPTER V

Science and Religion in the Enlightenment

Introduction: The Assault on Established Christianity

In his *Dictionary of chemistry* (1789), the entrepreneur James Keir (1735–1828) reported that a new spirit was abroad: "the diffusion of a general knowledge, and of a taste for science, over all the classes of men, in every nation of Europe." Even allowing for exaggeration, that growing appetite for science contrasts with the leaner fortunes of certain religious institutions. Fifty years earlier in England it had not been uncommon to hear Anglican clergy bewailing a widespread notion that Christianity had been discredited. The contrast is striking. Between 1660 and 1793 the scientific world established itself with more than seventy official scientific societies (and almost as many private ones) in urban centers as far removed as St. Petersburg and Philadelphia. In France alone there were thirty. The established churches, however, often perceived themselves to be in danger, both from dissenting religious movements and from a ground swell of rationalism and ridicule. But what connections were there between the popularization of science and rationalist movements that threw Christian theologians on the defense? In this chapter, which is focused on the eighteenth century, we aim to find out.

Attacks on the power of the Christian churches, and of the Roman Catholic Church in particular, were launched by deists, who denied the authority of doctrines supposedly derived from revelation; by materialists, who denied a duality between matter and spirit; and by agnostics, who, like David Hume (1711–76), argued that it was impossible to know

152

anything about the nature of God that need affect human conduct. The extent to which each of these critiques drew on the resources of science will be our principal concern.

Historians are in no doubt that there were connections between reverence for science and irreverence toward religion, especially among representatives of Enlightenment culture in France. Zealous in his popularization of Newton's science, Voltaire (1694–1778) was an irrepressible critic of the Catholic Church. Here is his caricature of an Old Testament text in which reference is made to human dung as the fuel over which Ezekiel would cook his breakfast:

Can I repeat without vomiting what God commands Ezekiel to do? I must do it. God commands him to eat barley bread cooked with shit. Is it credible that the filthiest scoundrel of our time could imagine such excremental rubbish? Yes, my brethren, the prophet eats his barley bread with his own excrement: he complains that this breakfast disgusts him a little and God, as a conciliatory gesture, permits him to mix his bread with cow dung instead. Here then is a prototype, a prefiguration of the Church of Jesus Christ.[1]

Or consider the sarcastic proposal of Jean d'Alembert (1717–83), whose *Preliminary discourse* (1751) to the *Encyclopaedia* of Diderot outlined a philosophy in which the rigor of geometrical reasoning set the standard for all pretensions to knowledge:

My idea would be to show great politeness to these poor Christians, to tell them that they are right and that what they teach and preach is as clear as daylight, that it is impossible that everybody should not ultimately agree with them, but that, in view of human vanity and obstinacy, it is a good thing to allow each man to think what he likes, and they will soon have the pleasure of seeing everybody share their opinion.[2]

An attitude of respect for the sciences, but condescension toward orthodox religion, was usually sustained by extolling the power of human reason. Newton's gravitational theory, with its solution to the problem of planetary motion, symbolized what the human intellect could achieve. Science was respected not simply for its results, but as a way of thinking. It offered the prospect of enlightenment through the correction of past error, and especially through its power to override superstition. An English deist, Matthew Tindal, saw such untapped wealth in human reason that he suggested in 1730 that men ought to be judged, and would be judged by the

Almighty, according to the use they had made of it. The most rational religion, as Joseph Priestley (1733–1804) later argued, was the one that would prevail, once all religions could compete on equal terms. A common grievance among eighteenth-century rationalists was that established churches were privileged only through arbitrary political power.

A further connection between reverence for science and irreverence toward religion was forged at an epistemological level. If scientific knowledge derived from reflection on ideas that arose ultimately from sense experience, it was tempting to generalize and say that no other mode of knowing was possible. Claims for knowledge based on revelation, divine illumination, or any form of intuitive access to a realm of divine grace could be dismissed as ill-founded. In this respect, the philosophy of John Locke (1632–1704) was a valuable resource for the deists, not least because, in his *Letter concerning toleration,* written between 1685 and 1686 during exile in Holland, he had argued that religious beliefs were of a kind that could not be legislated.

The concept of natural law provided another channel through which the achievements of science could be favorably compared with the effects of religious complacency. The physical sciences supplied a model of progress, as diverse effects were explained by single laws. Why should the search for laws not be undertaken in the study of human societies, and even human nature? Such a project proved attractive to many eighteenth-century philosophers, notably Montesquieu, who sought to correlate the character of a society with its political, legal, and social structure, the temperament of different peoples with the climate to which they were exposed. Once the laws governing the operation of the mind became a legitimate field of inquiry, the effect was to make men and women products of nature and of earthly society, rather than pilgrims bound for heaven. And if to be a pilgrim meant the renunciation of natural passions, so much the worse for Christian ethics. Enlightenment thinkers as diverse as Diderot and d'Holbach, Rousseau and Hume sought to liberate natural instincts from what the latter condemned as the "whole train of monkish virtues." For Diderot, the guiltless sexuality of the Tahitians made an inviting contrast with the sexual repression of the clergy.

There were connections, too, wherever scientific reasoning pointed to conclusions at variance with religious orthodoxy. During the eighteenth century, theories of the earth were proposed, challenging the customary view that human history and the earth's history had been coextensive. By the end of the century, especially in France, there was a growing aversion among natural philosophers to talk of divine intervention. Almost imperceptibly, confidence had grown that solutions to human problems lay with human effort rather than through the protection of the Church. In the seventeenth century, especially before 1660, religious exorcism had been common in tackling the insane; but, thereafter, satanic explanations faded away. In Britain, extreme religious enthusiasm was itself seen as a symptom of madness by new medical elites. By the end of the eighteenth century, with notable exceptions such as the evangelical preacher John Wesley, it had become less acceptable to ascribe illness to divine warning or punishment. A cold in the head, which Samuel Pepys had interpreted as divine retribution for illicit flirtation, would, by later generations, be attributed exclusively to the draught from the broken window against which he had been sitting.

Such shifting sensibilities are not easily explained, but they were often proclaimed by writers who saw in the sciences a vehicle of social and intellectual liberation. The polarity between "reason" and "superstition" was a recurring motif in the rhetoric of the Enlightenment, reinforced by claims for a rigorous methodology in the sciences that religious inquiry could not match. But to reduce the relations between science and religion to such a polarity, even in the age of reason, would be misleading. Science was popularized for many reasons having nothing to do with religion. In some cases, it was seen as a friend of Christianity, not a foe. Conversely, the motivation of those who pitted science against religion often had little to do with gaining intellectual freedom for the study of nature. It was often not the natural philosophers themselves, but thinkers with a social or political grievance who transformed the sciences into a secularizing force as they inveighed against clerical power. Such qualifications deserve special attention.

The Religious Utility of Science

Any suggestion that science was valued in the eighteenth century only, or even principally, as an antidote to "priest-craft" would be false. As propagandists for science found new social contexts in which to assert its utility, it was practical utility, in the sense of solving technical problems, that usually came to the fore. A classic example would be the determination of longitude at sea, a problem kept in the limelight by the association of mercantile interests with the safety of ships. One of Newton's popularizers, William Whiston, was instrumental in petitioning the English Parliament in 1714 to offer incentives for a viable solution. As repositories of specialized knowledge, scientific academies and societies were increasingly consulted. The Royal Society was approached in the 1770s when the British government wished to know whether a blunt or pointed lightning conductor would be the more effective. By opening new avenues for patronage, the scientific societies often provided a forum in which the rhetoric of utility could yield new dividends. The chemist William Cullen (1710–90) addressed himself to problems in agriculture, bleaching, and salt purification to please aristocratic patrons with whom he conversed at the Philosophical Society of Edinburgh. In England, the Lunar Society of Birmingham gave Joseph Priestley access to the wealth of entrepreneurs such as Matthew Boulton, James Watt, and Josiah Wedgwood. In return, he would offer advice on the effects of different gases in the steam engine or analyze clay samples for Wedgwood's pottery.

Studies of eighteenth-century scientific societies have, however, shown that the concept of utility carried broader meanings than the solution of technical problems. Even in a town such as Manchester, at the heart of Britain's industrial revolution, the men who gathered at the Literary and Philosophical Society (founded 1781) justified their allegiance to science with a multiplicity of arguments. The promise of technical application was dangled before manufacturers, even if relatively few benefited. The misfortune, declared one member, Thomas Henry, "is that few dyers are chemists, and few chemists dyers." But science was also commended as polite knowledge, a means of cultural expression particularly congenial to the medical membership. An acquaintance with natural knowledge could be urged as one mark of a gentleman. As founder of the Derby Philosophical Society in 1783, Erasmus Darwin (grandfather of Charles) announced that his society would seek "gentlemanly facts." Science was also promoted as rational entertainment, suitable for the youth of a burgeoning town who might otherwise be seduced by tavern or brothel. A "relish for manly science" was considered by one member of the Manchester society, Thomas Barnes, to be second only to religion in cultivating a mind likely to succeed in business.

Scientific knowledge could be prized for its supposed objectivity, in salutary contrast to the warring factions of political or religious parties. A dispassionate quest for objective knowledge was proclaimed a virtue in its own right, as by Fontenelle, the perpetual secretary of the Paris Academy of Science, whose eulogies of deceased members invariably praised their selfless commitment to human welfare. In the Manchester society, science was commended as a profession and an integral part of a reforming ideology that placed higher value on intellectual attainment than noble birth or inherited wealth. Finally, science was considered useful as a means of theological instruction. It gave content to arguments for God's power and foresight.

This belief that science had religious utility was widespread in the eighteenth century, particularly in Protestant countries. Certainly the God inferred from nature was not always the God of Christian orthodoxies; but in the minds of many it was. And, for them, the marvelous adaptations in the organic world underlined the wisdom of the God in whom they already believed. Christian, as well as anti-Christian, values could assist the popularization of science. Indeed, popular lecturers on science in Britain habitually displayed the powers of fire and electricity as impressive, almost theatrical, testimony to divine power. On a more elevated plane, the popularization of Newton's philosophy in England was due, in part, to the sermonizing of Anglican divines who, like Samuel Clarke (1675–1729), insisted that what was commonly called the course of nature was "nothing else but the will of God producing certain effects in a continued, regular, constant, and uniform manner."

When Robert Boyle had drawn up his will in July 1691, he had bequeathed fifty pounds annually for sermons to prove the Christian religion "against notorious infidels, viz. Atheists, Theists, Pagans, Jews, and Mahometans," without descending to controversies among Christians themselves. From the sermons of the first Boyle lecturer, Richard Bentley, it is apparent that certain features of Newton's science could fit the bill. The ambiguities in Newton's philosophy of nature, outlined in the preceding chapter, were systematically resolved in favor of a God active in both nature and history. Against materialist options, Bentley rejoiced that gravitational attraction was not an innate property of matter. He also argued that a mechanistic account of action-at-a-distance would require the emission of an effluvium from the attracting body and projected in every direction — a concept which he found repugnant. He also capitalized on Newton's point that gravity operated from the center of bodies, its magnitude depending on mass, not surface area. This seemed to exclude mechanical causation based on the pressure of corpuscles. For, if gravity could penetrate to the center of a solid body, it had to be a power "entirely different from that by which matter acts on matter." That was the opinion of Clarke who, like Bentley, inferred that the world depended "every moment on some Superior Being."

For Bentley, as for Newton, the elegance of the inverse-square law was proof of the divine mathematician, as was the fact that the planets had gone into closed orbits. Against those who derived the universe from a chance collision of atoms, Bentley had another Newtonian device. Committed to the ultimate unity of all matter, Newton had argued that the density of a body reflected the amount of empty space it contained. This porosity of matter was confirmed by the transmission of light corpuscles through as solid a material as glass. Such was the range of densities that most bodies must contain more space than matter. The implication, seized by Bentley, was the paucity of real matter in the universe. There were so few atoms compared with the dimensions of space that, given a uniform distribution, the odds were against any two colliding, let alone enough to make a world.

Bentley's arguments for the religious utility of science had political as well as philosophical dimensions. In the belief that atheism would encourage social instability, he was assuming that rational argument would be an effective counter. In other more specific ways, the political circumstances in England during the last decade of the seventeenth century created the space for Bentley's polemic. Following the Revolution of 1688, when William of Orange displaced James II from the throne, those clergy who sought to justify their allegiance to a new king could argue that a special divine mandate must take priority over a divine right to rule. Low Church Protestants warmed to the notion that God had willed the expulsion of

James, whose Catholic sympathies had produced unwelcome effects in as cloistered an environment as Cambridge.

During the 1680s the fear of popery had been a prominent theme in the thinking of those who would call themselves Whigs. For John Locke, the prospect of a Catholic monarch, with a threatening allegiance to a foreign power, had meant the abandonment of "the whole kingdom to bondage and popery." Newton himself had shared Locke's opinion that Catholicism could only appeal to uninquiring heads and unstable minds. Yet the beast had reared its head in Cambridge just as he was completing his *Principia* (1687), and in so ugly a manner that he had sacrificed his privacy to slay it. The occasion was a mandate from James II, asking Cambridge University to confer the degree of Master of Arts upon Alban Francis, a Benedictine monk, without requiring him to take the usual oath to uphold the Anglican faith. If James could catholicize his army, this looked like the thin end of the wedge in a bid to control the universities. Newton had encouraged the university to resist the royal will and was one of its representatives when the royal fury demanded an explanation.

Newton's science could be invested with political meanings because it was so easy to construct parallels between God's intentions for nature and for society. Newton's emphasis on the freedom of God's will, as disclosed in nature, could be useful when justifying the intervention of God's will in human affairs — even in the removal of a king. And, since God's activity in nature was mediated by physical laws, one could also argue for the moderating effect of Parliament in constraining an overzealous monarch. It would, however, be misleading to imply that such analogies provide the only explanation for the popularization of Newton's science. Prominent among those who promoted it were Huguenot émigrés from France and Scotsmen who, like the mathematician Colin Maclaurin (1698–1746), appreciated the quality of Newton's intellectual achievement. Among the first teachers of Newton's science at the University of Oxford were the high churchmen John Freind and John Keill, the latter having followed his mentor, David Gregory, from Edinburgh.

It would also be misleading to imply that Newtonian divines such as Richard Bentley and Samuel Clarke were representative of a dominant orthodoxy within the Anglican

Church. At least eighty percent of Anglican clergy in the period 1689–1720 were high churchmen who looked to parliamentary legislation rather than intellectual argument to suppress the atheist threat. By their fellow clergy, the Boyle lecturers were often perceived as heretics rather than upholders of orthodoxy. And not without reason. Newton privately disbelieved in the doctrine of the Trinity. His successor, William Whiston, lost his Cambridge chair by going public on that very issue. Bentley and Clarke, in the perception of their critics, were tarred with the same brush. To argue for the religious utility of the latest science was hardly a passport to preferment.

The fact that certain Anglican clergy assisted the popularization of Newton's science must not be allowed to conceal another important fact — namely that theological contentions were sometimes to the fore in critiques of Newtonian philosophy. This was especially true on the Continent of Europe where Leibniz, wielding a different metaphysical theology, would have no truck with Newtonian concepts. Leibniz was a Lutheran philosopher who, in wishing to do justice to the claims of theology, was also motivated by the desire to see Europe's Protestant churches united. Writing to his Catholic employer in 1679, he had even claimed that his philosophy would be accessible to the Jesuits. Rejecting the Cartesian definition that matter consisted only in extension, he insisted that every true substance had its own individuality, its own "soul." With the claim that he was restoring a concept of substantial form, he implied that he was creating space for a Catholic piety that had been compromised by Descartes.

During the second decade of the eighteenth century, Leibniz became embroiled in controversy with Newton's advocate, Samuel Clarke. Their exchange, which amounted to five letters each, began with a declaration from Leibniz that natural religion in England appeared to be in a state of decay. Writing to Caroline, princess of Wales, in November 1715, he observed that "many will have human souls to be material; others make God himself a corporeal being." Having cast aspersions on Newton and his followers, Leibniz excited a swift rejoinder from Clarke, who published the ensuing correspondence in 1717, following Leibniz's death. Theological issues dominated the debate, which was, however, colored and intensified by other matters. Antipathy had grown between

Newton and Leibniz because of a priority dispute over the calculus. Indeed, Newton once complained that Leibniz's critique of his gravitational theory had been designed to persuade the Germans that he (Newton) had lacked the wit to invent the new technique. National pride and personal jealousies were also involved. Having become court philosopher to the House of Hanover, and since on the death of Queen Anne there was a Hanoverian succession to the English throne, Leibniz cherished the hope that he would gain an official position at the new English court. For Newton and his supporters this was not a delightful prospect. Leibniz, for his part, was concerned that his royal pupil, Caroline, was falling under the spell of her Newtonian tutor, Samuel Clarke. The attitude that he struck may also have had deeper political roots, in that his aversion to systems of nature that stressed God's unrestrained power may have sprung from their use in sanctioning absolute powers claimed by an earthly sovereign — notably Louis XIV, whom he had long recognized as a threat to the peace of Europe, and of the German States in particular. In 1683 he had satirized Louis's military expansionism with references to Mars, the God of War.

An antipathy to voluntarist theologies is evident in Leibniz's remark that a secure foundation for law is to be found not so much in the divine will as in His intellect, not so much in His power as in His wisdom. Justice would be established "not so much by the Will as by the Benevolence and Wisdom of the Omniscient." The theological reasoning that underlay this position was concerned with the grounds on which God could be praised for His work in creation. If nature was deemed to be good merely because it was the product of God's will, this provided weaker grounds than if God could be shown to have deliberately structured the world according to standards of goodness, beauty, and wisdom that were independent of His will.

In the context of natural philosophy this meant that the world must reflect the rational constraints by which divine wisdom had been guided in the process of creation. Accordingly, theological arguments were pitted by Leibniz against the Newtonian vacuum:

To admit a vacuum in nature is ascribing to God a very imperfect work. I lay it down as a principle, that every perfection, which God could impart to things without derogating from their other perfections, has actually been imparted to them. Now let us fancy a space wholly empty. God could have placed some matter in that space: Therefore there is no space wholly empty: Therefore all is full.[3]

If there were a vacuum, there was no good reason why God should have stopped creating rather than enriching His work with further variety. Newton's atoms were also repudiated. It involved an unacceptable breach of continuity to suppose that at a certain stage of division one would suddenly encounter units not further divisible. The gravitational force was rejected as unintelligible. Because Newton had failed to explain its agency in mechanical terms, it had to be either an occult quality or a perpetual miracle. Leibniz had already set his face against perpetual miracles in his criticism of Nicolas Malebranche, who had responded to the Cartesian problem of how mind could interact with body by postulating a continuous intervention of God. Leibniz had preferred an alternative solution, sometimes called "psychophysical parallelism," according to which mental states correlated with physical states, but with each the result of independent causal chains, bound only by a preestablished harmony. There was nothing to be gained, in physiology or physics, by requiring constant miracles.

In Leibniz's theology there were grounds for praising God if His creation could be shown to be the best of all *possible* worlds. But such a world could hardly require the cosmic repairs urged by Newton, Bentley, and Clarke. Miracles, Leibniz insisted, were to supply the needs of grace, not to remedy second-rate clockwork. The controversy raged, too, on the nature of space and time, with Leibniz ridiculing Newton's reference to space as the sensorium of God. If the deity were possessed of organs, He could not be spirit but must be a corporeal Being. To crown his objections, Leibniz appealed to his principle of preestablished harmony. Once the motions of celestial bodies had been fixed, and living organisms created, everything that followed had been "purely natural, and entirely mechanical."

There was irony in this. For Leibniz's desire to uphold a category of the *purely* natural, the *entirely* mechanical, only confirmed the impression in English circles that it was his philosophy, not Newton's, that would benefit the deists. Leibniz would consider that a false impression because he took pains to distinguish between his kingdom of nature and kingdom of grace. The latter was the province of human minds, which were causally independent of the material world and not naturally destructible. Even the kingdom of nature was an empire within an empire. The natural philosopher could provide his mechanical explanations for the workings of an organism, but this need not exhaust the manner in which God's creatures were to be conceived. They also embodied His purposes. Teleological and mechanical explanation belonged to different modes of analysis; they were not mutually exclusive. Even the behavior of inanimate objects illustrated a purposiveness enshrined within nature. Leibniz was particularly impressed by the science of optics, which showed that light always seemed to travel by the "easiest" path. In the last analysis he believed, as strongly as the Newtonian divines, that science had a religious utility. It gave content to talk of perfection in the construction of the universe. His complaint was that Newton's universe was not perfect enough.

Some Nonscientific Roots of Religious Disenchantment

To assert an absolute polarity between science and religion in the Enlightenment would be to overlook such claims for the religious utility of science. But it might also conceal the fact that grievances expressed by critics of Christianity often had nothing to do with science. Nor was an appeal to science necessarily their most common or effective strategy. Because Newtonian science was susceptible to different interpretations, the particular gloss placed upon it could depend on prior religious convictions. These, in turn, had complex social and educational roots. Hostility to established churches, Protestant and Catholic, was often expressed through the complaint that institutionalized Christianity was a perversion of *natural* religion – a rational religion that would have been common to all humanity had it not been for the interference of priests.

The creed of this natural religion was formulated by Voltaire:

When reason, freed from its chains, will teach the people that there is only one God, that this God is the universal father of all men, who are brothers; that these brothers must be good and just to one another, and that they must practise all the virtues; that God, being good and just, must reward virtue and punish crimes; surely, my brethren, men will be better for it, and less superstitious.[4]

Such a creed appealed to many who believed that established churches had too much political power. Indeed, religious disaffection commonly had political rather than scientific roots. The priesthood was sometimes despised, as by Thomas Hobbes in the seventeenth century, for interfering with what might otherwise be a smoothly running monarchy. By the early eighteenth century it was commonly despised in France for lending support to a tyrannical monarch. In prescriptions for the moderation of the sovereign's power, the constitutional procedures recommended often had the effect of diminishing clerical power. And among the unfashionable few whose politics were republican, the priesthood might be deemed as dispensable as the monarch. As virulent a critic of Christianity as Voltaire was neither republican nor atheist. He even considered it a virtue of Newton's philosophy that it encouraged an awareness of a Supreme Being who had "created everything, arranged everything freely." Nevertheless, Voltaire's Supreme Being was not Newton's biblical God and emphatically not the God preached by the Catholic clergy.

The political grievance that united many deists and free thinkers concerned the threat to religious toleration posed by established churches and by all temporal powers claiming exclusive rights as interpreters of God's will. Discontent had been fueled in the later years of Louis XIV, who suppressed political and religious dissent in France with a severity that aroused sensibilities across Europe. The repeal of the Edict of Nantes in 1685 had forced Protestants out of France (including illustrious foreign members of the Academy of Science, such as Huygens), into Switzerland, Holland, and England, thereby creating nuclei of opposition to any regimes intolerant of toleration.

By the early eighteenth century, the political situation in England contrasted with that in France. Under the 1689 Toleration Act, religious dissenters enjoyed freedom to worship, provided their ministers complied with certain conditions, which included subscription to the doctrinal portion of the thirty-nine articles. But there was still scope for grievance. Dissenters were still discriminated against in that Test and Corporation Acts debarred them from educational privilege and access to the professions. And the more radical dis-

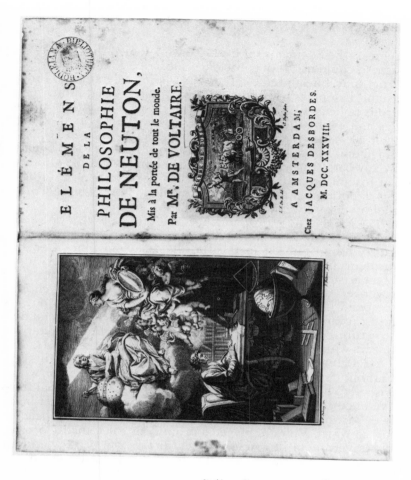

Figure V. 1. Frontispiece and title page from Voltaire's *Elémens de la philosophie de Neuton* (1738). Newton is here almost deified as a source of enlightenment. Unable to sympathize with the use of astronomy to secure the foundations of biblical chronology, Voltaire assured his readers that Newton had turned to such matters "to amuse himself after the fatigue of severer studies." Reproduced by courtesy of The Bodleian Library, Oxford; Shelfmark 50. b. 3.

senters, the Socinians for example who denied the divinity of Christ, could not with conscience subscribe to the thirty-nine articles. In England, the case for extending toleration to Protestant dissenters was often thwarted by fears that more generous policies would lead to the toleration of Roman Catholics and even atheists, the former (in Locke's words) owing their allegiance to a foreign prince, the latter (again in Locke's judgment) not to be trusted because they need not feel bound by any oath. Late in the eighteenth century when the Unitarian Joseph Priestley campaigned for the toleration of Catholics, he offended many of his fellow dissenters, who felt that their own case was being jeopardized.

The issue of toleration touched a sensitive nerve if one had suffered oneself. Voltaire, already imprisoned once for his satire, found himself in the Bastille a second time after an incident in 1726 when he had been mugged by the servants of an aristocrat with whom he had quarreled. His clamor for revenge was silenced by prison and an exile, which he chose to spend in England. On his return, he continued to nurse his grievance against the government and the offending structure of French society. Throughout his life he sympathized with those persecuted for their religious beliefs. In 1762 a Huguenot merchant was tortured and executed in the Catholic city of Toulouse for allegedly murdering his son, who wished to become a Catholic. Voltaire was convinced that the evidence pointed to suicide and that the wretched father had been victim of religious prejudice. Three years of campaigning, in which Voltaire played a prominent role, finally forced the government to rescind the verdict. Much of the venom that went into attacks on established Christianity was produced by a sense of political injustice rather than by the content of novel scientific theories.

Anticlericalism sometimes had social, as distinct from overtly political, roots − in that deists could draw on a reservoir of ridicule, which, in England at least, had been tapped since the restoration of the monarchy (1660) and increasingly directed against the sterner demands of an ascetic Christian ethic. Not all were as candid as one scoffer who had said that he would rather be in bed with his mistress than in heaven with Jesus Christ. But, among many who mocked the clergy or the sanctity of marriage, the sentiment was much the same. Bawdy Restoration comedies, with such titles as *The merry milkmaid of Islington, Love lost in the dark, The politic whore or the conceited cuckold,* had already poked fun at conventional morality and defined a set of attitudes (rather than an intellectual doctrine), which church leaders rightly recognized as subversive of their authority. A liberated wit became fashionable in early eighteenth-century England, a "social deism," as it has been called, paving the way for the more cerebral kind. Where Christian morality was perceived as a threat to human happiness, a common objection was that it was artificially sustained by the promise of a heavenly reckoning.

Thus the assault on Christian mores could owe more to the playhouse than it ever could to Newton's *Principia.* For as renowned a deist as the second earl of Shaftesbury, it was not a new and uniquely authoritative view of the universe that informed his outlook. What impressed him about natural philosophy was still its divisiveness, its lack of consensus. In fact it served him as a model for the diversity and inconsistency among those who sought to give Christianity a rational defense:

'Tis notorious that the chief opposers of atheism write upon contrary principles to one another, so as in a manner to confute themselves. . . . 'Tis the same in natural philosophy; some take one hypothesis, and some another.[5]

The Threat to Christianity from Deism

What weight *did* the deists attach to scientific authority in their crusade against "priestcraft"? It was always open to them to construct unfavorable comparisons between the apparently deductive methods of theological reasoning and the supposedly inductive methods of the natural sciences. If they agreed with Hobbes that ignorance of natural processes was one of the seeds from which religious opinions had sprung, they could find inspiration in scientific enlightenment. But their case for rejecting, or purging, Christianity was argued on other grounds as well.

Indeed, some of the more conspicuous rationalists had to modify Newton's understanding of matter in order to bring science on their side. An example would be John Toland, whose *Christianity not mysterious* (1696) is the most famous example of the suspect literature that flooded England once the Licensing Act had lapsed in 1694. In Ireland, where Toland was born, his book was burned by the common hangman. Some would have had Toland burned too. His thesis was that Christianity should be reduced to a set of teachings that were shorn of paradox and mystery. True to his purpose he had even turned the Virgin Mary into a rationalist:

The Virgin Mary, tho' of that sex that's least proof against flattery and superstition, did not implicitly believe she should bear a child . . . till the Angel gave her a satisfactory answer to the strongest objection that could be made.[6]

Toland's reductionism was accompanied by a vision of nature as self-sufficient and self-organizing. Whereas Newton's "active principles" had not been inherent in matter, Toland wished to have them so. Accordingly, he injected into Newton's world-picture the animism that he found in Giordano Bruno, a kindred spirit in the cause of free thought. Newton's laws, Toland declared, did not have to be interpreted Newton's way. They were "capable of receiving an interpretation favourable to my opinion." This is a clear indication that connections between Newton's science and radical critiques of religion were not straightforward. Everything hinged on how the new science was interpreted. It did not of itself generate deism.

The "bible" of the deists was Matthew Tindal's *Christianity as old as the creation* (1730). An examination of this work shows just how much science there was in the deist gospel. Tindal's aim was to show that revelation, whether in the person and teaching of Christ, or in the Scriptures, could add nothing to what had always been known about moral obligations. Discounting the possibility that God might have wished to reveal progressively more of Himself, or that Christ had been sent to redeem humanity, Tindal adopted the maxim that no addition was possible to a religion that was already perfect. This perfect, natural religion, to which all rational minds could subscribe, amounted to little more than obedience to civil laws. The duties of a truly religious person, Tindal wrote, and of a good citizen, are the same. The good citizens who read his book were spared a God of wrath and treated to One who was universally benevolent. Human happiness was obtainable by obeying the light of nature, which the Creator had placed in every human heart. The priesthood was therefore surplus to requirements, the Catholic clergy particularly reprehensible:

Figure V. 2. Fra Filippo Lippi (c. 1406–69), *The Annunciation*. The modest, submissive virgin in this classic portrayal helps to throw Toland's rationalist treatment into relief. Reproduced by courtesy of the Trustees, The National Gallery, London.

The Popish priests claiming a power by divine right to absolve people upon confession, have been led into the secrets of all persons, and by virtue of it have governed all things.[7]

Tindal was particularly severe on religious ritual. The use of crosses, pictures of saints, and other stimulants to devotion had led to idolatry like that of the ancient Egyptians, who had taken to worshiping their oxen, leeks, and onions.

Tindal did exploit scientific innovations that threatened the content of Scripture. The Copernican system, he claimed, was sufficient proof that, over natural phenomena, science and the Bible *seldom* agreed. This was, however, a generalization from one instance. For another example he could only resort to the fallibility of Christ and St. Paul when they spoke of the seed *dying* in the earth before bearing fruit. The very success of the sciences also showed that confidence in human reason was not misplaced. Tindal's God of infinite wisdom and goodness could be rationally demonstrated from nature's laws.

Science did have a place in the deists' bible. It contributed to a new tone of thought that made common religious practices appear more superstitious than before. But it was still a subordinate place. Tindal himself laid far greater stress on cultural relativism. Ever since the great voyages of discovery, there had been problems concerning the exclusivity of the Christian dispensation. If a culture as civilized as the Chinese had prospered without the Christian gospel, the implications could be disturbing. Particularly disturbing to Tindal was the thought that, if the primacy of reason were denied, one would succumb to the religious mores of one's native society. When he cast aspersions on the biblical miracles, it was not their incompatibility with scientific laws that he stressed, but rather the occurrence of miracle stories in every religious tradition.

The majority of Tindal's arguments had nothing to do with science. How were those to be judged who had lived before Christ, or had never heard of him? Only according to the use they had made of their natural reason. In their case it had to be justification by reason rather than faith. As for claims to revelation, there were simply too many for any one to be valid. Tindal's ploy was to inquire of defenders of revelation their reasons for accepting the authority of Scripture rather than the authority of reason. If they obliged with reasons for renouncing their reason, they confirmed that "men have nothing but reason to trust to." If the Scriptures were necessary to disclose the nature of God, they signally failed, because descriptions of the deity couched in terms of human characteristics had to be qualified if the uneducated were not to be misled. Scientific progress was not the mainspring of Tindal's attack. In fact it was sometimes used against him by orthodox divines, notably by Joseph Butler, who, in his *The analogy of religion* (1736), observed that obscurities in the meaning of Scripture (in which the deists reveled) might, with further research, be clarified – just as obscurities in nature had yielded to scientific research.

The Threat to Christianity from Materialism

Deists who wished to replace institutionalized Christianity with natural religion did at least affirm the existence of a Supreme Being whose laws had been implanted in nature. Other critics, however, went further, embracing doctrines of materialism and atheism. Materialism could mean several things: that the universe is a chance product of matter in motion, that all that exists *is* matter, and that the workings of the mind can be understood without reference to spiritual substance or agency.

The arguments of the heretical French priest Jean Meslier (1664–1729) show how a materialist philosophy could turn into a tirade against Catholic Christianity. His notorious *Testament* was widely, if clandestinely, circulated in eighteenth-century France before being condemned by the Paris Parliament in 1775. Meslier considered it pointless to ask who made matter and set it in motion. That question merely prompted another: Who made the being who supposedly did the job? And what was to be gained by inventing a perfect being when He would have to be held responsible for both good and evil? Atheists were just as capable as other men of practicing virtue. Religions in general were fabrications fostered by ruling elites. Meslier drew on relativist arguments to undermine revelation and moral arguments to brand the Old Testament conception of a favored people as unjust. Christian morality was indefensible because it encouraged the acceptance of suffering, submission to one's foes, acquiescence in the face of tyranny – just such tyranny as had been practiced by the kings of France, their police, censors, and tax gatherers. The earliest Christians had been exemplary in sharing their goods, but that ideal had long vanished. The horror for Meslier was that injustice and deformity could be so blithely explained away as the will of an all-wise Being. The doctrine that the wrongs of this world would be righted in the next was equally distasteful: It enlisted a fraudulent immortal soul and encouraged apathy towards social reform.

In common with other materialists, Meslier argued that matter could organize itself. But did contemporary science support that proposition? Newton's science was not a great help. But there were three discoveries during the 1740s about which a materialist could get excited. Matter within living systems was shown to have unsuspectd powers. One revelation was the evidence for spontaneous generation claimed by the English Catholic priest John Turbeville Needham (1713–81). A black powder obtained from rotten corn, when moistened, gave rise, so it appeared, to microscopic eels. Other vegetable infusions produced their own swarms of minute organisms. Needham's critics would soon be arguing that he had failed to sterilize his apparatus. To forestall such criticism, he conducted an experiment with gravy from roast meat, sealing the vessel and heating it to destroy any organisms already present. Still the microscopic creatures appeared. If Needham was right (and spontaneous generation proved to be one of those ideas extremely difficult to disprove), there was the disturbing possibility that all living things might have emerged from a primeval gravy. Voltaire reported that Needham's results were proving attractive to philosophers who supposed that matter could organize itself.

A second revelation came from the Swiss naturalist Albrecht von Haller. Knowing that movement persisted in the hearts of animals that had recently died, Haller inferred the existence of an unknown force in the fabric of the heart. This notion of an inherent force was given new vividness when he showed that muscle tissue, removed from the body, would automatically contract when pricked. Matter, it appeared, had its own power of movement, independent of an organizing soul. A materialist such as La Mettrie, whose *L'homme machine* (1747) was directed against the arguments for an independent soul, welcomed Haller's new force with open arms. The "soul" after all was affected by disease, sleep, drugs, food, age, sex, temperature, and climate. An obstruction in the spleen was all that was necessary to turn bravery into cowardice. The human body, in short, was a "machine which winds its own springs." It was the "living image of perpetual move-

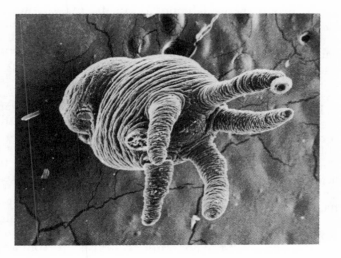

Figure V. 3. Scanning electron micrograph of a hydra. Reproduced by courtesy of Dr. Tony Brain/Science Photo Library, London.

ment." Even the intestines, La Mettrie observed, kept up their motions long after death. His materialist physiology was an integral part of a secular philosophy in which religious beliefs were considered irrelevant in the conduct of one's life.

A third disclosure was the most astonishing. In the early 1740s, news spread that a freshwater polyp, the hydra, could regenerate itself when cut into pieces. Abraham Trembley's discovery was so astounding that a wave of polyp chopping swept across Europe as scholars repeated the experiment – hundreds of times in the case of R. A. Réaumur who reported the news to the Paris Academy of Science. The commotion was not simply caused by matter reorganizing itself. If one polyp could become two by artificial division, then indivisible animal souls surely lost their credibility. It was a delight to the materialists, who wished to have done with souls altogether.

In France, by the late 1740s, a new-style scientific materialism was winning adherents. Between 1746 and 1749 Diderot shifted his allegiance from a fashionable deism to a more radical materialism. In a work of 1749 he used the device of a blind man to raise the question whether the designs at-

tributed to providence might not be in the eye of the be-
holder. Diderot was impressed by monstrosities and deform-
ities in the animal kingdom, which may have disposed him to
speculate on a history of organic forms in which defective
combinations of limbs and organs had automatically perished, giving the illusion that surviving combinations had been
designed. By 1753, when his *Interpretation of nature* was published, Diderot was speculating that, over millions of years,
organic matter might have passed through an almost infinite
number of organized states. Only one step remained toward
a full-blown materialism and he duly took it. The ability to
feel, the power of sensibility, was simply a product of the
organization of matter. There was no reason why a stone should
not have feelings.

It has been said of Diderot that he was the first modern
atheist – both in the sense of making matter the ceaseless
cause of all things, and of rendering the question of God's
existence a matter of little consequence. His shift from a position in which the *only* proof of a transcendent reality came
from what the sciences could say of the organization of nature, to a position in which that last proof was gone, marked
the end of the road for the God of the philosophers. The
question still remains, however, whether his materialism was
the only legitimate response to the biological discoveries. All
three could be interpreted in alternative ways that did little
damage to religious belief. As a Catholic priest, Needham
saw no great danger in his results. In fact Voltaire dismissed
them as the fraudulent miracle-mongering of an Irish Jesuit
– a bit hard when Needham was neither Irish nor Jesuit.

Similarly, Haller's discussion of his force of "irritability"
shows that he was able to resist a materialist gloss. He presented it as analogous to Newton's force of gravitation, known
by its effects and ultimately an instrument of the Creator. In
dispute with C. F. Wolff (1734–94) on the nature of embryological development, he considered that other biological
concepts were more likely to lead to atheism. The issue was
whether development of the embryo was the growth of an
already organized being, preformed in the egg, or whether
the process involved "epigenesis," with highly differentiated
tissues gradually emerging from less-organized material. As
so often in science, the empirical data were not decisive. When
Wolff claimed to have seen the formation of chick organs
from certain "vesicles" developing from a formless jelly, Haller

retaliated that the preformed organs *had* been present, but too minute to be seen through Wolff's microscope.

One's sympathies may lie with Wolff, for whom it made no theological difference whether one opted for preformation or epigenesis: "Nothing is demonstrated against the existence of divine power, even if bodies are produced by natural forces and causes, for these very forces and causes . . . claim an author for themselves just as much as organic bodies do." Haller, however, considered epigenesis the more dangerous doctrine. If material forces could produce living forms out of unorganized matter, it might be argued, by analogy, that no Creator had been involved in the ultimate origin of life. By contrast, the successful prepackaging of all organic structures could be used to argue for divine wisdom and foresight. They were "accommodated to foreseen purposes."

Even Trembley's polyp, though placed at the cutting edge of materialism, was hardly an incisive argument against a nonmaterial component of reality. It might embarrass the view that polyps had souls, but the applicability of the argument to human souls was not self-evident. It was even possible to place a conservative interpretation on the antics of the hydra. Its curious powers of locomotion suggested that it might occupy what had been an awkward gap in the great chain of being – an otherwise missing link in God's creation, between plants and animals.

Given that a speculative materialism was not demanded by biological data, what made it so attractive? One answer is that it could be applied to the human mind in such a manner that every person became a determinate product of their past experience. This association with determinism could then give zest to campaigns for educational reform. A common complaint against the Christian doctrine of original sin was that it checked any hope that human nature might improve. Once released from a state of doom, however, people would become pliable, open to instruction, which, if properly executed, would be a perfecting process. If, by rational means, it was possible to escape from hunger, cold, and disease, why should it not also be possible to escape from war, tyranny, and crime? The attraction of a determinist philosophy was that, given the right educational input, it seemed to guarantee a favorable result.

But how was the right input to be determined? The creed favored by many French philosophers was so antithetical to

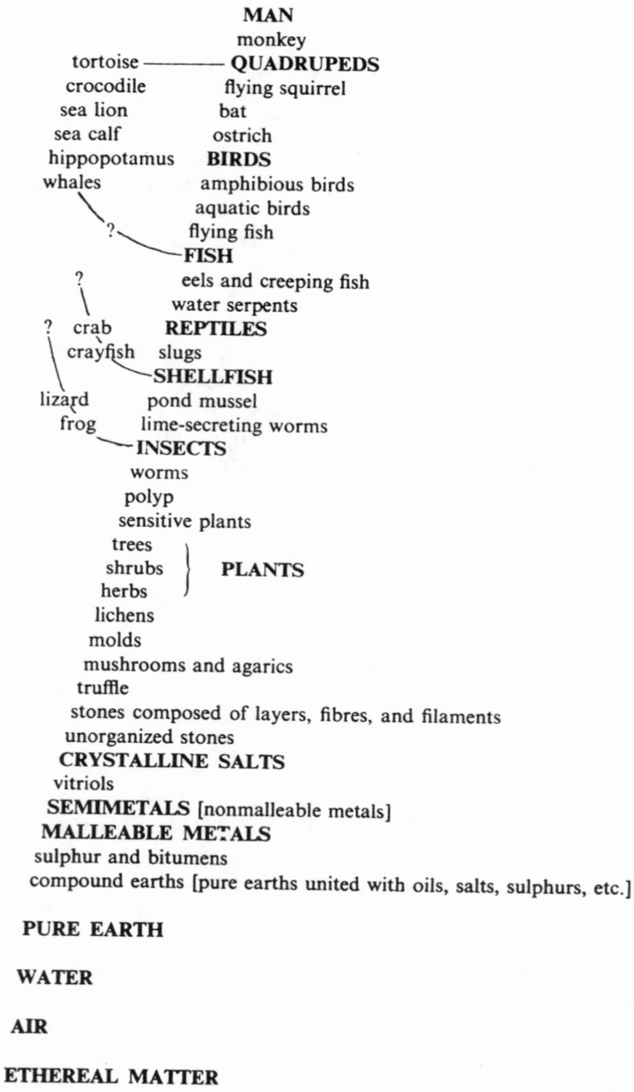

```
                              MAN
                             monkey
              tortoise ———— QUADRUPEDS
              crocodile       flying squirrel
              sea lion        bat
              sea calf        ostrich
              hippopotamus  BIRDS
              whales          amphibious birds
                   \          aquatic birds
                    ?\        flying fish
                       \—FISH
                 ?          eels and creeping fish
                  \         water serpents
         ?   crab    REPTILES
          \  crayfish   slugs
           \       \—SHELLFISH
         lizard     pond mussel
           frog     lime-secreting worms
                — INSECTS
                   worms
                    polyp
                    sensitive plants
                    trees      ⎫
                    shrubs     ⎬   PLANTS
                    herbs      ⎭
                    lichens
                    molds
                   mushrooms and agarics
                   truffle
                  stones composed of layers, fibres, and filaments
                  unorganized stones
                  CRYSTALLINE SALTS
                  vitriols
                  SEMIMETALS [nonmalleable metals]
                  MALLEABLE METALS
                 sulphur and bitumens
                 compound earths [pure earths united with oils, salts, sulphurs, etc.]
```

PURE EARTH

WATER

AIR

ETHEREAL MATTER

Figure V. 4. The construction of the Chain of Being as in Charles Bonnet's *Contemplation de la nature* (1764). Attempts to produce a natural system of classification based on a single linear series were to be frustrated, though the basic idea of a linear ascent, when reinterpreted as a process through time, was to regulate some of the earliest concepts of biological "evolution."

traditional Christian doctrines that it was almost a mirror image. But it was also possible to argue that the new values should be derived not from the rejection of Christianity, but from its purification. In Britain this task of purification was undertaken by the dissenting minister Joseph Priestley, who achieved eminence for experimental work on a variety of gases, including the "dephlogisticated air" that Lavoisier would rename "oxygen." Far from placing Christianity and materialism in opposition, Priestley welded them together. Part of his mission was to convince the philosophes that in rejecting Roman Catholicism they were only rejecting a corrupt form of Christianity, not the real thing. In Priestley's mind, science and true Christianity were fighting on the same side against superstition and political oppression. By superstition Priestley certainly meant popular but erroneous religious beliefs, such as the notion that God could directly intervene to change a state of mind, or that the mind itself was a spirit entity linked to an immortal soul. Reconstructing the history of Christianity, he tried to show that the common dualism between matter and spirit was a legacy not from the Bible, but from the contamination of biblical Christianity by Greek philosophy. Rather like Newton, he argued that the vocabulary in which the doctrine of the Trinity was formulated betrayed this corrupting influence of Platonism.

In his *Disquisitions relating to matter and spirit* (1777), Priestley produced scientific, philosophical, and theological arguments for collapsing the matter–spirit duality. The foundations of the scientific enterprise were best safeguarded, he suggested, if God were considered to work uniformly through a causal nexus of powers that were neither material nor immaterial as traditionally understood. One could associate these powers with matter, but in so doing one's conception of matter was changed. It was no longer the inert stuff of impenetrable atoms, but suffused with active powers that had so often been vested in some immaterial reality. Because he was collapsing the material–immaterial distinction, Priestley did not mind whether he was dubbed materialist or immaterialist – as long as it was clear that those labels no longer had their customary meaning.

A philosophical justification for his monism came from the consideration that matter and spirit had been habitually presented as such incommensurable things that it was inconceivable how they could ever interact. Moreover, it was facile to

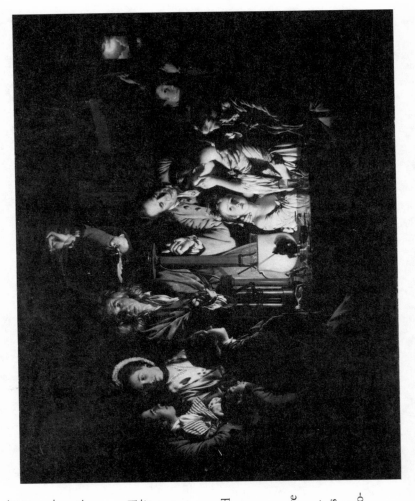

Figure V. 5. Joseph Wright of Derby (1734–97), *Experiment with the air pump* (1768). Withdrawing air from the vessel containing the small bird would cause its lungs to burst. The experiment is shown as part of a polite domestic scene. Until he developed a chemical test, Priestley would use mice to test the quality of "air" – that is, the extent to which it would support respiration. Priestley's "dephlogisticated air" [our "oxygen"] surprised him by being more supportive of combustion and respiration than common air. It has been said of Wright's composition that "emotions previously attached to religious themes by painters were transferred to wonder at the physical world." The concentration of light in this process of scientific enlightenment is dramatically conveyed. Reproduced by courtesy of the Trustees, The National Gallery, London.

define matter in terms of solid atoms because, if there were no attractive powers between the parts of the atom, those parts could not cohere and there would *be* no solid atom. Attractive and repulsive powers were ontologically prior to any kind of solid matter. He had no qualms in suggesting that mental phenomena could be explained by similar powers operating in the brain. By those who felt that no spirit meant no God, Priestley was vilified as an atheist. But behind his campaign were burning religious convictions.

The dualistic view, he decided, was clogged with religious difficulties. If the soul was immaterial and the body material, there was the old chestnut whether the two came together at conception, birth, or whenever. But more seriously, the doctrine of an immortal soul made the doctrine of the Resurrection superfluous. Priestley belonged to that mortalist tradition which proclaimed "when you're dead you're dead." But, by the grace of God, one had the hope of resurrection. That, for Priestley, was the biblical view and it was at odds with belief in the automatic survival of death or the Roman Catholic invention of purgatory. On this theological point he stood his ground because, although he denied the doctrine of original sin and of Christ's atonement, the doctrine of the Resurrection was central to his Christianity. Without it, and without the promise of rewards and punishments, there could be no social stability and no ultimate rationale for the reformation of character.

Priestley's theology expressed the egalitarian doctrine that happiness was intended by God for all humanity, not (as in the Calvinist theology of his childhood, which he had come to reject) for an already chosen elect. He appealed to egalitarian principles too in his attack on Anglican theology. A doctrine such as the Trinity was simply too abstruse for the common plowman. His pleas for greater religious toleration had similar inspiration. People were not treated equally if they were penalized for religious nonconformity. He himself found it irksome that, as a dissenter, he still had to contribute tithes to the Anglican Church he had renounced. His sympathies lay with the French Revolution and with America in the war over independence. Eventually emigrating to America, he wrote to Thomas Jefferson in June 1802 that he looked forward to living under the protection of a constitution "the most favourable to political liberty, and private happiness, of any in the world."

Priestley makes a fascinating study because he personified a set of values that allowed the integration of scientific and industrial progress into a process theology, which promised the eventual triumph of rational Christianity. Progress in science was to be "the means under God of extirpating all error and prejudice, and of putting an end to all undue and usurped authority in the business of religion as well as of science." Those same values were shared in late eighteenth-century England by other Unitarian families who, in expanding manufacturing towns, were often active in promoting closer links between scientific research and its practical application. Such families had a high profile in the Manchester Literary and Philosophical Society, but, without the Unitarians, *most* of the provincial scientific societies would have been significantly weaker. The driving force behind the Newcastle Literary and Philosophical Society, William Turner (1761–1859), was another apostle of rational dissent, for whom (it has been said) the industrial revolution was taking place not behind God's back but at His express command.

To set up a general antithesis between scientific materialism and religious values as a characteristic of the Enlightenment can result in caricature. What may have been true in France was not necessarily the case for Britain. In fact, Priestley nicely captured the contrast, as he reflected on a dining engagement he had enjoyed in France:

When I was dining at . . . Turgot's table, M. de Chatellux . . . said the two gentlemen opposite me were the Bishop of Aix and the Archbishop of Toulouse, "but," said he, "they are no more believers than you or I." I assured him that I was a believer; but he would not believe me.[8]

The Threat to Christianity from Agnosticism

Priestley's object had been to establish rational criteria whereby the historicity of facts, and the authenticity of doctrines based upon them, could be determined. The pure Christianity that remained would then be unimpeachable. He would only accept a biblical miracle if there had been many witnesses to it. A casualty was the Virgin Birth, because, in the nature of the case, there could hardly be witnesses to a miraculous conception! Yet, however much he jettisoned, he clung to the prevalent view that the organization of the natural world was proof of a caring deity. The high value that he placed on scientific activity stemmed in part from its disclosure of hidden connections between phenomena, which testified to the economy and benevolence of the system. There was economy, Priestley believed, in the fact that a single chemical principle, phlogiston, conferred common properties on all the metals. As for benevolence, the fact that vegetation could restore common air, contaminated by breathing, not only exemplified design but provided a scientific parable of the transformation of evil into good. In his experimental work, he actively sought the mechanism by which this restoration occurred.

During the Enlightenment, the most serious threat to this kind of natural theology came not from science, or deism, or materialism, but from a subtle form of agnosticism. Priestley knew he had a dangerous adversary in David Hume, the Scottish skeptic, whose most extensive critique of natural theology was published posthumously in his *Dialogues concerning natural religion* (1779). His arguments are commonly considered decisive against claims that a knowledge of the nature of God can be obtained by rational inference from the natural order. The relationship between science and the cultivation of an agnostic mentality was, however, far from straightforward.

In Hume's Edinburgh circle a life of civic virtue had come to mean the cultivation of an independent mind and a commitment to improving the economy and culture of Scotland. As a new republic of letters, Edinburgh had become *the* place in Britain where a public reputation could be gained for excellence in the arts and sciences. Its university, formerly little more than a seminary, had become increasingly responsive to the needs of a professional class. Unlike Oxford or Cambridge, there were no religious tests to control access. Its distinguished medical faculty and strong links with the Continent gave it a cosmopolitan flavor. The study of chemistry was to be a particular strength through the work of William Cullen and Joseph Black. But, for the Edinburgh literati, the proper study of mankind was man. To that study Hume brought his distinctive contribution. Justice, morality, politics, and religion were grounded not in reason, Hume argued, but in habit and custom. It was a point of view that encouraged a sociological inquiry into man and his history. Personality was shaped by social experience, not by the power of innate reason.

That the Christian religion could be given a rational defense became one of Hume's principal targets. He did not deny that the universe must have a cause. The question was whether anything could be known about it. By contending that the universe, or its constituents, resembled a human artifact, proponents of the design argument could infer that its cause must be intelligent and purposeful. Hume's strategy against such reasoning was to expose the limitations of analogical argument. Even if the world did resemble a machine, it would not follow that its cause was the transcendent God of Christian theology, or even the beneficent God of the deists. The more one pressed a similarity between the mind of a human and a divine contriver, the more one renounced – so Hume maintained – any claim to infinity in the attributes of the deity.

The design argument was fragile for a second reason. Analogical argument, as in science, might *suggest* possibilities. It could hardly prove them. In science the hypothesis would usually be open to experimental test. But the hypothesis of a mind behind creation was clearly not testable in the same way. No one had witnessed the creation of worlds. There was no direct test by which the inference could be established.

Hume's reference to the creation of *worlds* highlighted a further difficulty. We know only one – our own. If comparisons could be drawn between different worlds, it might be possible to say that *this* world is more like a machine than *that*. But, faced with the singularity of the universe, how can the choice of one analogue over another be justified? This problem was germane to Hume's critique. Such organization as was discernible in nature could justify analogies with plants and animals as readily as with machines – in which case, the world could have had its origin in a seed or egg. Hume was not arguing that the world really was a cosmic cabbage but merely that no one analogy was so privileged that inferences drawn from it were irresistible. Philo, who voiced the skeptical arguments in the *Dialogues,* did venture to suggest that "the world plainly resembles more an animal or a vegetable than it does a watch or knitting loom." With such reasoning, the god of the natural theologians could be superseded by a super seed. Even if the world were, after all, more like a machine, it could still have been the work of a stupid mechanic who had merely copied the work of another. Many worlds, Hume suggested, might have been "botched and bungled."

If this were not damaging enough, he identified a further respect in which natural theologians had themselves bungled. Even if parts of the universe resembled machines, it did not follow that the same could be said of the whole. In short, "a total suspense of judgment is here our only reasonable resource."

But suppose one replied that it did not matter whether the universe as a whole resembled a machine. Would not one example of contrivance — in the working of the human eye for example — be sufficient to establish a divine contriver? Were there not multiple instances of such adaptation of structure to function? Were not the beak of the woodpecker, the hump of the camel, sufficient proof in themselves? With such examples, William Paley would later argue that the case for design was cumulative. Each example *independently* implied a designer. Hume, however, had already developed an alternative perspective. Examples of apparent design might be illusory and, even if design were conceded, it often appeared malevolent.

Without anticipating Darwinian theory, Hume ingeniously drew on philosophical traditions that had nurtured the idea that one could have adaptation without design. An organism with a particular structure would adopt patterns of behavior for which it was equipped. It did not follow that the organism had been deliberately designed for that life-style. With a different combination of limbs and organs, it would have lived differently — or perished. The temptation to see a mind at work arose from the failure to see that structures could precede "ends" without being determined by them.

And if there had been ends, what of those facets of nature that bred misery, pain, and suffering? Conjuring up an image of perpetual war, he invited his readers to contemplate the "curious artifices of nature, in order to embitter the life of every living being." Every animal was surrounded by enemies, incessantly seeking its doom. This was a darker vision than even Darwin was to evoke. For Darwin would surmise that, on balance, there was more pleasure than misery in the world: If animal life were accompanied by nothing but misery, whence the drive to survive and propagate? But Darwin, too, would consider the volume of suffering a forceful argument against a beneficent God.

Hume's insistence that design arguments provided no route to a transcendent God was capped by reference to the principle that causes must always be proportioned to their effects. No more was to be ascribed to a cause than was necessary for the production of the effect. Since the world is a world of finite objects, their ultimate origin need not lie in a cause of infinite power. And if Hume did finally allow that "the cause or causes of order in the universe probably bear some remote analogy to human intelligence," it was still a limited concession. For the implication was that such a proposition "affords no inference that affects human life, or can be the source of any action or forbearance." While precluding a strict atheism, it gave free reign to a practical atheism.

For many of Hume's contemporaries, the image of the universe as a machine was so seductive that his alternative organic analogies appeared eccentric – even a throwback to a prescientific age. Philo's conjecture that a comet might have been the world's egg drew a predictably scathing reply from Priestley: "Had any friend of religion advanced an idea so completely absurd as this, what would Mr. Hume have said to turn it into ridicule." The contrast between Priestley's rationalism and Hume's skepticism may also be seen in their attitudes toward the miraculous. Despite Priestley's rejection of biblical miracles that were not well attested, he was prepared to admit those where the reliability of witnesses did not appear to be in doubt. Hume, by contrast, flattered himself that he had found an argument that would bring *all* reported miracles into doubt.

The gist of it was that no testimony was sufficient to establish a miracle unless the testimony was such that its falsehood would be more miraculous than the fact it purported to establish. Defining a miracle as a violation of the laws of nature, Hume insisted that there must be a strong antecedent probability against its having occurred. Human testimony, however, was known to be capricious and corruptible. It was not improbable that those who reported a miracle had been deceived or even deliberately deceptive. Weighing the probabilities, it was always more likely that the testimony was defective than that the reported miracle had occurred.

Hume's contention that the more unusual the fact, the greater the evidence required to establish it, has a thoroughly modern ring. The burden of his argument, however, fell not so much on the impossibility of miracles, as on the fallibility of the supporting testimony:

We entertain a suspicion concerning any matter of fact, when the witnesses contradict each other, when they are but few, or of a doubtful character; when they have an interest in what they affirm; when they deliver their testimony with hesitation, or on the contrary, with too violent asseverations.[9]

Add to this the many instances of forged miracles and their abundance among the more barbarous nations; add a love of gossip, a thirst for the spectacular, a desire for the prestige that comes from being the first to transmit sensational news; add the consideration that competing religions had their own stocks of competing miracles. What then was left of the fabric of reported miracle? For Hume it became a maxim that "no human testimony can have such force as to prove a miracle, and make it a just foundation for any system of religion."

For all his skepticism, Hume did construct a hypothetical example in which human testimony would have to be taken seriously:

Suppose, all authors, in all languages, agree, that, from the first of January 1600, there was a total darkness over the whole earth for eight days: suppose that the tradition of this extraordinary event is still strong and lively among the people: that all travellers who return from foreign countries, bring us accounts of the same tradition, without the least variation or contradiction: it is evident that our present philosophers, instead of doubting the fact, ought to receive it as certain, and ought to search for the causes whence it might be derived.[10]

This example has two interesting features. It shows that Hume believed that, in the recovery of the past, certainty was at least possible. In this he was more optimistic than a German critic such as Gotthold Lessing. And the role of science in such a case was not so much to reinforce the "usual course of nature" as to explain how the unusual might, after all, be possible.

What, then, were the connections between Hume's agnosticism and eighteenth-century science? The answer is not straightforward because Hume would, arguably, have been inconsistent if he had said that miracles were physically impossible. In his analysis of causality he had criticized the common view that effects were produced by their causes as a result of some necessary connection binding them together. He was concerned that terms such as causal *power* or *efficacy* were being bandied about without any clear idea of their sig-

nificance. His recommendation was that all such metaphysical conceptions of causality should be abandoned in favor of
an account that grounded the relationship of cause and effect
in the constancy of experience. If, in our experience, event
A has been invariably followed by event B, we speak of A as
the cause of B. When A recurs, our past experience raises in
us the expectation that B will follow. The idea of a causal
connection had its origin in human psychology rather than in
some perceived form of physical necessity. One could experience cause and effect, but not what bound them together.

If Hume was saying that there is no necessity binding effects to causes, if events in nature are in a sense "loose and
separate," if there are no grooves in the universe, then how
could miracles be declared impossible? For Hume himself
this was a relatively minor disadvantage because all his argument required was that miracles be exceedingly improbable,
and that was sufficiently guaranteed by the uniformity of past
experience. The minor disadvantage was more than offset by
new skeptical possibilities, which his analysis of causality released. If he could sustain the argument that we have no adequate idea of causal power, he could affirm, as he did in his
Treatise on human nature (1739–40), that "we have no idea
of a Being endowed with any power, much less one endowed
with infinite power." Although his critique of analogical argument was not dependent on his analysis of causality, it has
seemed to many commentators that it was reinforced by it. If
A may be said to be the cause of B only when A and B have
been repeatedly conjoined in our experience, there is a new
difficulty when B is the universe, not merely because we cannot experience its creation, but also because of its singularity,
its unrepeatability.

At the very least, Hume believed he had grounds for attacking that voluntarist theology of creation that saw God's
will behind the regularity of mechanical laws. To postulate
the constant activity of a divine power, as had Boyle, Newton, Bentley, and Clarke, was to exploit a concept that was
unclear because it referred to something beyond human experience. On this point, science did come to his aid, but in a
paradoxical manner that might be anticipated from our analysis in the previous chapter. Whereas Newton had wished to
keep the voluntarist option open, Hume invoked the authority of Newton to suppress it! He could do so because, during
the 1740s, Newton's conception of an ether had come to his

attention. It seemed to Hume that Newton had not vested causal efficacy in the divine will alone: "Sir Isaac Newton (tho' some of his followers have taken a different turn of thinking) plainly rejects it, by substituting the hypothesis of an aetherial fluid, not the immediate volition of the deity, as the cause of attraction." It had not been that plain or simple.

Although Newton's queries came to Hume's aid, it is less clear that Hume's philosophy came to the aid of science. In one respect his empiricism was arguably as subversive of science as of religion. He had no intention that it should be, but, taken to extremes, his philosophy would preclude the introduction of theoretical entities into science that were not directly perceivable. The issue is whether scientists would always have prospered in practice if they had outlawed concepts such as *atom* or *gene,* which were not verifiable when first introduced, and which have been used to bind effects to causes by the very kind of physical necessity that Hume apparently rejected. Later scientists having an aversion to religious language would sometimes express a similar aversion to scientific terms if they transcended their experience. The remarks of one of the great chemists of the nineteenth century, Marcellin Berthelot, show how religious agnosticism could foster a kind of scientific agnosticism: "I do not want chemistry to degenerate into a religion; I do not want the chemist to believe in the existence of atoms as the Christian believes in the existence of Christ in the communion wafer."

The threat to Christianity from agnosticism was not as straightforwardly science-based as is sometimes imagined. In Hume's *Dialogues* it was Cleanthes, the spokesman *for* design in nature, who placed science on a pedestal: "The true system of the heavenly bodies is discovered and ascertained. . . . Why must conclusions of [a religious] nature be alone rejected on the general presumption of the insufficiency of human reason?"

The nuances we have been examining raise the question whether the greatest threat to established religion might not have come from the human rather than the physical sciences. When the study of man was directed toward the origin of his religious beliefs, the effect could be more startling than inferences drawn from physical laws. Hume's most devastating blow was neither his demolition of arguments from design nor his assault on miracles, but his *Natural history of religion.* In this work, composed about 1750 but which he chose not to publish, he rejected the contention of the deists, shared by many orthodox divines, that monotheism was the "natural" religion of mankind. The origins of religious practices were to be found neither in the exercise of natural reason nor in the contemplation of the natural order, but in fear, ignorance, and attempts to placate local deities. The original religions of the world, Hume suggested, were polytheistic, as the contrary events of human life were ascribed to limited and imperfect deities. Nor was the historical movement from polytheism to monotheism the product of rational refinement. Men had either supposed that, in the distribution of power among the gods, their nation was subject to the jurisdiction of their particular deity or, projecting earthly models heavenward, had come to represent one god as the "prince or supreme magistrate of the rest." And it was again through *fear* of this supreme god that his praises were sung. Monotheism had been reached not by reason but by the "adulation and fears of the most vulgar superstition." Once the attributes of the deity had been extrapolated to infinity, he had become so remote and incomprehensible a figure that mediators had been invented, which, like the Virgin Mary, had become more accessible objects of devotion. When this process was overdone, a reaction toward a purer theism would again set in.

Hume applied this scenario to every religion, observing in them all a tendency toward fanaticism. Monotheistic religions, more than polytheistic, had been guilty of intolerance. But they had destroyed human happiness in other ways too. Originating in fear, they fostered fear, perpetuating the conditions in which they could flourish. A preoccupation with one's eternal salvation was apt to "extinguish the benevolent affections, and beget a narrow, contracted selfishness." Familiar with the psychology of a dour presbyterianism, he came close to treating all religious observance as pathological.

His allegation was that religion corrupted morality. In the Christian tradition, one was to adore a Being who had manufactured instruments of eternal torment. Whereas a responsible civic morality required that human actions be useful to society, the religious virtues were sterile:

Celibacy, fasting, penance, mortification, self-denial, humility, silence, solitude, and the whole train of monkish virtues; for what reason are they everywhere rejected by men of sense, but because they serve no manner of purpose; neither advance a man's fortune in the world, nor render him a more valuable member of society.[11]

One wonders how he would have reacted to the alternative view that contemplation could focus the mind on altruism — the kind of altruism visible, for example, in those religious institutions that promoted care of the sick. Many disparaged by Hume would doubtless have observed that it was not the goal of their religion to advance their fortunes in this world. Hume, however, was as cynical about otherworldliness as he was critical of Christian morality. In contrast again with Priestley, he would not allow that the doctrine of a future life had any use as a vehicle of social control. He acknowledged that, if it were seriously believed, it might be expected to have. But the reality was that men and women were too attached to present things for so uncertain a prospect to affect their actions. The threat posed by cynicism was perhaps the hardest to answer.

Conclusion: Taking Diversity Seriously

We have seen in this chapter that even where the methods and conclusions of science were deployed in attacks on established religion, the rhetoric of the rationalists often concealed relationships of considerable complexity. Deists and materialists interpreted the "laws of nature" (and the nature of matter) in their own distinctive ways; but it would be wrong to imagine that they succeeded in excluding other ways of relating God to nature. In order to emphasize the diversity of eighteenth-century thinking, it may be useful to conclude with another reformer: the evangelical preacher and inspiration of the Methodist movement in England, John Wesley (1703–91). In his *Journal* Wesley stressed an agreement between natural philosophy and religion in persons of "sound understanding." His attitude to the sciences is particularly instructive because it shows how fine distinctions are required if the texture of past thinking is to be recovered. In Wesley one sees how a deeply pious man could be an enthusiast for certain styles of science, and a sharp critic of others. Highly theoretical, mathematical science did not find favor with him. At Kingswood School, which he founded in 1739, Wesley listed Newton's *Principia* as a work to be studied during the second year of the curriculum. The evidence suggests, however, that he found elaborate theorizing presumptuous. It was the incomprehensible in nature that nourished his piety. Of certain electrical experiments he wrote: "How must these confound those poor half-thinkers who will believe nothing but what they can comprehend."

Faced with the choice, Wesley seems to have preferred the natural philosophy of John Hutchinson (1674–1737), whose critique of Newton's system enjoyed a persistent, if minority, following in Britain during the eighteenth century. In his *Moses' principia* (1724) Hutchinson, who had been steward to the duke of Somerset, offered an alternative philosophy of nature, ostensibly in closer harmony with the Hebraic Bible. He developed his own three-element theory of matter, with fire, air, and light corresponding to the three persons of the Trinity. He glimpsed the specter of pantheism in Newton's conception of space, complaining also that the Newtonian image of an interfering deity detracted from divine transcendence. In his critique of Newtonian science, Hutchinson had objected to the action-at-a-distance implied by gravitational theory, preferring to substitute an ether of his own devising. His contention that Newton had damaged the cause of revealed religion had been reinforced by the fact that republicans and deists, notably John Toland, had claimed Newton for their cause. Hutchinson's claims for a biblically based natural philosophy are known to have appealed to a succession of high church Tories and found favor for a while in the University of Oxford.

Wesley was attracted to Hutchinson's discussion because he, too, was sensitive to the exploitation of Newton by the deists. Nevertheless, Wesley's aversion to *systems* of natural philosophy also fostered a skepticism toward the Hutchinsonian view. Referring to an essay by one of Hutchinson's disciples, William Jones, Wesley noted that "he seems to have totally overthrown the Newtonian Principles, but whether he can establish the Hutchinsonian is another question."

From Wesley's suspicion of theoretical pretensions, it might be tempting to record a wholly negative attitude toward the sciences of his day. His credulity concerning the possibility of witchcraft has not helped his reputation. It would also be difficult to maintain that he showed much discrimination as a commentator on current scientific trends. But for a balanced view it is just as important to recognize those forms of scientific practice to which he was attracted. In 1760 he published a text having the subtitle: "Electricity made plain and useful." Passing through five editions by 1781, it was informed by a knowledge of Franklin's work and addressed such issues as the nature of lightning and of electrical conductivity. There was no contradiction for Wesley between the use of lightning rods and a respect for divine providence. On the contrary, Wesley was drawn to those aspects of science that did offer help to humanity. The fascination of electricity derived in large measure from its medical applications. Wesley believed it to be the "most efficacious medicine, in nervous disorders of every kind." Practical medicine was an abiding interest. He published on the subject and even had the insight to recognize that bodily malfunction could be caused or influenced by the mind.

A compendium of popular cures was not, however, Wesley's only compendium. In 1763 there appeared *A survey of the wisdom of God in the creation, or a compendium of natural philosophy.* In this work of natural theology, Wesley placed a high value on the study of nature precisely because it promoted a sense of awe and humility toward its divine author. He stressed, as before, that his aim was not to account for things, only to describe them. But there was clearly a religious interest in achieving accuracy of description. The propagation of heat, of light and sound, were given some theoretical treatment and the book as a whole gives considerable insight into the state of popular science in the second half of the eighteenth century. The lesson that Wesley's many readers would have absorbed was that a science of nature, not bedeviled by arrogant theorizing, could offer rational support for Christian piety — revealing, as it appeared to do, a marvelous organization and adaptation within the created order. For Wesley, there was even a biblical basis for promoting a natural theology. He often said that his purpose was not to "entertain an idle barren curiosity, but to display the invisible things of God, his Power, Wisdom and Goodness." This was a tacit reference to Romans 1:20 — a text commonly used by Christian writers who saw religious utility in popularizing the sciences. The fortunes — and misfortunes — of this type of reasoning will be the subject of the next chapter.

CHAPTER VI

The Fortunes and Functions of Natural Theology

Introduction

In 1802 there appeared what was to be one of the most popular works of philosophical theology in the English language: William Paley's *Natural theology*. Its author, an Anglican priest, claimed that there was proof of the unity of God. It came from the "uniformity of plan observable in the universe." This uniformity was both assumed and confirmed by scientific inquiry. Reaffirming a Newtonian argument for the religious utility of science, Paley underlined his point: "One principle of gravitation causes a stone to drop toward the earth and the moon to wheel round it. One law of attraction carries all the different planets about the sun." But Paley's argument was more ambitious than this. It aimed to establish a God with personality — one whose goodness, for example, could be deduced from the fact that, to the necessity of eating food, He had superadded pleasure. Paley was particularly struck by the exquisite mechanisms discernible in the structure of living organisms. The human eye, for example, was so remarkable an instrument that it was as certain it had been made for vision as it was that the telescope had been made for assisting it.

Paley's argument, that every part of every organism had been meticulously designed for its function, was not merely a piece of academic philosophy. It defined a way of looking at the world that was probably shared by the majority of his contemporaries. Some of his examples were presented in ways that strike us as decidedly quaint. So perfectly designed was the human epiglottis that no alderman had ever choked at a feast! His statement that "it is a happy world, after all," may have alienated those less privileged than himself. Yet, despite the threat to established Christianity from deism, materialism, and agnosticism, the union between science and religion forged by arguments from design proved remarkably resilient. Hume may have emasculated the argument for the benefit of fellow skeptics; but the opinion of his fellow Scotsman Thomas Reid (1710–96) must also be taken into account. According to Reid, the design argument had always made the strongest impression on thinking minds. With the advance of scientific knowledge, it had gained in strength. Paley exuded that same confidence.

The resilience of natural theology is the principal theme of this chapter. The idea that divine wisdom could be discerned in nature was attractive in different ways, both to Christian apologists and to deists. Christians found the argument useful in their dialogues with unbelief. It seemed to offer independent proof of a God who they believed had also revealed Himself in the person of Christ. On the other hand, deists also had reasons for promoting the design argument. The more that could be known of God through rational inference, the less perhaps was it necessary to refer to revelation at all. In his *Age of reason,* which appeared in three parts between 1794 and 1807, Tom Paine took that argument to its ultimate conclusion. Whereas the Bible had been written by men, nature was the handiwork of God. Whereas the Bible had suffered corruption through copying and translation, nature had an indestructible perfection. Whereas the Bible portrays a passionate God, changeable and vindictive, nature shows Him to be immutable and benevolent. The biblical revelation had come late in time and been vouchsafed to one nation only. The revelation in nature had been always, and universally, available. In the Bible God communicated with men through magic; in nature it was through their ordinary senses. Paine's triumphant conclusion was that theology was merely the study of human opinions concerning God, whereas science was the study of the divine laws governing nature. Paine, no less than Paley, was confident that nature disclosed a benevolent Creator. Where he differed was in his confidence that people had been equipped with all the natural faculties for working out their own salvation.

The fact that natural theology could be used both to attack and defend Christianity may be confusing, but that very ambivalence also helps to account for its resilience. Without additional clarification, it is not always clear to the historian (and was not always clear to contemporaries) whether proponents of design were arguing a Christian or deistic thesis. The ambiguity could itself be useful. By cloaking potentially subversive discoveries in the language of natural theology, scientists could appear more orthodox than they were, but without the discomfort of duplicity if their inclinations were more in line with deism. Consequently, those with firm Christian convictions would often try to differentiate their statements about the God of nature from those of the deists. By a curious irony – the converse of that which had earlier linked mechanical philosophies to deism – there were new scientific advances that would make their task easier. Thus the nineteenth-century Cambridge geologist Adam Sedgwick (1785–1873) would turn the fossil record to advantage. That there had been repeated, and seemingly *progressive* creation allowed him to reject the static, mechanical conception of nature beloved by the deist. Sedgwick's point was that a world that had seen the introduction of new species could not be a world in which God had lost interest.

The resilience of natural theology deserves attention for three special reasons. The first concerns its value as a defense of Christian theology. Long before Newtonian science had given new impetus to the argument for design, Pascal had warned that those who sought God apart from Christ, who went no further than nature, would fall into atheism or deism. To base arguments for a personal God on impersonal forces would lead to a bankrupt religion, because to neglect the person of Christ in dialogues with the atheist was to neglect the one mediator between God and humanity. It is ironic that attempts to establish God's existence should encourage atheism, but there is a sense in which Pascal's prophecy came true. There is an old adage that no one had doubted the existence of God until the Boyle lecturers undertook to prove it!

The point can be put this way. Because "atheism" often takes its character from the particular form of theism it rejects, to understand the origins of modern atheism it is important to locate the particular form of theism to which it could be an apposite response. The wisdom of looking at it this way is nicely illustrated by a remark of the nineteenth-century secularist Charles Bradlaugh (1833–91): "I am an

Atheist, but I do not say that there is no God; and until you tell me what you mean by God I am not mad enough to say anything of the kind." The problem for Christian apologists was this: In seeking to capitalize on the most accessible proof of God's existence, and one having the authority of the sciences behind it, they came close to saying that what they meant by God was the craftsman, the mechanic, the architect, the supreme contriver behind nature's contrivances. From this to atheism *could* be one short step. It only required an alternative metaphysics in which the appearance of design could be dismissed as illusory. As we saw in the last chapter, Diderot took precisely that step. If the *only* proof came from design, one was left with nothing on its collapse.

The point is not that science undermined the design argument — certainly not in the eighteenth century. Quite the contrary. It was rather that religious apologists were asking too much of it. A religious burden was placed on the sciences, which they were eventually unable to carry. This overburdening can be seen in contrasts between the style of natural theology represented by Paley and that to be found in earlier paradigms of Catholic and Protestant theology. Severe limits had been placed on the scope of natural theology by Thomas Aquinas and, at the Reformation, by Calvin.

For Calvin, any knowledge of God inferred from nature would be distorted, the defective product of a dimmed and fallen intellect. The image could only be rectified by reading nature through the spectacles of Scripture. Without that privileged source, just enough could be known of God to ensure that no man would be without excuse at the Day of Judgment. But that was relatively little. Nor would Aquinas have countenanced the facile procedure whereby divine attributes were gleaned from nature, independently of revelation. In fact he is associated with the saying that we know of God rather what He is not than what He is. This refers to his so-called negative way of approaching the nature of God. By successively denying Him the characteristics of finite things, such as materiality and mutability, a knowledge of His attributes (albeit in a negative sense) could be gained. It was a far cry from that position to Paley's claim that God's caring nature could be discerned in the hinges on the wings of an earwig. Indeed, Aquinas had warned against the anthropomorphic reasoning that simply conferred human attributes onto God. He had even excluded animate objects from his teleological proof on the grounds that it was the unconscious cooperation of *inanimate* things, in achieving particular ends, that spoke more clearly of God's existence. Animate objects, after all, could initiate their own purposive activity. The contrast with the eighteenth-century New England divine Jonathan Edwards (1703–58) is engaging. Formerly blind to the marvels of nature, he wrote that, once he had experienced divine grace, he could see evidence of higher purpose in a spider's web. And what a wonderful provision of "pleasure and recreation" there had been for the flying spiders of the New England coast. A spider's web was, however, a fragile thing on which to suspend proof of God's providence.

The second reason for considering the resilience of natural theology is that, as long as apologists both for science and religion linked scientific knowledge to the proof of design, the basis existed for connections between scientific and religious discourse to be made and remade rather than severed. The shift in the style of natural theology, to which we have just referred, had not been unconnected with the scientific movement of the seventeenth century. Mechanical images of nature had encouraged inferences to a divine clockmaker and they were still the mainspring of Paley's apologetic. But just as important had been the experimental and observational thrust of natural philosophy, which had revealed the amazing intricacy of anatomical structures. The seventeenth-century English naturalist John Ray had repeated John Wilkins's observation that natural objects, seen through a microscope, revealed a perfection altogether lacking in human artifacts. Design arguments pointing up the excellence of divine workmanship had even played a role in the liberation of natural philosophy from a lingering scholasticism. Reformist clergy in Spain, for example, pointed out that scholastic treatises on the nutritive and locomotive faculties of living things did little to generate an image of the Creator. Experimental philosophy, by contrast, had opened up a microscopic world that spoke volumes. According to the Benedictine monk Feijoo, when a dog's heart had been brought to his cell, his fellow monks agreed that "we had never seen or contemplated anything which gave such a clear idea of the power and wisdom of the supreme Craftsman."

Because such connections continued to be made well into the nineteenth century, it becomes important to analyze the bearing of natural theology on scientific thought and con-

versely the bearing of a changing science on the perception
of design. These issues will be addressed in later sections of
the chapter. In the long run, the religious burden placed on
the sciences did prove too much to bear. By the middle of
the nineteenth century, Charles Darwin's concept of natural
selection would evacuate the design argument as Paley presented it. But by then, natural theology in Britain had already
been transformed – in the process of rebutting secular forms
of science emanating from France.

The third reason for studying the resilience of natural theology derives from the fact that it was not equally resilient in
every European country. It was emphatically a European
phenomenon – not merely a quirk of the British. In Chapter
V we saw how Leibniz saw religious utility in science, even
though Newton's science, in his judgment, could not bear
the weight. In Germany, books with such titles as *Insect theology* and *Water theology* emerged, each celebrating divine
forethought. The fact that ice was less dense than water, enabling aquatic creatures to survive beneath a frost, was a favorite example of divine prescience. We have already seen
how, in Spain, the heart of a dog could stir the heart of a
monk. In Sweden, the taxonomist Linnaeus (1707–78) turned
the rhetoric of natural theology to advantage in raising the
profile of the natural scientist:

If the Maker has furnished this globe, like a museum, with the most
admirable proofs of his wisdom and power; if this splendid theater
would be adorned in vain without a spectator; and if man the most
perfect of all his works is alone capable of considering the wonderful economy of the whole; it follows that man is made for the purpose of studying the Creator's works that he may observe in them
the evident marks of divine wisdom.[1]

In the Netherlands, too, a tradition of physico-theology
was established, manifesting itself in over fifty theological texts
and claiming religious incentives for the study of nature. Even
the French deists did not all follow Diderot in rejecting a
designed and harmonious universe. One of the most popular
books in eighteenth-century France was Noel Antoine Pluche's
Spectacle de la nature (1732), which conjured up a vision of
nature not dissimilar to that later preached by Paley.

It is, nevertheless, true that natural theology, as an aspect
of scientific culture, was particularly prominent and persistent in Britain. Charles Darwin was later to testify how diffi-

cult it had been to emancipate himself from its presuppositions: "I was not able to annul the influence of my former belief, then almost universal, that each species had been purposely created; and this led to my tacit assumption that every detail of structure, excepting rudiments, was of some special, though unrecognised, service." To examine the forces that sustained this outlook in Britain, when it had largely disappeared among scientific practitioners in France, should help to show that perceptions of the relationship between science and religion have been contingent upon wider historical circumstances than the current state of science.

Natural Theology in Britain: The Makings of a Pervasive Tradition

If the English were peculiar in the determination with which they drew moral lessons from nature, the peculiarity may lie with the English landscape, deficient as it is in earthquakes and volcanoes. It was surely more difficult to rejoice over nature as the work of benevolent design in those countries wracked by natural disasters. Was there anything in England to compare with the dreadful famine afflicting Naples in 1746, when thousands of deaths gave new power to those seeking economic and agricultural reform? Was there anything in England to compare with the great Lisbon earthquake, which for Voltaire was a macabre mockery of the idea that this was the best of all possible worlds? Was there anything in England to compare with the earthquake that the young Charles Darwin was to experience at Concepcion in South America, which made so deep an impression upon him? If beneath England, he wrote, "the now inert subterranean forces should exert those powers, which most assuredly in former geological ages they have exerted, how completely would the entire condition of the country be changed!" In every large town, "famine would go forth, pestilence and death following in its train." A contrast indeed with the happy world of Paley's vicarage garden.

But this is only to say that there was little in England to generate a predisposition against the idea of beneficent natural laws. More positively, through the mediation of the Boyle lecturers, Newton's science gave both precedent and impetus to English physico-theology. It could be used to meet the challenge of atheism and of any philosophical system that detracted from God's freedom by implying that the universe could not have been other than it is. As we saw in Chapter V, useful parallels could be drawn between God's providence in nature and in history during the turbulent aftermath of the 1688 revolution. In the defense of a constitution in which the power of the monarch had to accord with the rule of law, parallels with a natural order, in which stability was also assured by the mediation of laws, could be attractive. A natural theology, in this wider sense of giving rational justification for a particular political system, continued to be a useful weapon whenever there were rumblings of a Stuart threat to the Hanoverian succession. In England, unlike France, the idea of the divine right of kings had been suppressed, even if it had not been fully dissolved.

A richer account of the context in which physico-theology could flourish would have to include reference to economic changes and the new values associated with them. Historians investigating the peculiarities of the English Enlightenment have suggested that the growth of a market economy may have helped to undermine religious confessions. In the early eighteenth century, a new preoccupation with money, an emerging individualism, and an explicit concern with how happiness was to be achieved ushered in what one historian describes as a "new moral order of refinement." This was one effect, as perceived by Voltaire:

Take a view of the Royal Exchange in London, a place more venerable than many courts of justice, where the representatives of all nations meet for the benefit of mankind. There the Jew, the Mahometan, and the Christian transact together as though they all professed the same religion, and give the name of Infidel to none but bankrupts. There the Presbyterian confides in the Anabaptist, and the Churchman depends on the Quaker's word. . . . And all are satisfied.[2]

As doctrinal and confessional divisions were eroded, so a natural theology would become more conspicuous as a bedrock of rational belief. As one of the admirable indications of divine management, William Derham, in his *Physico-theology* (1713) referred to the provision whereby the uniqueness of every individual had been ensured. A man's handwriting could always "speak for him though absent, and be his witness, and secure his contracts in future generations." Generations later, William James could not resist the observation that "a God so careful as to make provision even for the unmistakable signing of bank checks . . . was a deity truly after the heart of eighteenth century Anglicanism."

Contrasts between England and France, some of which we drew in the previous chapter, provide further clues as to why natural theology assumed cultural importance in England. According to Voltaire, all were satisfied in London – except Voltaire himself, for whom the contrast with his own country was a source of deep regret. In France there was no such continuum of tolerated belief. A far sharper polarization between Roman Catholic orthodoxy and religious dissent was the norm. A system of censorship had to be evaded. In England there was no need for enlightened minds to overthrow religion itself because there was no pope, no inquisition, no Jesuit order, no comparable grip of the priesthood on families through the practice of confession. Consequently, a critical mentality and faith in progress could thrive in England *within* piety. The peculiar resilience of natural theology can then be understood as a visible, and enduring, symbol of an Enlightenment goal – the pursuit of science – thriving within piety.

Natural theology flourished in England not because of a peculiar English mentality but because there were social and political circumstances that gave the English Enlightenment a distinctive character. From 1688, the constitution had incorporated demands – such as representative government, the sanctity of property, and a degree of religious toleration – that, in other European countries, remained on the agenda of reformers. England, early in the eighteenth century, was almost alone in that Enlightenment hopes were accommodated, rather than thwarted, by the existing order of state and society. The result has been described as a "drive towards inclusiveness" in English thought that was lacking in France:

Whereas militant French philosophes represented the world in contending opposites – light versus dark, body versus soul, humanity versus priestcraft – English thought went for comprehension: individual *and* society, trade *and* gentility, conscience *and* self-love, science *and* religion.[5]

Natural theology could prosper as a manifestation of that more general phenomenon of inclusiveness – especially when so many cultivators of natural history were themselves clergymen.

Figure VI. 1. Plate 5, figure 1 from the first edition of William Buckland's *Bridgewater treatise,* entitled *Geology and mineralogy considered with reference to natural theology* (1836). The question was whether the beast had been designed. Reproduced by permission of the Syndics of Cambridge University Library.

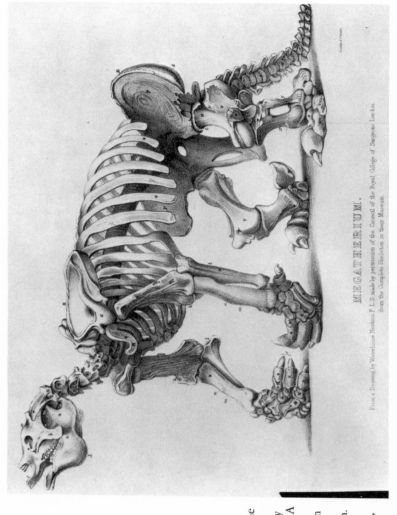

MEGATHERIUM.

From a Drawing by Waterhouse Hawkins F.L.S made by permission of the Council of the Royal College of Surgeons London.
from the Complete Skeleton in their Museum.

Figure VI. 2. Plate 6 from the 4th edition of Buckland's *Bridgewater treatise*, edited by Francis T. Buckland (1870). A later profile of *Megatherium* based on a complete skeleton in the Museum of the Royal College of Surgeons, London. Reproduced by courtesy of the Science Museum Library, London.

The high profile of clerics among students of natural history, before the life sciences became professionalized in the nineteenth century, meant that they were in an authoritative position to react if their science were ever attacked by fellow clergy. The vocabulary of natural theology served a useful function in meeting such a challenge. If one were accused of dangerous views, at variance with Scripture, one could always protest that, on the contrary, one's science only confirmed the power and wisdom of the Creator. The Oxford geologist and clergyman William Buckland (1784–1856) was to make precisely that move, early in the nineteenth century, as he strove to introduce geology into the Oxford curriculum. As improbable an item as an extinct giant sloth was transformed by Buckland into an argument for adaptation and design. When the newly founded British Association for the Advancement of Science came to Oxford in 1832, he detained his audience until midnight as he expatiated on the design of its grotesque forelimbs.

They had seemed grotesque to certain French commentators, but Buckland deduced that they had been used for the excavation of roots, for which they were the perfect tool. Christening the beast "Old Scratch," he showed how it would have dug quite a gutter in digging for its daily bread. In Buckland's public performances, science, natural theology, and revealed theology were kept in harmony. A few years earlier he had even suggested that a universal flood, such as that recorded in Genesis, was a necessary postulate to explain geological phenomena. Natural theology flourished in England because it had become an acceptable (if not always successful) means of fending off religious criticism. In a pre-Darwinian universe, the design that Robert Boyle had discerned in the minutest of God's creatures could be seen in monsters too. Natural theologians, such as Paley and Buckland, could exploit the intimate relationship between organic structure and function, apparently oblivious of a more subtle analysis that had been achieved in Germany.

Natural Theology in Germany: The Kantian Alternative

The intimacy between science and religion, so apparent in English physico-theology, was subjected to a searching criticism by the great German philosopher Immanuel Kant (1724–

1804). In certain respects has critique resembled that of David Hume, whose skeptical challenge we examined in Chapter V. Both endeavored to expose the limits of reason, though Kant was the more earnest in making room for a well-grounded religious faith. Like Hume, he attacked the argument from design by exposing the deficiencies of analogical argument. Thus, the most the argument could show was that the world has an architect, working on preexisting material. That was as far as the analogy with human manufacture could go. To establish the contingency of matter, the role of God as Creator, or the existence of a necessary Being, required a different form of argument. Hume had raised the objection that to explain the presence of order in the world by postulating mental order in a Creator was to invite an infinite regress since an explanation in terms of order still left the source of that mental order unexplained. The infinite regress could only be avoided if mental order in a god could be taken as self-explanatory. But that was surely an a priori assumption. For similar reasons, Kant argued that the design argument begged important questions. It assumed that a self-existent Being could be established as the First Cause of the cosmos. In his *Critique of pure reason* (1781), Kant showed that rational proofs of such a Being were unattainable.

In his critical writings, Kant disentangled the threads that had bound science and religion together through the agency of physico-theology. While he could still say that scientific investigation was only possible where nature was conceived as if its laws were the result of design, the emphasis fell on the "as if." That laws of nature were to be regarded as if they had been prescribed by a lawgiver was not sufficient to establish that they had. Only if no *other* explanation for the appearance of design could be found would the inference to a divine designer be secure. But one could not know that all other possibilities had been exhausted. In Kant's specialized philosophical terminology, a teleological principle was a regulative principle for the reflective judgment. It was not an a priori determining principle of things. Pure reason could not grasp the essence of nature. To affirm the reality of design would be to introduce into science a kind of causality beyond what was required.

Even in the study of living organisms, where it was effectively impossible to avoid references to teleology, Kant observed that the power of organization was not analogous to any other known causal agency. In his *Critique of teleological judgment* (1790), he contrasted the linear causal sequences that were postulated in the physical world with the peculiarity of living things, which were, in a sense, both cause and effect of themselves. The purposive causality of an organism could *not* be explained by analogy with a work of art. In the latter, the cause of the parts was simply the artist's idea of the whole. But the parts of an organism were not related by an idea external to them. The formative power was inherent in the organism itself. It did not follow that the ultimate origin of all bodies was mechanical in nature. But Kant maintained that only mechanical causes of material phenomena could be grasped by the understanding.

Such subtlety of reasoning makes a remarkable contrast with the somewhat glib inferences to design made by many of his contemporaries. Kant never denied that natural science had metaphysical foundations. But it was possible to separate what he called the "metaphysics of corporeal nature" from general metaphysical issues concerning God, freedom, and immortality. In his *Metaphysical foundations of natural science* (1786) Kant expressed the view that the very possibility of matter required that there be forces of repulsion and attraction; but the manner in which he expressed it suggested that the forces, determined by physical inquiry, were of the essence of matter. They were not, as it were, added by a spiritual Being. Thus his analysis of teleological reasoning, his concern with the boundaries of science, and his insistence that "an original attraction belongs to all matter as a fundamental force appertaining to its essence" could all be read as severing the bond between Newtonian science and the theological inferences drawn from it.

The supreme deficiency of physico-theology, according to Kant, was that no matter how much wise artistry might be displayed in the world, it could never demonstrate the *moral* wisdom that had to be predicated of God. To postulate a Being who was holy, good, and just, one had to transcend the anthropomorphism of the design argument, for no man could claim to be holy, even though he might be a superb mechanic. But how could one reason one's way to the existence of so holy a Being? Kant's answer was that one could not. There was no "proof" available in the usual sense. There was, however, a form of moral argument that would make *faith* in such a Being a rational foundation for one's conduct. While denying that knowledge of the divine Being could be attained through the use of speculative reason, or through science, he insisted that the *idea* of a God who was holy, just, and good was a necessary guiding principle. Otherwise there could be no adequate account of our moral experience.

Kant's argument runs like this. It is our moral duty to promote the highest good. This highest good must be attainable since "ought" implies "can." But finite humans lack the power to achieve it. They can achieve virtue, but they cannot ensure that happiness will be added to virtue, which is what the highest good requires. That requirement can only be met by postulating a rational and moral Being who, as Creator and sustainer of the world, has the necessary power to make happiness proportional to virtue. Two ancillary postulates were also necessary: the freedom of the will and the immortality of the soul.

Kant did not pretend to have established the objective truth of any of these postulates. They were simply the precondition of a rational account of moral experience. It was inappropriate to say that *it is* morally certain that there is a God. One could, however, say that *"I am* morally certain. . . ." The rationalist proposition that one needs to know whether God exists or not before deciding whether to take a religious stand was precisely that which Kant rejected: "The critically aware mind comes to view the entire theoretical approach to God as a development made out of a fundamental, moral and religious belief in Him, rather than the converse." There was no proof for the objective existence of God. Indeed, Kant is often credited with the "deobjectification of religion." God was no longer an object of knowledge in the sense in which human beings were, or in which extraterrestrial intelligences might turn out to be. But, for religious thinkers, there were gains as well as losses, particularly from his stress on the personal nature of faith.

The style of Kant's critique can be illustrated by his essay *On the failure of all attempted philosophical theodicies* (1791). One of the assumptions of much eighteenth-century natural theology was that a sufficient reason could always be found for those features of the world (pain, suffering, natural disasters, evil) that were not suggestive of a benevolent God. By showing that the blemishes were only apparent, that they filled some ulterior purpose, the path was smoothed. Science sometimes lent a hand in the smoothing operation. John Ray had argued that the apparently excessive amount of water, in the shape of the oceans, was no more than was required to generate, by evaporation, sufficient rainfall to ensure that habitable lands would be fertile. Even the most noxious insects, William Derham protested, served the purpose of teaching us watchfulness. And the rattlesnake, for all its venom, did oblige with a warning before it struck!

This was a type of special pleading, which, in the moral sphere, Kant found offensive. A man of moral and religious sensibility would have to acknowledge that all was not right with the world. There was palpable injustice in the vicissitudes of life. How could the holiness of the Creator, as legislator, be squared with moral evil? How could the goodness of the Creator, as governor, be squared with the countless woes of humanity? How could the justice of the Creator, as judge, be squared with what Kant described as the "unbecoming disproportion between the impunity of the guilty and the gravity of their crimes"? Kant examined the reasons habitually given, but in every case found them wanting. There was no license for optimism, for it would require omniscience to recognize such perfection in any world that one could say of it with certainty that no greater perfection was possible. To blame moral evil on man, not God, was unsatisfactory; for, if God could not prevent it, it must be a product of the nature of things, of human limitations, for which mankind could not be held responsible. If the toil and suffering of this life were rationalized as a necessary means of perfection for some future existence, the proposition was easier to state than understand. With reference to the justice of God, it was sometimes said that no criminal ever escaped punishment. The malefactor would always be plagued by conscience. But this would not do, Kant replied, because it projected onto vicious men the sensitive conscience of the virtuous.

To invoke a future life, in which all would get their just deserts, was quite gratuitous. In common with Hume, Kant pointed out that it was impossible in *reason* to find a justification for any different dispensation in a future world than that experienced in this. As long as the power of reason in such matters remained undemonstrated, all attempts to fathom the ways of divine wisdom must be abandoned. Instead, the question of justifying the ways of God to man was one of faith, not of knowledge. As his example of an authentic stance in the face of adversity, Kant chose Job, who had been stripped of everything save a clear conscience. Submitting before a divine decree, he had been right to resist the advice of friends who had sought to rationalize his misfortune. The strength of Job's position consisted in his knowing what he did not know: what God thought He was doing in piling misfortune upon him. The moral wisdom of the Creator had to be postulated, but it could not be understood, even less proved.

It was not Kant's intention to abolish religion, rather to reconstruct it in terms of moral faith. It was a human-centered religion in the sense that its focus was what man must hope and believe if he is to fulfil his sense of duty, and if cultural achievements are to have lasting worth. But they were hopes and beliefs compatible with, and arguably informed by, a Christian ethic. For Kant, one of the strengths of Christ's moral teaching was that it tempered self-conceit and self-love with humility. The precept of a Christian ethic, that a man should love his neighbor as himself, was, however, so pure and uncompromising that it destroyed one's confidence in one's capacity to obey. Nevertheless, Christ's teaching also helped to reestablish that confidence "by enabling us to hope that, if we act as well as lies in our power, what is not in our power will come to our aid from another source, whether we know in what way or not."

Christian theologians have often complained that Kant's Christology was deficient in that Christ emerged simply as an authoritative moral teacher. Kant, however, believed that he had found a way of preserving a personal faith, and a way of articulating it, which distinguished it from knowledge, on the one hand, and an uncritical willingness to assent to unproved propositions on the other. Faith was "the moral attitude of reason in its assurance of the truth of what is beyond the reach of theoretical knowledge." It was a steadfast principle of the mind, according to which "the truth of what must necessarily be presupposed as the condition of the supreme final end being possible is assumed as true in consideration of the fact that we are under an obligation to pursue that end."

The fact that so much popular religion was based on, or had its origins in, fear and superstition had, for Hume, been a good reason for abandoning religious comforts altogether. For Kant it was a good reason to erect an enlightened theology that would bar all attempts to probe into and manipulate God as if He were an object of nature. His moral theology, he believed, would serve just such a purpose. In Kant's world, man could assert his autonomy, but he still had to assert his faith in a divine author of the moral law.

The Continuing Functions of Natural Theology

The effect of Kant's critique was not to bring science and religion into opposition but to separate them in a manner that has often been imitated since. Science could proceed with the quantification of natural forces without colliding with theology, whose province was that of morality. In Germany, where Kant had his greatest impact, there were physiologists who clearly welcomed that separation. They felt free to discuss the integration of different levels of organization in organic systems without investing their teleological language with theological meaning. In Britain, however, the association of science with the argument for design continued, largely unscathed by the criticisms of either Hume or Kant. Even among relatively sophisticated thinkers, having some understanding of Kant's position, it was assumed that Christian theology still gave the best *explanation* for the appearance of design in nature. In the 1830s, the Cambridge philosopher William Whewell (1794–1866) would still be describing a hymn to the Creator as the "perpetual song" of the temple of science. The beneficial interaction between so many laws of nature indicated an intelligence behind the system.

How is the survival of natural theology to be explained? One element of an explanation had been foreseen by Hume. He had acknowledged that religious preconceptions were so strong that it was difficult to clear them from the mind. Religious believers, educated to see the world as the work of a wise and merciful Creator, would find it hard to remove their spectacles – or, as Hume preferred, their blinkers. To be told that the design argument only works on the supposition that mental order in a god is self-explanatory would hardly move the believer who was perfectly happy with that assumption. Where God was already assumed to be good, natural theology could be conceived as supplying both empirical data to corroborate the assumption and theoretical considerations to show why other data need not falsify it. A more modest orientation of natural theology might still be possible – and still attractive if the argument from design could win assent from persons of different religious persuasions.

Here it becomes important to explore the functions of natural theology within religious communities. To treat the traditional "proofs" of God's existence as if they were exclusively proofs (and, as such, necessarily failures) is to miss other roles they played in the religious cultures that had nurtured them. An awareness of these nondemonstrative functions is vital for a balanced assessment. Not even Kant had considered the arguments valueless. As *proofs* they were without meat. But they still had a function in the context of analyzing conceptions of God. They could function, nondemonstratively, in a clarification of what it would be for God to be God. In his *Critique of pure reason,* Kant proposed that otherwise discredited proofs could still regulate the ideal of a Supreme Being to ensure that it remained a flawless ideal. The arguments did not work as proofs, but they retained a function in purifying the concept of God and in ensuring that it was self-consistent.

Within religious communities, references to design could help to evoke a sense of awe and wonder. When the ancient Psalmist had declared that "the heavens are telling the glory of God; and the firmament proclaims his handiwork," he had not been constructing an argument but expressing a sense of the divine presence in poetic form, hoping to elicit a similar response in others. Natural theology as evocation rather than proof was unlikely to be disturbed by the philosophical critiques. A similar nondemonstrative function was to confirm to the understanding what was already a matter of faith. It helped to reassure the faithful if beliefs taken on trust could be rendered plausible on independent grounds. There was poignancy in this if doubts had been actuated by scientific discovery. In his pastoral capacity, Adam Sedgwick tried to assuage the doubts of a friend with the argument we noted earlier: Geology testified to successive acts of creation, not to the eternity of living forms. Ten thousand creative acts, he wrote, had been recorded on stony tablets.

Natural theology had another recurrent function in the context of intellectual engagement between representatives of different religious traditions. In the seventeenth century, Jesuit missionaries had used the proofs of Thomas Aquinas to establish common ground with other cultures. It was not necessary, for example, to convince thinking Hindus *that* God exists. They believed that already. Rather, natural theology had been the vehicle for showing them that their God and the Christian God were the same. From that platform one could then proclaim what was distinctively Christian. Natural theology, as a prelude to evangelism, continued to appear in many guises. In the conclusion to his *Bridgewater treatise,* written in the 1830s, the Scottish divine Thomas Chalmers (1780–1847) willingly conceded that the design argument, as a proof, was not compelling. But one thing it did was raise *questions* about the origin and purpose of creation – questions that infidels ought to be asking, and to which the only satisfying answers were to be found in the Gospel.

In explaining the survival of a modest natural theology, we have concentrated so far on functions of the design argument that may be said to be intrinsic to theology. There were, however, functions associated with the promotion of *science* that kept design arguments in the picture. One was the unifying function for scientists who were either clerics or deeply committed to a religious interpretation of nature. Not only did William Buckland work wonders with the limbs of the sloth; he actually *defined* geology as the "knowledge of the rich ingredients with which God has stored the earth beforehand, when he created it for the then future use and comfort of man." As a member of a nation undergoing rapid industrial expansion, Buckland was impressed by the resourcefulness of the Supreme Anglophile who had ensured that coal, iron ore, and limestone would coexist in convenient localities.

The second of these additional functions was the role of natural theology as a mediating agent between different theological positions, when the object was to avoid religious and political discord. Belief in a beneficent designer could serve as a lowest common denominator for men whose primary interest was the pursuit of science. This function continued to be important, in Britain and America, until well into the nineteenth century. When the British Association for the Advancement of Science was founded in the early 1830s, its desire to project a united front and to cultivate a favorable public image created the perfect environment for design arguments to flourish. They flourished in Adam Sedgwick's address before the 1833 meeting when he specifically warned against the distractions of religious and political discord. Natural theology could be a way of avoiding doctrinal disputes and the political issues so often associated with them.

This does not mean, however, that the content of natural theology was politically neutral. It was certainly not perceived to be by the more radical exponents of social reform in Britain, who associated it with the protection of privilege in the alliance between the established church and the universities of Oxford and Cambridge. And they had reason to do so. For the scientists of those universities had found in natural theology a useful resource when the social order was most under threat. In his Easter sermon for 1848, the year of revolution in Europe, the dean of Westminster was still reiterating an analogy between the fixed laws of nature and the fixed laws of society: "Equality of mind or body, or of worldly condition, is as inconsistent with the order of nature as with the moral laws of God." The remark is interesting both from the point of view of the time when it was made and by whom it was made. In 1848 the dean of Westminster was none other than the erstwhile geologist William Buckland. And the timing is interesting not just because it happened to be 1848. A fortnight before, steps had been taken to defend Westminster Abbey from a potentially unruly gathering of workingmen, claiming the right to self-organization.

Earlier in the nineteenth century, natural theology had featured in scientific discourse because it had provided an explicit means of dissociating scientific discoveries from revolutionary implications. The critical point is that, in Britain, this particular function of arguments for design had become more urgent in the period following the philosophical critiques of Hume and Kant than before – and for three quite different reasons. First, by the end of the eighteenth century, the evangelical revival (associated with John Wesley) was well under way. Second, there were also more potentially subversive theories and cosmologies coming to light – particularly those emanating from France, where a secular model for the origins of the solar system was proposed by Laplace, and for the origins of living systems by Lamarck. Third, during the last decade of the eighteenth century, there was such a conservative backlash against the horrors of the French Revolution that any revolutionary scientific conjecture was liable to start the warning lights flashing. It was during the 1790s that Joseph Priestley had his house and chemical laboratory destroyed by a Birmingham mob because of his known sympathies with events in France.

This change in sensibility, which allowed arguments for design to meet a greater rather than lesser need, can be illustrated by the changing fortunes of one hypothesis of organic development – that of Erasmus Darwin (1731–1802). Darwin's departure from religious orthodoxy had occasioned little comment in reviews of his earliest works. Even his *Zoonomia* of 1795 got off lightly, though the *British Critic* attacked him for depriving man of his soul. But during the next few years his reputation was assailed on all sides. By 1803, when his *Temple of nature* was published, even those organs that had once been sympathetic now condemned him. Damned by the *Critical Review,* his work had as its "great fault" its "unrestrained and constant tendency to subvert the first principles and most important precepts of revelation." It was no coincidence that Paley's *Natural theology* (1802) was published when it was. And he, too, took a swipe at Erasmus Darwin, along with others who had suggested that organic structures could develop by natural means in accord with the needs of the organism. In these years following the French Revolution, the contrast between England and France is particularly striking because it would have been difficult, in France, for scientists to revive the arguments of natural theology without appearing to be supporters of the ancien régime.

The Relevance of Natural Theology to Science

The fact that natural theology was sustained by a series of functions that can be analyzed in social and political terms raises the question whether arguments for design, emanating from scientists themselves, might have been little more than a convenient veneer on their science. It is impossible to generalize about this since religious beliefs differed from one naturalist to another. Nor does it follow that because an argument can fulfill a social function it thereby loses its integrity. It is, nevertheless, pertinent to ask what bearing, if any, natural theology had on the practice of science. It is one thing to say that the results of a piece of scientific work were susceptible of a providentialist interpretation. It would be another to suggest that religious belief structured scientific theory and practice.

Among the roles for religious belief *within* science, identified in Chapter I, were three that might be expected to be visible if conceptions of design were doing any work. Belief in nature as a designed system might regulate scientific thinking both in the choice of problems and in the construction of acceptable solutions. It could also play a selective role in conferring greater plausibility on one theory rather than another if it happened to be more congenial to religious interests. And, in certain contexts, it could play a constitutive role, as when religious dogmas were placed at the summit of deductive explanation. Because the argument from design presupposed a Creator, one would expect creationist models to be visible in theories about the geographical distribution of species and preeminently in discussions of the origin of species. Natural theology undoubtedly had a bearing on science in each of these respects.

One science in which the relevance of natural theology can be discerned is that of human anatomy. A recent study of German anatomists of the late seventeenth and early eighteenth centuries suggests that the desire to construct a physico-theology created a predisposition in favor of mechanical analogues for human organs. Representatives of this "Anatomia Theologica," as it was called in the early eighteenth century, included Friedrich Hoffman (1660–1742), professor of medicine and physics at Halle; Georg Albrecht Hamberger (1662–1716) of Jena, and his successor Johann Friedrich Wucherer (1682–1737), who was both physicist and Lutheran theologian. In their works the intricate machinery of the human body was displayed to evoke amazement and deference to an incomparable designer. And if physico-theology affected the content of their science, it also affected its promotion. To appreciate just how wonderfully humans had been made, it was necessary to dissect their corpses. But against that practice there had long been taboos of many kinds. By providing a theological rationale for the practice of dissection, the exponents of "anatomia theologica" found a way of challenging those taboos. Lorenz Heister (1683–1758) of Helmstedt and Albrecht von Haller would advertise dissections to which the public were invited. How could it be wrong, the argument ran, to practice dissection, if divine craftsmanship was thereby more fully revealed?

Another example of the regulative role of natural theology occurred in the previous chapter, in the work of Joseph Priestley, whose determination to find the mechanism for the restoration of common air coincided with his conviction that, if nature were a rational and viable system, there had to be such a mechanism. A similar teleological principle regulated the geological system of Priestley's contemporary James Hutton (1726–97), whose Theory of the earth (1795) enjoyed notoriety for a cyclic view of earth history that promised no vestige of a beginning, nor prospect of an end. A solid body of land, Hutton explained, "could not have answered the purpose of a habitable world." Soil was the crucial requirement to allow the growth of plants. But the same processes of erosion that produced soil from mountains would, in time, destroy the land. Accordingly, if the world was the work of "infinite power and wisdom" there *had* to be restorative forces, which he went on to locate in the earth's central fire.

The Huttonian cycles of elevation, erosion, deposition, and consolidation were grounded in teleological reasoning, which assumed that nature (and/or God) purposed the maintenance of plants and animals. The insertion of that ambivalence is deliberate because, whether or not he was familiar with the Kantian critique, Hutton seems to have perceived that teleological reasoning could play a regulative role without having to be attached to a theological base. His own references to "infinite power and wisdom" were helpful to those who sprang to his defense, but he was equally at home when denying the "possibility of anything happening preternaturally or contrary to the common course of things." A role for natural theology is visible in Hutton's science, but in precisely that ambivalent manner that we identified at the beginning of the chapter.

During the nineteenth century, conceptions of design continued to regulate scientific theory in several fields of study. One was paleontology where the assumption that an organism was a well-adapted structural whole could regulate the reconstruction of fossil species from fragmentary remains. If only certain combinations of bones were compatible, then useful constraints could be placed on the reconstruction of the whole from its parts. In conjuring up the plesiosaurus, British paleontologists designed the beast on the assumption that it had a previous master designer. Buckland's account of "Old Scratch" also shows how teleological reasoning could penetrate technical work. The feeding habits of extinct species were inferred on the assumption that they had been designed for their habitat.

In anatomy it even became fashionable for a while to suppose that all vertebrates had been modeled on a single structural archetype. This view could constitute a challenge to the argument from design as formulated by Paley. Organic structures might be explained by their conformity to the type rather

8. The intent and ends of anatomy are various: Its end
the primary one is an acquaintance with, and an and in-
admiration of, the work of the Creator in the tents.
human frame: a ferious contemplation of the
ftructure of this amazing fabrick, of the appropriated figure of the feveral parts of it, their connections, communications, actions and ufes, is one
of the ftrongeft of all arguments againft atheifm:
it carries a proof not only of the exiftence of a
Deity, but at the fame time of his amazing great-
nefs and wifdom; and leads the obferver immediately to the adoration, as well as the acknowledgment, of a God. The glory of the Creator may,
therefore, be very juftly declared to be the great and
primary end of anatomy. The fcience, treated in
this light, may therefore be called philofophical,
phyfical, or theological anatomy, and is highly ufe-
ful to every one who ftudies true wifdom and
theology.

B 2   9.

Figure VI. 3. Part of the text from page 3 of Lorenz Heister's *A compendium
of anatomy translated from the Latin* (1752). Heister's presentation of his
"theological anatomy." Reproduced by permission of the Syndics of Cambridge University Library.

than by an *intended* function. But, for the distinguished British anatomist Richard Owen (1804–92), the archetype had
been an idea in the mind of the Creator who had adapted it
to the needs of individual species. In his *Anatomy of vertebrates,* published in the 1860s, Owen was still proclaiming
unity of structure as a regulative principle of scientific inquiry. In this text the connection with natural theology was
explicit. One of his aims had been to disclose "the unity which
underlies the diversity of animal structures; to show in these
structures the evidences of a predetermining will, producing
them in reference to a final purpose."

The most palpable effect of natural theology in the fabric
of science was the resistance it produced to theories of organic transformation. The exponents of the design argument
whose names have recurred in this chapter – Paley, Buckland, Sedgwick, Whewell – were united by an opposition to
theories that turned God's creatures into products of nature.
This was not merely a clerical prejudice. As we shall see in

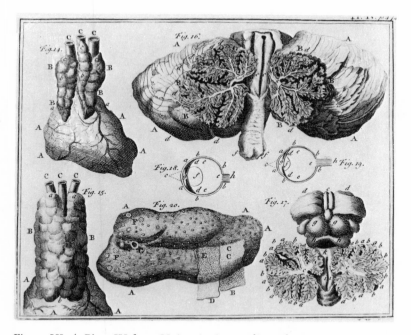

Figure VI. 4. Plate IV from Heister's *Compendium of anatomy* (1752). The intricacy of what Heister called "this amazing fabric" was revealed through dissection. The figure in the top right is of the human cerebellum showing a division of lobes that, according to Heister, was commonly overlooked by anatomists. Also depicted (on the left side) are two examples of the thymus gland adhering to the heart. The ascending branches of the aorta are clearly visible. Reproduced by permission of the Syndics of Cambridge University Library.

the next chapter, an anticlerical geologist, Charles Lyell, reacted just as strongly to the evolutionary theory of Lamarck. The creationist assumptions behind the design argument did, however, manifest themselves in both overt and subtle ways. Overtly, they were manifest in Whewell's remark that it was not within the competence of science to solve the mystery of the origin of species. In subtler ways creationist assumptions intruded in reconstructions of the fossil record. An effective strategy for disposing of Lamarck, and his few British admirers, was to eliminate any linear trends toward increased complexity on which they might capitalize. Paradoxically, in

Figure VI. 5. Illustration from Richard Owen's *On the nature of limbs* (1849). Owen's skeletal "archetype" and the relationship it bears to actual vertebrates, supposedly modeled on it, are suggestively displayed. This Platonist anatomy was developed by Owen to counter more secular and materialistic conceptions of "unity of composition" developed by Etienne Geoffroy St. Hilaire in France and adopted, during the 1830s, by certain medical reformers in Britain. Reproduced by courtesy of the Syndics of Cambridge University Library.

even subtler ways, natural theology left its mark on Charles Darwin, even after he became an evolutionist. We have already had occasion to note his confession that he had found it almost impossible to annul the idea, taken from Paley, that every detail of structure must have some purpose.

The Relevance of Science to Natural Theology

Just as certain scientific theories were shaped by the assumptions of natural theology, so the form of the design argument was itself shaped by scientific innovation. The existence of this reciprocal relationship meant that science could have a direct bearing on the fortunes of natural theology. With the growth of scientific specialization and its institutionalization in the nineteenth century, this increasingly came to mean a direct bearing on its misfortunes. On one level, natural theology was not so much destroyed by science as eased out of scientific culture by a growing irrelevance. The questions that flowed from viewing nature as the product of design simply became too blunt to yield precise information at the rock face of research. This was a point Charles Darwin came to appreciate as he reflected on the fact that each island of the Galapagos archipelago had its distinctive species, closely resembling those on neighboring islands and bearing a general similarity to those of mainland South America. This was simply too tantalizing a puzzle to solve by invoking the will of God.

A growing resistance to natural theology *within* science was also prompted by the excessive zeal of religious apologists, some of whom created doubts about the logic of their enterprise by the inanity of their examples. Darwin's cofounder of the theory of evolution by natural selection, Alfred Russel Wallace (1823–1913), lost his patience when he encountered the claim that the soft scar on a coconut was a "wise contrivance" without which the embryonic shoot beneath would be trapped. This was degrading as well as inane, since it imputed to the Supreme Being "a degree of intelligence only equal to that of the stupidest human beings." It was like praising an architect for remembering to put a door in his house!

These references to Darwin and Wallace are a pointer to the fact that there was another level on which science was to have a direct and more devastating relevance to the argument from design. Philosophers of the eighteenth century, such as Diderot and Hume, had suggested that the appearance of design in the organic world could be illusory. It was not surprising that existing species were well adapted to their environment. Had they not been, they would not have survived in the first place. With far greater sophistication, Darwin and Wallace would show exactly *how* a perfectly natural mechanism could counterfeit design. We shall examine Darwin's reasoning in subsequent chapters, but his conclusion would be that "the old argument from design . . . as given by Paley, which formerly seemed to me so conclusive, fails, now that the law of natural selection has been discovered."

But this is to anticipate. For the period from 1800 to 1850 it is more accurate to say that science prompted the diversification of natural theology rather than its demise. The pressure to diversify came both from the physical and life sciences, and especially as they acquired a historical dimension. Because the historical sciences were to transform perceptions of man's place in nature, their history will be the subject of the next chapter. Their impact on natural theology can, however, be traced in outline.

By the close of the eighteenth century it had already become difficult to defend one of the assumptions on which much of natural theology had been based — that God had created every creature it was possible to create and that they had all existed together since the original days of creation. A century earlier, it had been possible for John Ray to say that what he meant by the works of the creation were "the works created by God at first, and by Him conserved to this day in the same state and condition in which they were at first made." This assumption had allowed Ray to argue that animals had been preadapted to preadapted conditions. The excellence of the provision was confirmed by the fact that no species had ever become extinct. Even Ray had been perplexed by certain fossils, which, if they were admitted to be remains of once-living creatures, were sufficiently different from extant species to raise the specter of extinction. He had, however, hit on the neat solution that they were, after all, relics of still living forms that would eventually be found in some remote corner of the globe. In the 1770s it was still possible for Thomas Jefferson to say that no species had ever been lost. Such confidence was, however, soon shaken by the remains of giant mammoths, which impressed the French naturalist

Buffon (1707–88), and later by the quadrupeds of the Paris basin, which Georges Cuvier (1769–1832) showed were different from living forms. Natural theology had to readjust, if it was to survive.

The shift from a static to an episodic creation did not defeat the argument from design. It was always possible to contend that each species had been well adapted to the conditions prevailing during its life-span, that changes in those physical conditions were precisely those that enabled the Creator to introduce new species, thereby achieving a plenitude of created beings, even if they had not all existed concurrently. Variants of this idea of *progressive* creation were to be extremely popular in the nineteenth century. Buckland, Sedgwick, and Whewell each adopted it. But a world in which there had been a succession of worlds was no longer the single clockwork world that had so appealed to Paley.

Further diversification, and a further shift from Paley, would occur in response to another threat emanating from France – from the comparative anatomy of Etienne Geoffroy Saint-Hilaire (1772–1844). According to Geoffroy, teleological principles could be misleading in the analysis of living systems. A better guide was a principle of structural unity that could be affirmed without teleological trappings. An example discussed by Richard Owen illustrates the threat. Evidence for design had often been seen in the fact that, in viviparous animals, ossification of the head began at several centers, allowing a degree of compression that facilitated birth. It was just the sort of contrivance that Paley most admired. There was just one snag:

Our view of this provision is disturbed, when we find that the same mode of the formation of the bony framework takes place in animals which are born from an egg. . . . In this way, the admission of a new view as to unity of plan will almost necessarily displace or modify some of the old views respecting final causes.[4]

In acknowledging that displacement, Owen was admitting that if natural theology were to survive, it would have to change its format. He masterminded the change by locating the evidence for design in the modification of a structural archetype in special ways for special purposes. But Owen's move, even though it won the support of Whewell, signified a weakening of the conventional argument. Whewell admitted as much, even before the impact of Darwin was felt. With regard to the design that was seen in the organs of living things, Whewell observed that "though we can confidently say we see it, how obscurely is it shown, and how much is our view of it disturbed by other laws and analogies."

One last axis along which natural theology diversified can be illustrated by those scientists who simply reduced the design to the wisdom enshrined within scientific *laws*. This was something of a fall-back position. It appealed to the Oxford mathematician and philosopher Baden Powell (1796–1860) because it preempted objections arising from the further expansion of scientific territory. It did not matter to Powell if the development of the solar system, or even of organic systems, could be explained by natural laws. The fact that those laws had conspired to produce such a world as this was sufficient testimony to divine wisdom and foresight. Powell's position was ultraliberal among the clergy. It reflected years of frustration and disillusionment as he sought to advance the cause of science at Oxford during the 1830s and 1840s. It also had an embarrassing consequence. It gave tacit support to the materialist thesis of a book that appeared anonymously in Britain in 1844.

The work of a Scottish publisher, Robert Chambers (1802–71), the book had as its title *Vestiges of the natural history of creation*. Its message was that the transformation of species was not merely a fact but a law. Chambers drew on developments in astronomy, evidence from fossils and comparative embryology, the popular science of phrenology (which promised character determination from the topography of bumps on the head), even experiments purporting to show the in vitro production of microorganisms – all to make the point that the scientific elite who were denying organic evolution were missing the wood for the trees. All that was required for the emergence of new species was an abnormally long period of gestation in the development of an embryo. Widely considered a recipe for disaster, his book sold like hotcakes. Yet, for all its subversive implications (a "foul book" Sedgwick called it), Chambers presented it as a treatise on natural theology. If Newton's law of gravitation could be a reflection of divine activity in the physical domain, why could not the same be said of a law of organic development? If a law of development were denied, one would be left, according to Chambers, with the degrading image of a diety who had separately produced every new species throughout the universe, like so many conjuring tricks.

The fact that a theory of organic transformation could be framed in this way provides yet another instance of that ironic pattern we have discerned in earlier contexts. The association of natural theology with the extension of natural law had an effect rather like the Trojan horse. It smuggled a full-blown naturalism into territory that upholders of a more conservative natural theology, such as Whewell, still considered holy ground. Yet Whewell had many times said that laws of nature, and the effects resulting from their combination, provided as fitting a testimony to divine activity as the appeal to divine intervention. The irony is palpable, for Darwin was to place a quotation from Whewell to that effect at the front of his *Origin of species* (1859).

This, in outline, is the challenge that would be posed to Paley's natural theology by developments in the historical sciences. Before exploring those developments in detail, however, we should briefly return to our earlier question: whether natural theology, by claiming so much of its authority from science, might not have dug its own grave. From Newton, Bentley, and Clarke through to Paley, Buckland, and Whewell, had Christian apologists not placed too great a burden on arguments from design? They certainly had critics who thought so. In fact it would be a great mistake to imagine that when the kind of physico-theology we have been considering finally collapsed, it came as a shock and an embarrassment to every sector of the Christian Church. The vicissitudes of a science-based natural theology were arousing anxieties in Britain from the earlier years of the nineteenth century. The philosopher and romantic poet Samuel Coleridge (1772–1834) protested that he was weary of the word *evidences* of Christianity. Men and women should be made to feel the need of the Christian religion, not bludgeoned into it by reason. With some perspicacity, he identified one of the loopholes in a Newtonian natural theology, through which one could easily slip from theism into atheism:

It has been asserted that Sir Isaac Newton's philosophy leads in its consequences to Atheism; perhaps not without reason, for if matter by any powers or properties *given* to it, can produce the order of the visible world, and even generated thought; why may it not have possessed such properties by *inherent* right? And where is the necessity of a God?[5]

During the 1820s and 1830s some, such as the liberal Anglicans Richard Whately and Edward Copleston, were critical of the rational theology Paley had stood for. Their concern was that the social realities of deprivation and distress could give *proofs* of God's beneficence a hollow ring. It was better not to give the impression that Christian apologists were insensitive to social inequality. Another common complaint was that rational arguments based on the details of the natural world were ineffective in a pastoral context. In an extended critique of the *Whole doctrine of final causes* (1836), William Irons bemoaned the tendency in works of natural theology to degrade religious feelings by treating them as simple responses to material phenomena.

Even before Darwin altered the ground rules, the limitations of conventional arguments for design were widely canvassed. As sensitive a proponent as William Whewell was fully aware of their shortcomings. Even if the mutual adaptation of physical and biological phenomena could prove the existence of God, the proof would be valueless since the real requirement was an "active and living appropriation of the belief in God." As a philosopher of science and natural theologian, Whewell continued to argue that the best explanation for the mind's capacity to discover scientific truth was that it had been designed for the purpose. As priest and preacher, however, he stressed that the way back to God was not through such rational considerations. For one thing, that would leave God out of the conversion process; for another it would take insufficient account of the fact that design arguments were only really compelling to those who already believed.

In a sermon of 1839, the leader of the Oxford movement, John Henry Newman (1801–90), added his authority to the view that design arguments would only convince those with a preexisting faith. It is a "great question," he declared, "whether atheism is not as philosophically consistent with the phenomena of the physical world, taken by themselves, as the doctrine of a creative and governing power." A clue as to how that great question might be answered was provided by the anonymous *Vestiges,* which was read by many critics as an atheistic tract masquerading as natural theology. Whatever reasons religious leaders had for their reservations, they were compounded by the realization that references to designed laws might be just too facile. If they could be exploited by the author of a book that made humanity a material derivative of the animals, their integrity as well as their credibility was compromised. It was in response to *Vestiges* that another churchman delivered a telling blow. The historian of the Oxford movement R. W. Church pulled no punches: "The *Vestiges* warns us, if proof were required, of the vanity of those boasts which great men used to make, that science naturally led on to religion."

CHAPTER VII

Visions of the Past: Religious Belief and the Historical Sciences

Introduction

In his frontispiece to Gideon Mantell's *Wonders of geology* (1838), the painter John Martin conjured up a terrifying picture of ferocious dinosaurs fighting each other in an ancient, prehuman, and tormented world. It was a far cry from the benign world of eighteenth-century natural theology, a nightmare that also shrieked against a straight reading of Genesis where the implication was that pain and death had not entered the world until Adam's Fall. The integration of concepts drawn from astronomy, geology, paleontology, and evolutionary biology was to result in reconstructions of the past, which not only transformed perceptions of man's place in nature, but also contributed their own leaven to the growth of biblical criticism. And, irrespective of their implications for natural and revealed theology, the historical sciences could be perplexing, even wounding, to popular belief. Harmonious images of nature were shattered by new discords to which Cuvier had drawn attention in his *Preliminary discourse* (1812). Life on earth had often been disturbed by calamitous events, which, in the beginning, had penetrated the very depths of the earth's crust. Countless beings had been the victims of these upheavals — some annihilated by floods, others perishing of thirst as the seabed suddenly rose. Entire species had vanished forever, leaving their enigmatic traces.

No wonder John Ruskin was to wish he could escape the din of the geologist's hammer. No wonder the poet Tennyson associated the historical sciences with a sense of deprivation. Faced with the loss of a dear one, there had always been the consolation that nature, though careless of the individual, was careful of the type. But if entire species had vanished, even that solace was denied. No wonder the artist William Dyce, in his painting of Pegwell Bay (1858), chose to depict the futility of human life against the gloomy backdrop of comet and cliff, each of which spoke of aeons of time, dwarfing one's ephemeral existence into insignificance. And such melancholy thoughts as these were often as nothing compared with the revulsion experienced by those who shared Ruskin's horror of "filthy heraldries which record the relation of humanity to the ascidian and the crocodile."

The object of this chapter is to locate the considerations that led a succession of naturalists to construct historical models for the earth at variance with popular belief and with customary interpretations of Scripture. In addition to Darwin, we shall examine Linnaeus, Buffon, and Laplace; Lamarck, Cuvier, and Lyell. These figures have been selected not because they each made a durable contribution to the natural sciences (though that would be true of them all) but because their respective visions of the past raised further questions for the relationship between scientific and religious belief. Nor have these figures been selected because they were, in some sense, precursors of Darwin. There was no linear succession of scientific insights, inexorably culminating in Darwin's theory of natural selection. Even to look for precursors of Darwin among proponents of organic transformation can be simplistic; for many of the earliest speculations lacked a full historical dimension. As with the eighteenth-century naturalist Charles Bonnet, they were often based on transformation up a predetermined chain of being, with every link defined by an *extant* form. Not until fossil forms were proved to belong to *extinct,* but once viable, species was the space created for theories of organic evolution that were thoroughly historical, in the sense that the course of history could be locked in a fossil record, which revealed a unique, unrepeatable sequence of events in which novelties had appeared and disappeared.

The development of the historical sciences was riddled with controversy, not least because theories of organic transformation, such as that of Lamarck, could be seized by political radicals to support their schemes for the transformation of society. Recent research has revealed the depth of feeling among medical reformers in London who, during the 1830s, welcomed the transformism of Lamarck and the anatomy of

THE COUNTRY OF THE IGUANODON, RESTORED BY JOHN MARTIN ESQ. K.L.
FROM THE GEOLOGICAL DISCOVERIES OF GIDEON MANTELL, ESQ. LLD, FRS.&c.

Figure VII. 1 (*above*). Frontispiece to Gideon Mantell's *Wonders of geology* (1839 edition). Despite inaccuracies, Martin's restoration of "the country of the iguanodon" conveyed a sense of primeval violence long before humanity had appeared. Reproduced by courtesy of the Science Museum Library, London.

Figure VII. 2. Henry de la Beche's reconstruction (c. 1830) of a Liassic scene, used as frontispiece in Francis T. Buckland, *Curiosities of natural history*, 2d ser., 2 vols. (1860). For his half-whimsical sketch of life in Ancient Dorsetshire, de la Beche drew on William Buckland's research. As with John Martin's reconstruction, the reader is confronted by a spectacle of great ferocity, in this case of a plesiosaur being eaten by an ichthyosaur, and much other eating besides. Reproduced by courtesy of the Science Museum Library, London.

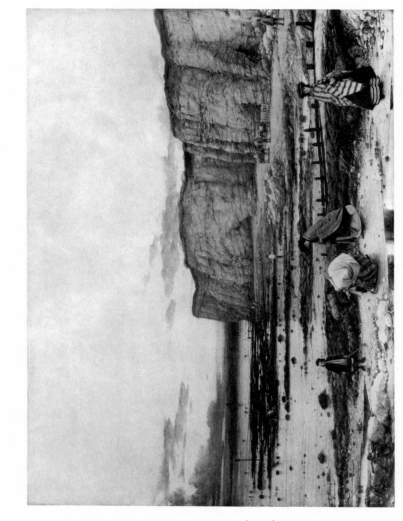

Figure VII. 3. William Dyce: *Pegwell Bay: a recollection of October 5th 1858.* Though barely visible in reproduction, Donati's comet is a striking feature of the original painting. Astronomy and geology ("Terrible Muses," Tennyson called them) combined in lending gloom to desultory, meaningless human activity. Reproduced by courtesy of the Tate Gallery, London.

Geoffroy St. Hilaire as they challenged the nepotism and the Oxbridge privileges perpetuated by the official medical corporations. It was the fact that theories of organic transformation had already acquired subversive moral and political connotations that allowed Darwin to say that admitting the mutability of species was like confessing to a murder. There were special reasons why Darwin's own theory was to prove so controversial once his confession was made public in 1859. Although the idea of organic transformation was by then familiar, the particular mechanism that Darwin proposed, with its vivid portrait of nature red in tooth and claw, was not only unpalatable to many but also questionable in its pretensions to scope and adequacy. The public debate was particularly animated because it was set within a larger controversy concerning the authority of Scripture. With the emergence of more sophisticated historical scholarship, particularly in Germany, it had already become clear to many Christian intellectuals that adherence to the literal inerrancy of Scripture was no way to present the credentials of Christianity in the modern world. But the general public was confronted with the implications of history as a science and the sciences as history more or less simultaneously. In the concluding sections of the chapter, we shall therefore consider the convergence of these two forms of criticism.

Linnaeus: The Fixity of Species and the Seeds of Doubt

The great taxonomic system for which Linnaeus is renowned was erected in the eighteenth century on the assumption of a fixed and rational creation. Because every plant displayed some affinity with others, Linnaeus was convinced that nature must conform to a plan. By subsuming every species under genera, orders, and classes, he hoped to capture something of the order imposed on the world by a mind that had preceded his own. In his *Philosophia botanica* (1751) species were defined as primordial types created by divine wisdom and perpetuated from the beginning to the end of the world. The idea of spontaneous generation was rejected, for if such a process were possible it would render superfluous the ingenious apparatus by which plants broadcast their seed. In his early writings, Linnaeus took as an axiom what Genesis seemed to say – that all God's creatures reproduced after their own kind.

The quest for an orderly world, typified by Linnaeus in his natural history, was a characteristic of Enlightenment culture, discernible for example in the classical structure of a Haydn symphony, in the symmetry of Georgian architecture, and in the fashion for formal gardens. Linnaeus himself redesigned the botanical gardens at Uppsala in accordance with his taxonomy. Plants of each taxa were sown in separate beds, which were themselves laid out in impeccable order. Often described as a second Adam, because of his propensity to give new species their names, he designed his own garden of Eden.

How did nature maintain this ordered, fixed creation? The answer was simple: Every creature was both predator and prey. Carnivores controlled birds, which controlled insects, which controlled plants. Far from being blind to a struggle for existence, Linnaeus knew that every species produced more offspring than could possible survive. But the resulting competition was merely a mechanism for preserving an overall balance and harmony. From this neat scheme the historical sciences eventually had to depart. But the scientific authority that Linnaeus conferred on a doctrine of fixed species was to have long-term implications. Because he is still cited with approval by latter-day creationists, it is important to appreciate that, in two respects at least, Linnaeus lets them down. Although his binomial classification was conceived as a psalter for divine worship, his account of creation was already a departure from a strict biblical theology. His idea that species had been created in pairs, in different climatic zones, on a paradisaical mountain, was effectively a conflation of the creation and flood narratives: These same species had migrated as the primordial waters subsided. It was a flood without Noah, for was it credible, he asked in 1744, that the Deity "should have replenished the whole earth with animals to destroy them all in a little time by a flood, except a pair of each species preserved in the Ark?"

The other respect in which this second Adam fell from grace consists in the doubts he began to harbor concerning the fixity of species. These were precipitated in part by the discovery of an unusual plant form by an Uppsala student, Magnus Zioberg. The deviant form, now thought to have been obtained by gene recombination (*Peloria* from *Linaria*), was rationalized by Linnaeus as a kind of hybrid. But it was a

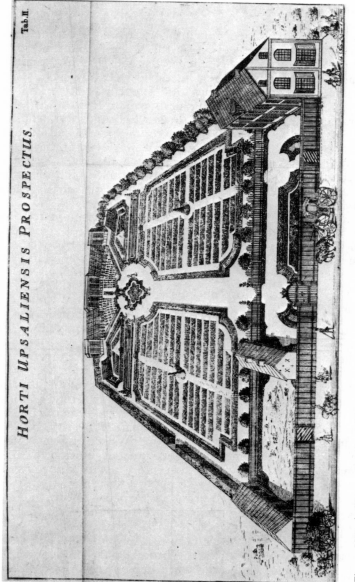

Figure VII. 4. Illustration from Linnaeus's thesis in *Hortus Upsaliensis* (1745). Linnaeus's design for the botanical gardens at Uppsala. Reproduced by courtesy of the Syndics of Cambridge University Library.

hybrid with a difference; for, unlike the proverbial mule, it was fertile. Further reflection on the scope of hybridization encouraged Linnaeus in the belief that species, like truth, might be daughters of time. When he developed his "System of Vegetables" he suggested that God might have created only one plant of each order, arranging intermarriages to form the genera, and leaving nature, by further hybridization, to pro-duce the species. On such a view, both species and genera became plastic, the former molded, he was even prepared to conjecture, by the influence of climate and geography. In a dissertation on the sexes of plants he would boldly declare that "a genus is nothing else than a number of plants sprung from the same mother by different fathers." Species and gen-era had a *history*. This was no great embarrassment to Lin-naeus's natural theology; for the upper levels of the taxo-nomic hierarchy could still be ascribed to God. But it was a departure from the biblical axiom of "each after its kind."

Buffon: A History for the Earth

When a vacancy arose at the Jardin du Roi in Paris in 1739, the successful candidate was asked to prepare a catalog of plant and anatomical specimens, which had been collected for their medical interest. From that enterprise, and from an empire-building operation in which the collection was con-tinually enlarged, Buffon was to launch his encyclopedic *His-toire naturelle*. Growing from a fifteen- to a forty-four-volume work, it was one of the most impressive monuments of the French Enlightenment. As a writer on natural history, Buffon was an unusually controversial figure. Where others would marvel at the taxonomic genius of Linnaeus, Buffon became his most conspicuous critic. To classify plants according to their sexual anatomy had generated a system that Buffon considered artificial in at least three respects. By abstracting the sexual criterion, Linnaeus had violated the need to con-sider the organism as a whole and the full range of its resem-blance to others. The very attempt to subsume species under a hierarchical scheme of classes, orders, and genera was arti-ficial in that these categories were human constructs, violat-ing both the continuity of nature and another cardinal prin-ciple — that nature only knew individuals. And Linnaeus was guilty in a third, more obvious, sense; for one consequence of his method had been the assimilation, in some cases, of species that were palpably dissimilar. But it was not merely as a critic that Buffon excited controversy. His reconstruction of the earth's history, together with a hypothesis for its origins, exposed him to the charge of impiety from ecclesiastics and recklessness from other naturalists.

In an early essay on the ass he had addressed the question that had come to occupy the mind of Linnaeus: whether species were permanent. In particular he inquired whether the horse and ass had descended from a common stock. He had no illusions about the importance of the question; for, if one species had ever been produced by the degeneration of another, no bounds could be fixed to the powers of nature. Why, in the course of time, might nature not have produced from a single individual all the organized bodies in the universe? Buffon's conclusion, however, went against such a prospect. He firmly rejected the notion that the ass was a degenerate horse, finding it contrary to Scripture, reason, and experience. It was not only that hybrid forms like the mule were sterile. There was also the problem of intermediate forms, of which no traces were known.

Despite that resolute answer, the question would not go away. Conscious of what had been achieved with domestic animals, Buffon was haunted by the possibility that, even in the natural state, some variation might be induced by external circumstances. Might not gradual, cumulative changes take place in an organism, with time as the great workman? The rate of change would depend on the degree to which the environment was altered and the rate at which the organism reproduced. Eventually he considered that the original constitution of a species might change, much as the different breeds of dog had been derived from an original wolflike species.

The possibility of organic change in response to changed environmental circumstances proved particularly engaging when Buffon compared the quadrupeds of the Old and New Worlds. Of two hundred species, only seventy were known in the Americas. Of these, forty were unknown in Europe and Asia. There was a possible solution to the riddle if quadrupeds from the Old World could have migrated to the Western Hemisphere and had subsequently degenerated under the influence of a less auspicious environment. Once again a historical dimension was being introduced.

Buffon's solution was by no means perfect. There was the

problem that some of the American quadrupeds were larger and more prolific than their Old World equivalents. There was also the difficulty as to how the migration could have occurred. But as long as one rejected creation in situ for the American species (which would, of course, be a contravention of the Genesis Eden), some kind of historical reconstruction was required. Buffon conjectured that America and Asia might still be joined in the north. If southern species were also derivatives then there must once have been a bridge in the south. The physical map of the earth also had its history.

In his *Epochs of nature* (1778) Buffon tried to give hypothetical content to that physical history. Suggesting that the earth had its origins in the material ejected from a collision between a comet and the sun, he required no less than seven epochs to bring it to its present condition. During the first, the incandescent matter from the sun gradually assumed the shape of the earth, the center of which began to solidify. During the second, solidification extended to the crust, with the formation of mountain chains. Only during the third was a temperature reached at which water could begin to condense on the earth's surface. In the hot seas, marine life was possible, but in forms that subsequent cooling rendered extinct. The fourth epoch witnessed a retreat of the waters, the smoothing out of valleys, and the exposure of hills. Further cooling still and the first animals could begin to appear at the poles, among them the giant elephants whose remains had been found in Siberia. Because similar remains had been found in the North American continent, Siberia and Canada had certainly been joined at that stage. If fracture had occurred during the sixth epoch, it was that which also saw the large mammals migrating toward equatorial regions in order to escape the encroaching cold. Not until the seventh did humans make their appearance. Looking to the future, they too would perish in a deep freeze.

In this work Buffon preferred spontaneous generation to organic transformation as nature's most universal method of introducing new species. But the effect was still a striking departure from biblical literalism. Human history was emphatically not coextensive with the history of the earth. Moreover, the age of the earth was greatly increased, Buffon citing experiments of his own in self-justification. Having heated various spheres (of iron, sandstone, marble, and glass) to their maxima, he measured the time it took for them to reach hand temperature and then to a colder level that he took to be the temperature of the earth. Elaborate calculations, eventuating in spurious precision, allowed him to gauge the length of each epoch (2,936 years for the first), the grand total far exceeding a conventional biblical chronology. His unpublished estimate was to be 3 million years. That the earth *had* cooled was surely confirmed by the remains of tropical animals in polar regions.

Not surprisingly Buffon was accused of having contradicted Scripture. After all, in Genesis the earth was created before the sun, not out of it. Where Genesis had plants created on the third day and fish on the fifth, Buffon was assigning both to the same epoch. The theological faculty at the Sorbonne had other grievances. In Buffon's reconstruction, the first humans had lived in a state of terror, fleeing from wild beasts. Where was the paradise described in the Bible and where man's domination over the beasts? And how could one possibly reconcile St. Peter's prediction of a final conflagration with Buffon's congelation? In extending the domain of naturalism, Buffon got the predictable response that it was a poor sort of God who needed a comet's tail to get the solar system organized. As for Buffon's hypothetical posture, it was, according to the Abbé Royou, a transparent evasion: "At first leave is asked for a hypothesis, and once granted it becomes transformed into a demonstrated truth." Such, he continued, was the way of young philosophers in cafés and of women in society. Not that all the opposition came from hypersensitive priests. Voltaire decried Buffon's system as the work of one who had tried to create a universe with a stroke of the pen. As for the generation of the earth from the sun, it reminded him of an old fable in which Minerva emerged from the brain of Jupiter.

The reader may well be reminded of something else. One cannot but be struck by the parallel between his successive epochs and the successive days of Genesis. For all that it was a countertheology, Buffon's vision was still structured by vestigial religious concerns. His critics were entirely correct in saying that the two schemes were incommensurable, but the homology was pronounced enough for Buffon to exploit. Meeting the theologians on their own ground, he undertook an exegesis of the Creation narrative in which days were expanded into epochs. His argument was that in a text such as "darkness was upon the face of the abyss and the spirit of God moved upon the face of the waters," the verbs were placed in the imperfect tense, allowing a long duration. It was a stratagem that popularizers of geology would adopt in the nineteenth century, many of them devout believers. That such an option was available means that a question mark must be placed against the common view that practicing naturalists were constrained in their speculations by the power of religious orthodoxy. It is not even clear that there was a single orthodoxy on such unprecedented questions as the historical models were raising.

Laplace: A History for the Solar System

In constructing his history of the earth, Buffon had resurrected a question that seemed to require an answer in naturalistic terms. Why did all the planets revolve around the sun in the same direction and almost in the same plane? This was the question that, in the seventeenth century, Descartes had answered with his theory of vortices, but which Newton found unanswerable in terms other than God's aesthetic choice. However one looked at it, the odds against such an arrangement were enormous if it had been a purely chance affair. For Buffon there had to be a root cause, which he identified with his solar collision, from which the earth and the other planets took their origins. A less catastrophic explanation was, however, also possible – one that soon took shape in the nebular hypothesis of Laplace. Calculating the odds against the combination of motions in the solar system being the result of chance, Laplace, in common with Buffon, pressed the antithesis between chance and mechanism rather than that between chance and design. The solar system must have had a prior state from which it emerged through purely physical processes.

In his attempt to reconstruct that prior state, Laplace was motivated by a positive desire to eliminate teleological considerations from physical explanation. If he could show that the order in the system had derived simply from the operation of physical laws, references to purpose and contrivance could be excluded. Accordingly, he postulated a solar atmosphere, which, by progressive condensation, threw off rings, the parts of which eventually coalesced to form the planets. As the rotating atmosphere contracted, critical points would be successively reached at which the material at the periphery would break away owing to the increase in centrifugal force. That the planets revolved in the same direction was a direct consequence of the atmosphere's original rotation. Supportive evidence came from the rings of Saturn and from the asteroids. The latter were indicative of a planet that had simply failed in the making, the ejected bits and pieces having refused to coalesce.

It should be emphasized that Laplace was not constructing a grand scheme of cosmic evolution in the later nineteenth-century sense. But he was giving a historical account of the origins of the planetary system. Though speculative, it gained in plausibility from the theorizing of William Herschel (1738–1822), who had been busy introducing a time dimension into stellar astronomy. In the fourth edition of his *Exposition du système du monde* (1813) Laplace replaced his original solar atmosphere with a nebulous cloud in order to take advantage of Herschel's claim that stars were themselves made from the condensation of nebulas over immense periods of time.

Herschel's principal aim had been to construct a classification of natural objects in the heavens, much as other eighteenth-century naturalists had sought to classify plant and animal species. At one end of his nebular series was the Orion nebula, which, because of its apparent change of shape, was interpreted as a formless, gaseous cloud. At the other extreme were the many nebulas resolvable into stars by Herschel's telescopes of unprecedented power. Intermediate species displayed both an astral and nebular composition. But Herschel had encountered a problem that his telescopes could not resolve: how to classify the so-called planetary nebulas, in which a central star was surrounded by a uniform and symmetrical halo of milky appearance. It was through the contemplation of this anomalous species that Herschel had injected a temporal dimension into his classification scheme. In observing such planetary nebulas, was one perhaps seeing the last stages of star formation? Was the nebular envelope in its final stage of condensation?

As with Laplace, Herschel was not proposing a complete evolutionary cosmology. The analogies through which he expressed his vision suggest that he thought in terms of the growth of species, not their subsequent transmutation. But the growth of a star was a historical process and the stars themselves were so many millions of light-years away that the universe groaned with age. The integration of Herschel's model with that of Laplace required a reorientation of the imagination, which could be acutely uncomfortable. In the context of natural theology, however, the reorientation was perfectly possible. One simply placed further back in time the role of divine wisdom in creating a set of initial conditions from which an orderly system would emerge. Such a move was anticipated in 1804 by the Edinburgh professor of natural philosophy John Robison. Despite a disposition against all things French (in 1797 he had published an anti-Jacobin tract with the title *Proofs of a conspiracy against all the religions and governments of Europe*), he was prepared to give Laplace his due. Robison was aware that Lagrange and Laplace had shown how the solar system could stabilize itself, without that correcting hand of God that Newton had seemingly required. For Robison, here was evidence that the solar system was even more perfectly designed, whatever its mode of production. His complaint against Laplace was simply that the French mathematician had so blindly refused to draw the correct theological conclusion.

During the nineteenth century the secular science of Laplace was frequently assimilated, and reinterpreted, within a framework of natural theology. Indeed Laplace's own antithesis between chance and mechanical law was seized by religious apologists, in both Britain and America, who liked the premise that the solar system could not be the result of chance. Although it was probably a minority view, there were even scientific figures, such as the Scottish physicist David Brewster, who claimed that Laplace had confirmed the Mosaic cosmogony. Such, however, was his reputation for atheism that Laplace did create difficulties for British commentators. He certainly created a difficulty for William Whewell, who liked to believe that the most significant advances in science had been made by the godly. Contributing to the series of works on natural theology known as the *Bridgewater treatises,* Whewell still found it necessary in the 1830s to show that a self-developing and self-correcting universe need not rule out providential design.

Lamarck: An Evolutionary History of Life

Through the patronage of Buffon, a young naturalist, Jean-Baptiste Lamarck (1744–1829), secured a position at the Jardin du Roi, and subsequent election to the Paris Academy of Sciences in 1779. Developing an expertise in the study of shells, Lamarck was struck by the existence of marine species, which in more than forty instances resembled fossil shells – not exactly, but sufficiently closely to place them in the same genus. The fundamental question was how to explain the differences between fossil and living forms. Had the former been destroyed, modified, or simply transported from one climate to another? With an aversion to global catastrophes, and an overriding commitment to the harmony of nature, Lamarck was predisposed toward the second of these possibilities. The strong resemblance between fossil and extant shells implied a gradual transformation, bypassing the uncomfortable concept of extinction. But did all fossil forms have their living counterparts? Lamarck considered it premature to suppose otherwise, because new species, particularly of marine invertebrates, were still being discovered. And there was always the possibility that the living counterparts of some nautical fossils were simply inaccessible, fathoms deep in the ocean.

In arguing for organic transformation, Lamarck was assisted by an analytical approach to classification and by a disregard, bequeathed by Buffon, for any restriction on the earth's age. Time, he insisted, was "never a difficulty for nature." Given enough of it, transformation could occur without the penalty of extinction. In his approach to classification he retained linear series as an ideal frame of reference, even though he was forced to fragment the traditional chain of being. Taking the animal series as a whole, he displayed a striking gradation between complex and simple forms. From the summit, occupied by human beings with many faculties, a skeleton, and a vertebral column, one could descend through animals of intermediate complexity down to those bereft of all the characters by which complexity was judged. As each member of the series could be shown to exemplify all the characters of the one beneath it, it was tempting to suppose that germane to life itself was a power driving toward an increase in complexity. Could the descending scale of complexity be reversed to represent the history of an ascending animal kingdom? Lamarck took this view, transposing natural history into a history of nature.

For the details of the historical process one had to look to the effects of environmental change, which, by inducing responses from the organism, disturbed what would otherwise be a smooth ascent. By what mechanism could organic transformation occur? Lamarck suggested the operation of four laws. First there was the "constant tendency for the volume of all organic bodies to increase and for the dimensions of their parts to extend up to a limit determined by life itself." Second, "the production of new organs in animals results from newly experienced needs which persist, and from new movements which the needs give rise to and maintain." Third, "the development of organs and their faculties bears a constant relationship to the use of the organ in question." Fourth, "everything which has been acquired . . . or changed in the organization of an individual during its lifetime is preserved in the reproductive process and is transmitted to the next generation by those who experienced the alterations."

We recognize here that doctrine of the inheritance of acquired characteristics for which Lamarck is often belittled. But this doctrine was by no means peculiar to him; nor was it the main feature of his theory. It was accepted by Darwin himself and only became controversial later when August Weismann (1834–1914), for example, attacked it with his principles of the insularity and continuity of the germ plasm. In contrast to caricatures of his position, Lamarck nowhere suggested that new organs could arise from the exercise of volition. It was simply that changes in the environment created new needs and therefore new habits. They were felt needs, but the mechanism producing change was entirely physical, depending only on the stimulation and movement of interior fluids. These fluids could enlarge, lengthen, divide, and build up bodily canals and organs by the materials that separated out from them. Certainly, it was a highly speculative scenario. Even today, however, the establishment of new habits still has its adherents as a creative force in evolution. The current, albeit controversial, version of the argument is that mutations favoring an *already learned* and beneficial habit will tend to be preserved, thereby accelerating the pace of evolution.

Because nature, through spontaneous generation, was constantly feeding the lowliest organisms onto the escalator, the process of transformation that Lamarck envisaged was continuous and, in broad outline, capable of repetition. The humblest organism, spontaneously generated today, could in principle run a similar course to those that emerged in the distant past. If this is the correct interpretation, it confirms how misleading it would be to speak of Lamarck as an evolutionist in anything like the sense in which that word may be applied to Darwin. Their respective visions of the past were historical in quite different senses. For, in Darwin's theory of common descent, the evolutionary process was unique, irreversible, nonlinear, and utterly unrepeatable: Species once extinct were extinct for ever. For Lamarck, they had never even been extinct in the first place.

There was, however, a sense in which Lamarck's theory did anticipate that of Darwin: in its thoroughgoing naturalism and inclusion of mankind. Lamarck had no doubt that the mental capacities of human beings arose from the physiology of their brain and nervous system. He was happy to pinpoint the similarities between man and the orangutan, ascribing human thought and will to the flow of nervous fluid. It was time, which, by adding progressively more complex levels of organization, had allowed the emergence of human intelligence. Because the brain appeared to be the organ most responsive to constant use, Lamarck had no inhibitions in declaring intelligence to be a product of habit — the habit of thought. Impatient with Christian accounts of man, he repudiated a mind–body dualism. All of living nature, including the human mind, was the product of historical processes.

Lamarck was not strictly an atheist since he distinguished between the productive powers of nature and the creative power of God. But by interpreting plants and animals as natural products rather than creatures, he was effectively pushing the Creator beyond arm's length. A vestigial deism is apparent in his conviction that nature exhibits harmony, secured by an ulterior plan. It was this conviction that had predisposed him against extinction. It may also explain the absence of references to struggle and competition, which were to be the hallmark of the Darwinian theory. In keeping with Enlightenment assaults on established religion, he regarded the idea of God as a human invention. The most he would allow was that ideas fostered by religion, a caring God and personal immortality, should be tolerated because they could be of comfort to a species supremely and uniquely aware of its mortal coils.

Cuvier: An Alternative History of Life

Lamarck's vision of the past had to compete with a rival interpretation that was, in certain respects, structured by its oppositional stance. When mummified specimens, brought back from Napoleon's visit to Egypt, were discussed at the Paris Museum, Lamarck took in his stride the fact that they showed no differences from contemporary specimens. It was precisely what one would expect given the relatively short passage of time. For Cuvier, however, four thousand years without a hint of modification were sufficient to cast doubt on an evolutionary scheme. Rival interpretations were possible of the same set of data. This was, however, only one aspect of a more comprehensive rivalry between the two men, which was aired in public lectures, a medium Lamarck was prepared to use in attacks on Cuvier's system of classification. In retaliation, Cuvier would marshal arguments against Lamarck's transformist creed. Because Lamarck was *denying* the reality of extinction, Cuvier had motive as well as grounds for *asserting* it. By diverting attention from Lamarck's marine fossils to the large quadrupeds of the Paris basin, he found a promising means of doing so; for huge mammals, unlike sea shells, were unlikely to survive a catastrophic inundation. The differences between fossil and extant species could also be thrown into sharper relief. And whereas it might be argued that living counterparts of marine fossils were yet to be discovered, it was inconceivable that mammals the size of elephants had passed undetected. His strategy was therefore to show that there were fossil species that had no living isomorphs.

Cuvier's "proof" of extinction may be illustrated by his reconstruction of a *Paleotherium* species, fragments of which had been fossilized in the gypsum of Montmartre. The shape of the teeth implied a herbivorous pachyderm. One group of molars was associated with the presence of canines, another without. Assembling the dental evidence, Cuvier was faced with a jaw containing twenty-eight molars, twelve incisors, and four nonprotruding canines. The crucial point was that the shape of the molars implied a resemblance to the modern rhinoceros, but the number of the teeth was in line with the tapir. The remains were indisputably of a distinctive species no man had ever seen. And the same was true of another skull, which lacked the four canines and which he christened *Anoplotherium*. The difficulties he encountered in reconstructing the whole are apparent from the presence among the heads of two kinds of feet. Was it the two- or three-digit foot that belonged to *Paleotherium?* Cuvier acknowledged the

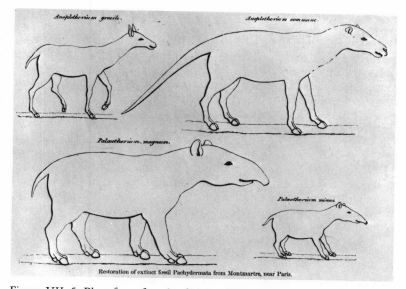

Figure VII. 5. Plate from fourth edition of William Buckland's *Bridgewater treatise,* edited by Francis T. Buckland (1870). Extinct species of *Anoplotherium* and *Paleotherium* are shown here, reconstructed from fossil remains investigated by Cuvier. Reproduced by courtesy of the Science Museum Library, London.

dilemma, resorting in the end not to an a priori deduction from a principle of correlation, but to the relative abundance of heads and feet. Consistency was achieved by matching the larger jaw with the smaller foot.

It was not merely the fact of extinction that informed Cuvier's vision of the past. As fossil species were reconstructed in greater numbers and from different strata, it gradually became clear that they had not all been contemporaneous; nor had they all been extinguished together. Fossils from the more superficial strata displayed a clear resemblance to modern species; the further one descended the stratigraphical column, the more the resemblance faded. How was the pattern to be explained? A succession of catastrophes, so repugnant to Lamarck, gave Cuvier his answer. His major point of reference was an extensive deposit of rough limestone, which stretched across Europe and which he attributed to the effects of a long-standing sea. It was only above that line that terrestrial mammals appeared, whereas shells, fish, and oviparous quadrupeds had left their mark beneath it. That each stratum had its characteristic tenants implied a geological succession, punctuated by catastrophes. The catastrophes that had preceded the rough limestone and modern alluvial deposits were certain; others only probable.

Because Cuvier has been maligned for his catastrophes, it is important to clarify their nature. He certainly did not believe that, in interpreting the past, he should restrict himself to current forces acting with their current intensity. Rain, snow, and ice might gradually erode mountains, but how could they account for the existence of mountains in the first place? He was willing to countenance more intense and dramatic events in the past. But his catastrophes had a conservative aspect in that they were reduced to the inundation of land by water. They were neither supernatural nor universal. Their effects could be sudden and violent. A species could perish in an instant. But it was no part of his vision that *all* species were annihilated by every catastrophe. Furthermore, he was uncomfortable with the view that a new creation had been required to produce the species existing today. It was rather that modern species had not always occupied their present location: They had migrated from elsewhere as part of a wider pattern, not of special creation but of mutual exchange. Just as Lamarck had accused Cuvier of drawing premature conclusions when he admitted both catastrophes and extinction, so Cuvier could reverse the charge. On his historical model, it was premature to postulate transformism when the effects of migratory exchange of species had been so little studied. His critique of Lamarck was reinforced with his principle of the correlation and interdependence of parts. There simply could not be a gradual accumulation of variation in any one part, unless all could change in concert. And that, for Cuvier, was simply too fanciful.

It is possible to see elements of a Christian Platonism in Cuvier's concept of distinct animal types under which species were subsumed. He certainly did not deny that God had created an original pair for every species. Because there was this respect in which he could be described as a creationist, his reputation was easily tarnished by later commentators. But, arguably, it was because he favored both catastrophes and the fixity of species that he was able to lay the foundations of paleontology in a manner that eluded Lamarck. By denying organic evolution, Cuvier had conferred on the earth a history, which, embracing the fact of extinction, helped make the Darwinian synthesis possible.

Cuvier's stature as a naturalist has sometimes been diminished with the imputation that he was fighting, albeit unconsciously, on the side of religion. Conversely, it has been implied that so rational a paleontologist could not possibly have held serious religious beliefs — that he was a hypocrite whose remarks on the noncontradiction between science and religion were a compliant response to the Catholic religious revival of the Empire and Restoration in France. The trouble with this hypocrisy charge is that Cuvier did not acquiesce in the Catholic revival. As an active member of a Protestant minority, he was caught between two prospects, neither of which appealed. He did not wish to see the Catholic Church gain a complete control of education; nor did he welcome the idea of a secular state, which could be equally subversive of religious freedom. His own conviction, not surprisingly, was that Protestant Christianity was more conducive to scientific inquiry than Catholic, which too often had the effect of polarizing faith and reason. As for the charge of veiled biblicism, it is completely contradicted by Cuvier's contention, foreshadowing that of Lyell, that geology had been retarded by the intrusion of biblical motifs. It was a science whose autonomy had to be respected. Speculations about a universal flood had distorted the direction of geological research. Even the attempts devised to explain the event had been misconceived, given that the biblical flood was supposed to be a miracle — a point on which Buffon had also dwelled in his dismissal of English sacred theorists. The great flaw in popular diluvialism was that it referred all fossil remains to a single cataclysm when, as Cuvier himself had shown, there was need for more. His time-scale for the history of the earth far exceeded the few thousand years estimated by those biblical chronologists who had studied Old Testament genealogies on the assumption that earth history and human history had been coextensive.

Recent research suggests that Cuvier, as with many of the scientific figures we have considered, was not a typical representative of the religious community with which he was affiliated. The absence of positive evidence for an inner spirituality is not decisive, since it is not clear in what the evidence would consist. The scattered remarks of contemporaries, however, suggest a man with some reputation for irreverence, whose outlook bore the marks of an Enlightenment deism, and whose conversion was prayed for by his daughter. Although his vision of the past was to be reconstituted in more overtly Christian terms, especially by clerical geologists in Britain, he himself regarded the story of Noah's flood as just one among many diluvial legends.

Lyell: Historical Explanation in a Historical Science

During the first half of the nineteenth century, the dialectic between transformist and antitransformist positions continued: New controversies flared up; social, political, and religious concerns fed into contending points of view; every controversy left its mark on a malleable and imperfect fossil record. In France, Cuvier became embroiled with a former collaborator, Geoffroy St. Hilaire, who emerged as Lamarck's successor in arguing the transformist case. The debate was politicized from the start by Geoffroy's accusation that Cuvier was a political conservative who was using his power in the Paris Academy of Sciences to suppress the freedom of scientific inquiry. A religious metaphor might seem to muddy the waters, but it was Geoffroy's own. Innovators, he wrote, like Christ himself, must wear a crown of thorns. Religious rhetoric was also infused into the dispute with Cuvier's claim that he was confronting a dangerous pantheistic system.

In Britain, the case against Lamarck and St. Hilaire was put by Charles Lyell (1797–1875), whose reconstruction of the past might be described as a form of history in depth. Whereas the kind of history epitomized by Cuvier's work has been labeled antiquarian, in that the reconstruction of extinct mammals from fragmentary bones bore a resemblance to an antiquary's reconstruction of a classical temple, Lyell's history was explicitly concerned with causal sequences. Just as the German historian Barthold Georg Niebuhr had traced a sequence of stages through which the Roman legal system had developed, each dependent on a preexistent state of affairs, so Lyell saw his task as one of connecting the present configuration of the earth with its past by identifying the successive (though not necessarily progressive) states through which it had passed.

Germane to this program of causal analysis was Lyell's insistence that only those forces known to be in action today could be invoked to explain the past. In his *Principles of geology* (1830–3), he greatly expanded the repertoire of what those current forces were. His insistence that they had acted with no greater intensity in the past was intimately bound up with his proposition that nature had had aeons of time at its disposal. Small gradual changes could have cumulative effects, which a superficial analysis might ascribe to sudden catastrophes.

Because Lyell deliberately excluded cataclysmic events from his science, and because such catastrophes were, for some contemporaries, a sign of providential intervention, his theorizing did have a radical edge. And because churchmen sometimes asked for a succinct statement as to why the earth had to be so old, it may be helpful to consider the kind of answer Lyell could give. Actually, no brief statement was possible because it would have been an oversimplification to claim *either* that immensity of time was directly deducible from "facts" or that it was merely an assumption imposed on the data. It was rather that the assumption and the interpretation of the data stood in a symbiotic relationship, which permitted a coherent and plausible account of causal sequences.

In Sicily, where he had the opportunity to study the slopes of Mount Etna, Lyell found a way of calibrating the past. The layered structure of the volcano constituted an argument against any paroxysmal origin for the mountain. It appeared to have grown from the gradual accumulation of many lava flows. Because Etna had a circumference of some ninety miles, it would have needed ninety flows, of a mile in width, just to raise the sides of the mountain by the thickness of one flow. But eruptions of Etna were recorded within the span of human history. One therefore had a rough guide to their frequency. The striking conclusion was that there had not been enough eruptions, during the time records had been kept, even to achieve that overall buildup of one thickness. The implications for the time required to produce a mountain of over ten thousand feet were staggering. And yet Lyell had evidence that beneath the mountain were fossil-bearing rocks belonging to geologically recent Tertiary strata. There were obvious consequences for any assessment of the time during which life had existed on earth. The fact that these same Tertiary strata were to be found in the center of Sicily, elevated to a height of some three thousand feet, was welcome evi-

View from the summit of Etna into the Val del Bove.

The small cone and crater immediately below were among those formed during the eruptions of 1810 and 1811.

Figure VII. 6. Illustration from page 93 of the third volume of Charles Lyell's *Principles of geology* (1830–3). Lyell's own sketch from the summit of Etna, which he made on 1 December 1828. Reproduced by permission of the Syndics of Cambridge University Library.

dence to Lyell that ordinary earthquakes were sufficient to account for such elevation – given that they had had the same immense time to operate as Etna had in the making. In his reasoning, Lyell was also guided by the principle that one could date fossil-bearing strata by reference to the proportion of extant species they contained.

Lyell's strongly held belief that clerical geologists should not try to do two jobs at once, his deliberate attempt to exclude biblical preconceptions from geological reasoning, and his dependence on none other than known natural causes has generally resulted in his being portrayed as a secular hero of science who championed the uniformity of nature against the theologically laden catastrophes of his day. Accordingly, his aversion to the idea of progressive creation used to be explained as the response of a deist, or crypto-Unitarian, to the overt Christianizing of the fossil sequence practiced by Buckland or Sedgwick. There may be truth in this, but scholars have recently had to consider an alternative source of his aversion: the progressive transformism of Lamarck. The argument is that Lyell's opposition to progression was triggered not so much by a distaste for the manner in which clerics were making theological capital out of the fossil record, as by the realization that to admit progression at all was to play into the hands of the transformists. If this interpretation is correct, Lyell sensed that the clerical geologists, in their bid to turn progression into an argument for providence, were actually setting a time bomb for themselves. And in a sense he was right; for, as he later admitted, it had been the progressive creationists who had come closer than he to the construction of the record required by Darwin. That admission is an embarrassment to those popular works in the history of science that either persist in seeing Lyell as a transformist at heart or which imply that all Darwin had to do was to apply Lyell's "uniformitarianism" to the organic domain.

Lyell had not always denied a sense of direction in the fossil record. It was only after he read Lamarck in 1827 that he abandoned progression in favor of a "steady-state" history, according to which species had appeared and disappeared in piecemeal fashion. But what was so abhorrent about the Lamarckian scheme? For Lyell it affronted human dignity, reducing mankind to a glorified orangutan. It is not clear whether his sensitivity was the product of religious belief or of humanistic concern, but throughout his career he emphasized the uniqueness and "high genealogy" of the human species. One manifestation of this was the exception he made in his *Principles of geology:* Man alone was a recent addition to creation. Another was his reluctance, much later, to go the whole way with Darwin even after he had been converted to species transmutation. He would never concede that natural selection was a sufficient explanation for the powers of the human mind. Utopian fantasies of the perfectibility of man were simply the compensation offered by the transformists for having naturalized the human soul.

Lyell's arguments against Lamarck were largely modeled

AWFUL CHANGES.
MAN FOUND ONLY IN A FOSSIL STATE——REAPPEARANCE OF ICHTHYOSAURI.

Figure VII. 7. Caricature by Henry de la Beche (1830), used as frontispiece in Francis T. Buckland, *Curiosities of natural history,* 2d ser., 2 vols.
(1860). Lyell's denial of directional change in the fossil record allowed him
to speculate in the first volume of his *Principles of geology* (1830) that, given
the steady-state and essentially cyclical nature of the earth's physical history, appropriate environmental circumstances might recur, permitting
genera similar to *extinct* forms to reappear in the future. It is this notion of
cyclicity in the history of life that de la Beche cleverly caricatures. The
sketch has the title "Awful Changes: Man found only in a Fossil State –
Reappearance of Ichthyosauri." The caption reads: "You will at once perceive," continued Professor Ichthyosaurus, "that the skull before us belonged to some of the lower order of animals; the teeth are very insignifi-
cant, the power of the jaws trifling, and altogether it seems wonderful how
the creature could have procured food." Reproduced by courtesy of the
Science Museum Library, London.

on those of Cuvier. Variation within a species was admitted
but strictly limited. Advocates of transformism might appeal
to the plasticity of species under domestication, but, despite
the great variety of dogs produced by breeders, they were
still dogs. To the selective breeding of plants, nature had also

set its limits. Individuals descending from the same stock could be modified in various directions, but experience showed that there was always a point beyond which the character in question could not be further altered. Lyell also argued that human interference was required to bring about even these modest changes. The propagation of a plant by buds, grafts, or cuttings was "obviously a mode which nature does not employ." Varieties cultivated under domestication even reverted to type when reintroduced into the wild. And as for the Lamarckian mechanism: Use and disuse might *modify* organs, but that was a long way from giving a convincing account of their *origin*.

To scotch the scheme altogether, Lyell stripped the fossil record of any progressive pattern. The persistence of certain marine genera throughout the fossil record showed that there had "never been a departure from the conditions necessary for the existence of certain unaltered types of organisation." Pursuing the same strategy, he would even suggest that the absence of mammals from lower strata did not prove that they were not already in existence. Fossilization required specific physico-chemical conditions. It was an error on the part of the progressionists when they reasoned as if nature had planned to leave a complete record of its works. One of Lyell's significant contributions to science was the theoretical justification he gave for regarding the fossil record as imperfect. It is not always appreciated that his insights were geared to, even if they did not arise from, the case against transformism.

The example of one controversial fossil, "Didelphis," may help to focus discussion. A jaw recovered from the Stonesfield slate near Oxford had a dental formula implying resemblance to an opossum. That, at least, was the mammalian diagnosis of Buckland, which also found acceptance among his colleagues at the Geological Society, including Lyell and Sedgwick. The Stonesfield marsupials were of peculiar significance because, if they were of Oolitic age, they were not only the earliest mammals but anomalously early according to standard progressionist accounts. To the progressive *creationists* this mattered relatively little, since they supposed no genealogical continuity between successive species. But to a transformist such as Robert Grant (whom the young Charles Darwin encountered in Edinburgh) it mattered a great deal, because a mammal located in Secondary strata disrupted the neat serial transition from Secondary reptiles to Tertiary mammals. Accordingly, Grant was the first to oppose the mammalian interpretation, succeeded by Henri de Blainville (also a transformist) who, by comparing the crowns and roots of the Stonesfield molars with those of the reptile *Basilosaurus,* showed that these features were not exclusive to mammals. Buckland, enlisting the help of Richard Owen, sought to defend his original slant, in the full knowledge that it would rupture the Lamarckian chain.

What of Lyell? Bent on erasing all semblance of progression, he welcomed these early mammals with enthusiasm: "The occurrence of one individual of the higher classes of mammalia, whether marine or terrestrial, in these ancient strata is as fatal to the theory of successive development, as if several hundreds had been discovered." With such exaggeration, Lyell differentiated his position from that of the clerical geologists too, for whom the ancient opossum was still something of an exception. That Lyell was wishing to turn an exception into a rule is clear from the critical remarks of W. D. Conybeare, who accused him in 1841 of blindness toward such progressive complexity as the fossil record still displayed. Surely, Conybeare wrote, "one cannot consider the wretched little marsupials of Stonesfield to counterbalance the general bearing of the evidence." Evidently what the progressive creationists relished was a few mavericks to defeat the serial progression of the transformists, but a broad panoply of progression in which to see the hand of providence. So, according to the preconceptions of three very different models of earth history (the steady-state, the progressive creationist, and the transformist), three different constructions were placed on the find. Later, there was to be a fourth. Operating with a more flexible model of evolution, Alfred Russel Wallace was to challenge Lyell's strategy:

Lyell says the Stonesfield mammal of the Oolite is fatal to the theory of progressive development. Not so if lowly organised mammalia branched out of *low* reptiles or fishes. All that is required for the progression is that *some* reptiles should appear before mammalia & birds or even that they should appear together. In the same manner reptiles should not appear *before fishes,* but it matters not how soon after them.[1]

The image of a *branching* tree, which appealed to both Wallace and Darwin, was to prove more accommodating than one of serial progression.

Darwin: The Historical Explanation of Speciation

An emphasis on *divergent* lines of speciation, extending from a common ancestor, allowed both Darwin and Wallace to transcend the parameters of the earlier debate between proponents of serial transformism and discontinuous progression. As with the growth of a tree, every new fork in every new branch was contingent upon the branching that had preceded it. Extant species were equivalent to the recent growth at the tips of those branches that had survived as the tree had grown. The image was delineated in a famous diagram, publicized in the *Origin of species* (1859). It was an image that would prove difficult to reconcile with the notion that the precise course of evolution had been foreordained.

The distinctiveness of the Darwinian theory was not confined to the pattern it imposed on the fossil record. In the mechanism that Darwin proposed for the transmutation of species, a historical process of "natural selection" was of paramount importance. The basic idea was that, in the struggle for existence occasioned by a disproportion between reproductive fecundity and the means of subsistence, the members of a given population most likely to survive and leave offspring were those that varied sufficiently from the norm in a particular direction to have a competitive advantage − in relation to the circumstances of their environment. Intraspecific competition throughout succeeding generations would lead to a situation in which the favorable variation would be preserved at the expense of the original norm. Given sufficient time, the accumulation of variation could conceivably lead to such a departure from the original species (as each minimally advantaged form effectively exterminated its predecessor) that a new species could be said to have come into existence − one that would not have been capable of producing fertile offspring were it crossed with a member of the original species.

The idea is deceptive in its simplicity. But it was neither self-evidently correct nor directly provable from available data. Darwin's achievement, rather, was to show how it could be used to inject coherence into an enormous range of biological data, from technicalities associated with classification to the geographical distribution of species, from the existence of vestigial organs to the nature of animal instincts. Above all, a theory of evolution by natural selection provided a his-

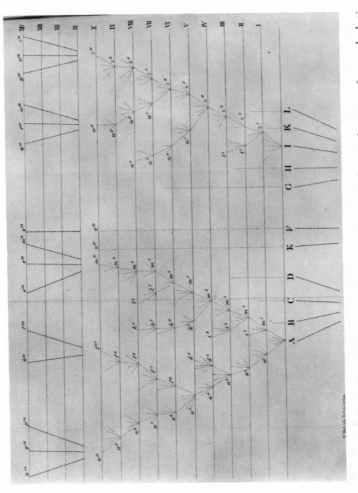

Figure VII. 8. Diagram from Charles Darwin's *On the origin of species by means of natural selection* (1859). The image of divergence superimposed upon divergence was one of the distinguishing marks of Darwin's theory that made it difficult to harmonize with concepts of a foreordained plan. Reproduced by courtesy of the Science Museum Library, London.

torical explanation for what Wallace in 1855 presented as nothing more than an empirical law, namely that new species, wherever they had arisen, had been created on the plan of preexisting ones in that locality. Wallace already knew that this would not be surprising if they were derived from them, but another two or three years were still to elapse before he finally hit on essentially the same mechanism as that which Darwin had been refining for the past twenty years.

The flora and fauna of the Galapagos islands, which Darwin encountered in 1835 on the voyage of *H.M.S. Beagle,* constitute the most commonly cited source for his early speculations. That several islands possessed their own species of tortoise, mocking thrush, and finches, he later reported, had filled him with wonder. He was also struck by a gradation in the size of beaks, discernible among the Galapagos finches: from one as large as that of the hawfinch to another no larger than that of the chaffinch. One might really fancy, he later wrote, that "from an original paucity of birds in this archipelago, one species had been taken and modified for different ends." Because he stated in his private journal that the species of the Galapagos had marked the origin of all his views, popular accounts of Darwin's conversion have conjured up the image of a patient fact collector suddenly bowled over by what he had found — a creationist suddenly propelled toward an exciting heresy. The trouble with such accounts is that they can trivialize the logic of discovery. They assume that the "facts" were somehow there, waiting at the Galapagos for Darwin to process. Darwin himself knew better than that. One of the things that had worried him earlier in the voyage was whether he was noting the right facts. What his experience at the Galapagos rather embarrassingly showed was not that some new facts pointed unequivocally toward a new theory, but that the constitution of a relevant fact depended on prior expectation.

It has long been known (since Darwin acknowledged it himself) that he was indebted to a chance remark of the vice-governor, Nicholas Lawson, to the effect that he could tell merely by looking at a tortoise from which island it came. Hitherto Darwin had not registered the "fact" that separate islands might possess their own representative species. And, to his subsequent consternation, he had been mingling his specimens, including his finches. He later implied that Lawson's clue had come in the nick of time, but recent reconstructions suggest that Darwin need not have interpreted the clue *at the time* as evidence against separate creation. If the tortoises were merely different varieties, as he continued to suspect the mockingbirds of being, there was little to be excited about. Not until he was back in England when, helped by authoritative taxonomists, he reconstituted his data, did the transmutation of species become a seductive and irresistible possibility. Thus he learned from his advisor John Gould that three of his mockingbirds could, after all, be considered separate species. Not until his return from the voyage did he also learn of the close relation between Galapagos birds and those on mainland South America. The route to the instability of species was not one of straight inductive inference. Rather, he eventually realized that *if* his finches, like the mockingbirds, were confined to separate islands, it might be possible to rationalize their peculiarities in terms of transmutation. But it was an "if," a suspicion, which he had to try to substantiate by reestablishing the lost localities from which each specimen had been obtained. An incipient evolutionary hypothesis allowed him, in retrospect, to solve the riddle of the finches. It had not been implicit in his data.

For the legends concerning the origin of the *Origin,* Darwin himself was in part responsible. His presentation of his theory as a response to hard facts concealed the extent to which he had been indebted to the theoretical structures of others. He did, however, admit that Lyell's *Principles of geology* had made such a deep impression on him that, during the *Beagle* voyage, he could not help but look at geological phenomena through Lyell's eyes. Staggered by the sheer numbers of South American mammals that had become extinct, he had nevertheless rejected catastrophic explanations. Lyell's principles, and especially his views on the geographical distribution of species, were one of the major resources on which Darwin drew. Here, however, we have to contend with another legend. Taking their cue from the fact that Lyell expressed a preference for the role of secondary causes in species production, some writers have taken the view that Lyell was an evolutionist at heart who directly paved the way for Darwin.

Such a view is incompatible with the analysis of Lyell's opposition to Lamarck that we have already given. But it needs further correction. Darwin did contemplate the geographical distribution of species with Lyell's precepts in mind, but in key respects Lyell let him down. It was the *failure* of Lyell's biogeography that prompted Darwin to explore the advantages of an alternative scheme based on species transmutation. Some of the basic facts of distribution were certainly hospitable to an evolutionary explanation. The fact that distinct, but closely related, species were found in adjacent geographical zones could point toward variability in the past. Similarly, the restriction of some genera (such as the South American sloth) to one continent, coupled with the concentration of congeneric species in particular locations, could easily be rationalized with evolutionary genealogies. Lyell, however, resisted the temptation. His assumption was that particular species had originated in those regions that, at the time of their appearance, offered a receptive environment. There was even a legacy from natural theology in such a view, Lyell affirming that the synchronization between the earth's physical history and the insertion of new life forms was capital evidence of a "Presiding Mind."

Lyell's explanation, based on adaptation, did, however, carry certain implications. It was these that were thwarted on Darwin's reading of the data. If Lyell was right, one would expect to find a living population in every habitable environment — and yet Darwin encountered propitious islands devoid of life. One would expect to find similar species in similar environments — and yet in similar climatic zones to the east and west of the Andes, Darwin discerned a marked contrast in occupants. One would expect to find dissimilar species in contrasting environments — and yet it eventually came to light that the birds of the Galapagos resembled those of mainland South America, despite the dramatic environmental divergence. One would expect indigenous species to flourish, immigrant species to be at a disadvantage — and yet Lyell himself had drawn attention to the success of domesticated mammals such as cats, dogs, and horses in colonizing other regions of the globe. Darwin gradually began to see that an explanation for the geographical distribution of species based on transmutation and migration carried greater explanatory power than one premised on prior adaptation.

A third set of resources on which Darwin drew related specifically to the nature of humankind. His speculations concerning transmutation, which began in earnest during 1837 and 1838, were accompanied by a serious inquiry into the alleged uniqueness of human reason. Traditional accounts omitting this dimension are defective because they overlook a point that Darwin grasped from the start: An all-embracing evolutionism would only look plausible if human mental faculties were of a kind that could have been derived from humbler species. A willingness to entertain that possibility undoubtedly owed something to the freedom of thought that blossomed in his family. His brother Erasmus was an atheist; his grandfather Erasmus had established a pedigree of radical religious dissent. A willingness to admit an animal ancestry may have been encouraged by his close study of the barbarians he encountered on the *Beagle* voyage. The sight of natives from the Tierra del Fuego, gesticulating on the shoreline, left an indelible impression: "I could not have believed how wide was the difference between savage and civilised man; it is greater than between a wild and a domestic animal, inasmuch as in man there is a greater power of improvement." That reference to a power of improvement is of particular interest since, on board ship, there had been three Fuegians who had been "civilised" in England, to be returned by Captain Fitzroy to their homeland. The contrast between the dandy Jemmy Button and his untutored brothers, the rapidity with which the Anglicized trio relapsed, might have made it easier for Darwin to strip the veneer from civilized man and lay bare the civilized animal. One's mind, he wrote, "hurries back over past centuries, and then asks, could our progenitors have been men like these? — men whose very signs and expressions are less intelligible to us than those of the domesticated animals."

What, after all, was supposed to be the difference between the mind of humans and of animals? In his second transmutation notebook, he recorded the opinion of his cousin, Hensleigh Wedgwood, to the effect that the only difference was "the love of the deity and thought of him or eternity." But Darwin already had his doubts. How faint were such sentiments in a Fuegian or an Australian! "We could never discover," he later reported, that they "believed in what we should call a God, or practised any religious rites." If the quintessential difference could be dropped, why not admit gradation?

The relationship between Darwin's science and his religious beliefs has often been discussed in terms of the impact of his science on a religious mentality that passed from a nominal Christianity through deism to agnosticism. The reality was more complex, the interaction taking place in both directions. A willingness to adopt a materialist account of the relation between mind and brain was a concomitant of his evolutionary speculations, not a sequel. "Oh you materialist," he chuckled to himself, as he speculated that love of God might itself be the "result of organisation."

A particular way of looking at the relationship between animal and human reason was one of the prerequisites of his theory. He realized that they might be assimilated if it could be shown that humans and animals were capable of expressing the same range of emotions – hence his visits to the London Zoo to observe the expressions of orangutans and baboons. It is not impossible that his crucial decision to reread Thomas Malthus's *Essay on population* in the autumn of 1838 was prompted by a curiosity to learn what Malthus had said about human passions. Looking for one thing, he found another – a way of calculating the consequences of superfecundity, which transformed his current understanding of the mechanism by which transmutation could proceed. If we are to accept Darwin's own testimony, it was while reading Malthus that he saw the light. The reality, the inexorability of intraspecific competition was brought home in a vivid manner as he encountered the arithmetic by which Malthus purported to show that population increase would always have a tendency to outstrip even an increasing food supply.

Because Darwin said that he had already been well prepared to appreciate the struggle for existence, there will always be doubts concerning the extent of his debt to Malthus, and doubts too about the experiences that had prepared him. On one interpretation his predisposition, and that of Wallace later, simply stemmed from an awareness of tough industrial competition, which Britain was experiencing under a laissez-faire capitalism. But there is another possibility if one is looking for such a projection. Darwin had a more immediate experience of European colonial conquest than he had of cut-throat commercialism at home. In the Pampas of South America he had witnessed the war between land-hungry Argentinians and the native Indians whose plight was intensified by the massacres of General Rosas. Images of colonial warfare were as real as images of industrial competition. Darwin may well have been more absorbed by Malthus's discussion of ancient empires than by any legitimation of the new industrial order.

By the autumn of 1838 Darwin had a theory with which to

work. He had seen that in an unavoidable struggle for existence, favorable variations would tend to be preserved, unfavorable variations destroyed. But twenty years were still to elapse before the full implications of that insight were to be made public. It is often supposed that once the mechanics of natural selection had been conceived, it required little further adjustment before belatedly appearing in the *Origin*. The delay is commonly attributed to Darwin's fear of persecution. Anxiety, if not fear, became real enough. He was painfully and poignantly aware of his wife's religious sensibilities. But it is too easy to project back to 1838 a set of assumptions that Darwin did not yet hold. It took him several years to shake off a legacy from natural theology, which had persuaded him that evolution was nature's method of preserving *perfect* adaptation when a physical environment changed. The ideas of relative and differential adaptation were only gradually given greater scope. Moreover, it was not until the late 1850s that he really satisfied himself *why* natural selection would favor evolutionary divergence. Not until September 1857, in a letter to Asa Gray, did he assemble the relevant considerations into a coherent argument. The same spot, he suggested, would support more life if occupied by very diverse forms. Because every organism could be said to be striving to increase its numbers, it followed that the varying offspring of each species would try to seize on as many and diverse places in the economy of nature as possible. The greater the dissimilarity between varieties derived from one and the same species, the more readily would each find a distinct ecological niche. The greater the divergence between newly incipient species, the less would they compete with each other, for the competition would always be severest between closely allied forms.

At last the branching and subbranching was explained, "like a tree from a common trunk; the flourishing twigs destroying the less vigorous – the dead and lost branches rudely representing extinct genera and families." Because Darwin was still resolving scientific difficulties in the late 1850s, it must not be supposed that his reticence over publication arose only from anxiety over religious reaction. In his relatively late clarification of how each step in evolution created new evolutionary opportunities, in his explanation as to why natural selection favored divergence, he did finally taken advantage of an analogy between the economy of nature and an industrial economy. The seizing of different ecological niches, through the diversification of form, was not unlike the division of labor in the process of manufacture. In both cases there was a premium on specialization.

We shall see in the next chapter how Darwin presented his case to the public. In both its development and its presentation he stressed the analogy between the effects achieved by breeders who selected for particular characteristics among domestic birds and animals, and the effects allegedly achieved by nature without the intervention of intelligence. This was the analogy that gave content to his metaphor of natural *selection*. He was impressed by the fact that varieties induced by the breeders were often so divergent that, to the uninformed, they would appear as distinct species. Bizarre varieties of the pigeon were a favorite example. Darwin counted as many as twenty forms, which, if presented to an ornithologist as wild birds, would probably be ranked as separate species. Although the analogy was to prove extremely controversial, it was a powerful expository device. If humans could achieve such marked effects in so short a time, what might not nature achieve during countless millennia?

In the last chapter of the *Origin*, Darwin summarized arguments that in his opinion gave evolution by natural selection the edge over separate creation. By setting up an antithesis between these two concepts, he appeared to many readers to be launching a frontal assault on an essential component of Christian theology. Darwin, however, never pretended to account for the origin of the first few living forms, still less for the origin of the universe. The concept of creation he was attacking was not the general doctrine that Christians affirm when they say that the universe depends for its existence (and continued existence) on God, but the much narrower concept that every species had been *separately* created by some means, by divine intervention or otherwise. It was, nevertheless, an attack with which a younger generation of naturalists was soon to identify, especially if they were looking for an autonomous and historical science of biology, resistant to religious intrusion.

History as a Science: The Wider Context of Biblical Criticism

The issues that Darwin's science raised for religious belief will be analyzed in the next chapter. The blow to a literal reading of Genesis was an obvious one. It hit the public,

however, at a time when Christian thinkers were already having to come to terms with what was, in many respects, a more
serious challenge to traditional concepts of biblical authority.
A literal interpretation of the creation narrative had already
been compromised by developments in geology. But the most
radical challenge to biblical authority came not from the history of science but from the science of history.

The importance that Christian apologists attached to the
gospel miracles had been challenged long before new historical methods called for a reassessment. The seventeenth-century Dutch and Jewish philosopher Benedict Spinoza (1632–
77) had been an inspiration to Enlightenment rationalists with
his insistence that the Bible should be approached as any other
book – as a work of human authorship. Rejecting a traditional duality between nature and supernature, he had developed a monistic conception of the universe in which it was
erroneous to construe God's activity in terms of intervention. He had even argued that what the Bible properly meant
by divine activity could be identified with the lawlike course
of nature. Claims for the miraculous were to be viewed with
suspicion because they depended on the assumption that the
occurrence could not have come about through natural causes.
And was it not presumptuous to suppose that the limits of
the power of nature were already known? With similar subtlety, as we saw in Chapter V, David Hume had called into
question the authenticity of all reported miracles. Eighteenth-century rationalists were wonderfully resourceful in
constructing alternative scenarios for the gospel events. Although he refrained from publishing, the Hamburg teacher
H. S. Reimarus (1694–1768) had conjectured that Christ's
Resurrection was a put-up job. The disciples had stolen the
body from the tomb, had carefully waited fifty days before
announcing the resurrection appearances (by which time it
would be impossible to *disprove* their claim by identifying the
rotten corpse as Christ's body), and all to ensure that they
might continue to live off charity by having something still
to preach.

The publication of such radical conjectures was invariably
seen as a political act. During the 1770s, Gotthold Lessing
chose to reveal Reimarus to the German public. The severity
of the reaction was such that the government of Brunswick
banned all of Lessing's theological works. Lessing himself was
arguing that the truth of Christian faith could be affirmed

without verification of the gospel texts. Faith could withstand, even gain from, an open inquiring mind. But to use that as a justification for publishing such subversive conclusions as those of Reimarus incensed the representatives of a more conventional piety. It was like saying that the true nature of chastity, and the duty to preserve it, could only be established when every piece of pornography had been printed. Thus argued Lessing's Lutheran opponent J. M. Goeze, who made much of the political implications. What, he asked, would Lessing reply to someone who says "the system of government practiced by the best and most just of rulers does not deserve allegiance until every conceivable libel and insult directed against the person of the ruler, has been set out in print." More sophisticated forms of biblical criticism than that of Reimarus were about to appear, often authored by devout scholars seeking a better understanding of the relationship between the sources on which the gospel writers had drawn. But sixty years later, in the Germany of the 1830s, the young critic David Friedrich Strauss (1808–74) would create an even greater outcry at a time when an orthodox piety was again in league with a monarchy harboring aristocratic rather than democratic ideals.

It was during the late eighteenth and early nineteenth centuries that historians became self-consciously critical about the methods they were employing in reconstructing the past. Through the scholarship of J. G. Herder, attention was focused on the different audiences for which each of the evangelists had written. But in their bid to answer the question What really happened? scholars faced a serious problem, which another German philosopher, Hegel, had underlined. If, during the unfolding of human history, ideas about nature and destiny had changed from one period to the next, how could the texts of a former age be related to the needs and assumptions of the modern world, without first identifying the prevailing, and in many cases obsolete, thought forms of that bygone age? The problem was particularly acute in the context of biblical scholarship because the prevalence of miracles in the gospel narratives could be taken as a paradigm case. Given the beliefs and expectations of Christ's contemporaries, which were themselves historically contingent, his followers would have expressed their religious experience in terms that were full of meaning to them, but which, to the nineteenth-century scholar, would have the character of myth.

This category of "myth" was not to be confused with falsehood. Important religious truths had been expressed in these historically remote forms. Nevertheless, through the work of German biblical critics, a question of far-reaching significance was given new urgency. Was it possible to divest the gospel narratives of their mythical character and see, behind the veil, what really happened? Or were the recorded "facts" of Christ's life so molded by the interpretation of his followers that it would prove impossible to recover the historical Jesus?

The distinctively historical style of the new criticism can be appreciated by contrasting it with two other approaches to the gospel narratives, both of which were attacked by Strauss in his *Life of Jesus* (1835). One approach, the more conservative, had been developed in response to apparent discrepancies of detail between the four gospels. Was it not possible to harmonize the various accounts of Christ's ministry, avoiding the alleged contradictions? An example that Strauss himself considered arose from divergent accounts given by Matthew, Mark, and Luke of Christ's healing in Jericho. Matthew had two blind men cured, Mark and Luke one. Moreover, Luke placed the miracle at Christ's entry into Jericho, Matthew and Mark at his departure. The ingenuity with which the harmonizers had tried to preserve consistency was, for Strauss, matched by its absurdity. Harmonists had broken the miracle into two, suggesting that Matthew had combined them in his account of Christ's departure. But, Strauss protested, if so much weight were allowed to Matthew's ascription of locality, why not give equal weight to his statement about the number cured there? In which case, the accounts of Mark and Luke would remain anomalous.

In painstaking detail, Strauss addressed himself to many such examples, berating the harmonists for their misplaced zeal. Only by multiplying the occasions on which Christ performed a particular act could they save the verbal inerrancy of Scripture. On the sensitive issue of the Resurrection appearances, Strauss was merciless. If Matthew was to be believed, then the first appearance, described by Luke and John as having occurred in Jerusalem, could not have taken place since Matthew stated that the eleven disciples had gone to Galilee for their first and last encounter with their risen Lord. Strauss tried to make the harmonists see that their approach was misconceived, relying as it did on their own ad hoc interpolations. It would be far better to acknowledge the historical process whereby reports of the original incidents had been affected by the passing of time, by the accretion of legend, and by human interpretation.

The second, less conservative, approach was that of the rationalists who tried to save the biblical miracles by showing how they could have happened by natural means. Among the rationalists, psychological considerations were called into play to explain how Christ cured those whose illnesses were described as demonic possession. In the more difficult case of an organic disease, such as leprosy, Christ would be credited with medical knowledge — despite the fact that his behavior was consistently depicted as that of a faith healer. His Resurrection would be rationalized with the suggestion that he was not dead, but in a coma when taken from the cross.

Strauss objected to such rationalization that it was not licensed by the gospel accounts. The rationalizers effectively turned Christ into a deceiver. This problem of deception arose in another context — that of the second coming of Christ. To propose that Christ had spoken as he had in Matthew 19:28 and Luke 22:30 merely to encourage his disciples, but without really sharing their belief, was to charge him with duplicity. Playing off the rationalists against the harmonists, Strauss showed how neither the naturalism of the former nor the supernaturalism of the latter did justice to the gospels as historical documents.

That an alternative, more genuinely historical approach was possible he knew from the lectures of the Old Testament scholar Wilhelm Vatke, whom Strauss had heard in Berlin and who became a personal friend. Vatke not only exposed poetic, legendary, and mythical strata in the books of the Old Testament, but also developed the thesis that the Pentateuch had been written in a much later age than that of Moses. Tracing the historical development of Judaism, he showed how, from having been a tribal war God, Yahweh, by the time of the prophets Jeremiah and the second Isaiah, had become a far less provincial deity. He had also become more concerned to establish a relationship with individuals. A strong emphasis on His transcendence was only to be found in the literature that followed the period of exile. Not until the Maccabean period had Messianic expectations begun to appear: The hope then was for the coming of an ideal king, modeled on David, who would bring liberation. Other Messianic types had emerged during this late period: the suffering servant of Isaiah and the Son of man as described in Daniel. The idea of the resurrection of the body also featured in this later stage of Judaism. It was clear both to Vatke and to Strauss that early Christian writers had drawn on all the available Messianic imagery in their interpretation of Christ.

In his *Life of Jesus,* Strauss made much of the generation gap separating the earliest gospel from the events of Christ's ministry. An oral tradition had been embellished with miracles not as a deliberate fabrication (as Reimarus had imagined) but as a perfectly understandable outcome of the interpretation placed on Christ's life and death by his disciples. Because the prophetic literature with which they were familiar had associated the Messianic era with signs and wonders, so those expectations were retrospectively woven into the tapestry of Jesus' life. The evangelists knew how the Messiah would have acted and wrote accordingly. It was not that, in some supernatural scheme, Christ had been an objective fulfillment of Old Testament prophecy – rather that the predicted attributes of the Messiah had been *transferred* to him. The whole process was natural and comprehensible once the canons of critical, historical scholarship were applied.

Not all advocates of the "higher criticism," as it became known, were led to such radical conclusions. Indeed much of the motivation for historical scholarship came from a desire to protect the idea of God's revelation by showing how human understanding of the deity had been progressively refined. To cling to an outmoded view of the verbal inerrancy of Scripture – a view that was, in any case, of relatively recent origins– was increasingly seen as a form of obscurantism that could only bring Christianity into disrepute. Strauss himself claimed that he was trying to preserve the moral insights of Christian doctrine by ensuring that they were not dependent on a dubious historicity of miracles, which his modern world could no longer accept. In reply to criticism, he repudiated the charge that he had turned Christ into an entirely mythical figure. He had denied neither his existence nor his charismatic personality – nor the extraordinary recovery of hope among his disciples, following his death. Nevertheless the effect of his book was shattering. It showed that one could give a plausible account of the gospels without admitting the historicity of the miracles to which they referred. That some of the early disciples believed in the Resurrection did not mean that it had happened.

For Strauss the Resurrection was a delusion, but not a fabrication. And delusions could contain a great deal of truth. In their deepest spirits, he wrote, the disciples had presented their master as revived, for they could not possibly think of him as dead. This was, however, too subtle a conception of truth for his orthodox critics. Other features of his book were to make it a much prized symbol for the political radicals of his generation, Karl Marx among them. In 1843 Ludwig Feuerbach would say that, for Germany, theology was currently the *only* practical and successful vehicle for politics. For those imbued with democratic ideals there was a powerful resonance in the shift of focus from Christ himself to the consciousness of a human community. Scholars have spoken of the replacement of Christ by humanity in Strauss's analysis. Humanity is portrayed as the miracle worker. The human species, rather than a unique individual, became the God-man. Whether Strauss intended or not, the displacement could sabotage the analogies still being constructed between Christ and the German monarch. Five years before Strauss published, there had been a further revolution in France. King Friedrich Wilhelm III had specifically warned that the influence of theology professors who did not feel bound by the dogmas of the Protestant Church was "extremely dangerous for the state." Strauss was perceived to be precisely that. He had to forfeit his academic career at Tübingen, and such was the commotion that when a chair of dogmatics fell vacant at Zurich in 1839, the Swiss government was petitioned by some forty thousand signatories that he should not be appointed.

Strauss provides but one example of a wide-ranging and increasingly specialist endeavor that created a rift between the populace and professional scholars — a rift that still exists within the Christian churches today. The Christian layman who was prepared to listen had to adjust to the idea that the Bible was a library whose books had many authors. He had to adjust to the idea that it showed a historical evolution. He had to contemplate a new and blurred image of Christ — one in which he was divested of miraculous power. He would learn that the Christianity preached by St. Paul had been informed by personal idiosyncracies. Above all, he would learn that one could understand the Scriptures only by regarding

them as the work of ordinary men whose beliefs and aspira-
tions were products of the period and society in which they
lived. And even if he could adjust to these alternative perspectives, he was faced with a dilemma. He could see that it
was a *presupposition* of the scholars that Christ be stripped of
superhuman qualities; and yet these peculiar qualities (his
consciousness of being God's eternal son, his preaching of a
new kingdom and prediction of a second coming, above all
his Resurrection) were apparently essential to the New Testament witness. The science of history had created a wa-
tershed. One set of presuppositions took one toward a human, but historically elusive Christ. The other — more
traditional — allowed the retention of the Christ of faith, but
at the cost of severing one's ties with what Strauss called "our
modern world."

Conclusion: The Convergence of Scientific and Historical Criticism

In the English-speaking world of the mid-nineteenth century,
it is possible to detect among the intelligentsia what has often
been called a crisis of faith. Men and women with the highest
religious sensibilities found they could no longer believe what
they had formerly believed or what, in retrospect, they discovered (with some relief) they had never really believed in
the first place. It was a crisis with many roots. Disenchantment with Christianity, as in the case of Darwin, could spring
from a moral repugnance toward some of its doctrines, notably the idea of eternal damnation for those who (like so
many of Darwin's relatives) were beyond the pale. In a particularly poignant passage Darwin once wrote that he could
not see how anyone could *wish* Christianity to be true when
it enshrined so damnable a doctrine. Whatever its roots, the
crisis was often compounded by the implications of the historical sciences and by the new higher criticism. Although
the conclusions derived from each stemmed from two quite
different kinds of inquiry, they were ultimately shaped by the
same set of presuppositions — those of a fervent naturalism,
out of tune with the faith of an earlier age.

When Strauss spelled out the presuppositions of his modern world, they included the conviction that "all things are
linked together by a chain of causes and effects, which suffer
no interruption." The totality of finite things might owe its

existence to a superior power, but it suffered no intrusion from without. This conviction had become so much a habit that the belief in a supernatural manifestation or immediate divine agency would automatically be attributed to ignorance or imposture. Darwin belonged to that modern world. In his *Autobiography* he insisted that "the more we know of the fixed laws of nature the more incredible do miracles become." And of the gospel writers: "the men at that time were ignorant and credulous to a degree almost incomprehensible by us." When, as a student, he had made notes on Paley's *Evidences of Christianity* (1794), he had looked sympathetically at the arguments against Christ's having been a misguided fanatic. But he could not refrain from adding that "there must have been a certain degree of imposture in his miracles and prophecies," which no argumentation could gainsay.

The convergence of scientific and historical criticism was most pronounced in the thesis that Darwin gave the world. When Strauss had articulated the assumption that the chain of causes and effects was never interrupted, he had not seriously considered whether the origin of species might not be an exception. For many of Darwin's contemporaries, it certainly was. But Darwin obligingly closed the gap with a naturalistic account of speciation. Not surprisingly, Strauss saw great value in Darwin's achievement. It not only destroyed any last pretensions to historicity in the Genesis creation story, it also sanctioned the welcome view that man had risen, not fallen:

Humanity has much more reason to be proud of having gradually worked itself up by the continuous effort of innumerable generations from miserable, animal beginnings to its present status. We prefer this to being the descendants of a couple, who, having been created in the image of God, were kicked out of Paradise and to knowing that we are still far from having reached the level from which they fell in the beginning. Nothing dampens courage as much as the certainty that something we have trifled away can never be entirely regained. But nothing raises courage as much as facing a path, of which we do not know how far and how high it will lead us yet.[2]

Strauss's *Life of Jesus* had become available to the English speaking world in 1846 through George Eliot's translation. It has been said that it aggravated and disturbed an entire century. One facet of that disturbance became visible in Britain with the publication of a volume entitled *Essays and re-*

views (1860), hard on the heels of Darwin's *Origin of species* and creating, if anything, an even larger furor. For the majority of the essayists were Anglican clergy who, even if they kept their distance from the more radical German critics, nevertheless roundly rejected the verbal inspiration of Scripture, replacing it with a historical approach. One of the assumptions that the essayists held in common was that the Bible should be read like any other book. To admit that it had been written by fallible men or that it contained an obsolete understanding of the natural world did not, however, vitiate its moral authority. A convergence of scientific and historical criticism is once again apparent, not merely from the coincidence of timing, but also from the content of the essays.

The message of one contributor, C. W. Goodwin, was that it was far better to admit that the Bible contained dated and false science than try to preserve its literal meaning by harmonizing it with the latest theories. It was to geology he turned to demonstrate the bankruptcy of this kind of harmonization program. During the nineteenth century, several attempts had been made to reconcile Genesis and geology. Thomas Chalmers and William Buckland had postulated an enormous gap in time between the initial creation and the first Genesis "day," thereby keeping the possibility of more or less literal days, while giving the geologist all the time he needed for the earth's physical history. But that scheme had foundered once it had become clear that living things had been created at intervals widely separated in time. An alternative scheme, mooted earlier by Buffon, was to stretch the Genesis days, treating them as symbols of long and consecutive epochs. A degree of concordance with the order of creation, as disclosed in the fossil record, had thereby been achieved. But even this scheme had foundered because in Genesis the birds preceded the reptiles, whereas in geology the reptiles came first. An even more elaborate concordance had been proposed by the Scottish evangelical Hugh Miller, who had equated the Genesis days with successive days on which the author had received a divinely inspired vision as to how the earth would have appeared in each of its primitive epochs.

Goodwin was critical of all such strategies. In the first place, they were in competition with each other: divisive and mutually exclusive. Second, they led to constant retreat as science advanced. Furthermore, they usually failed on points of detail. Buckland, for example, had glided over the difficulty that, in Genesis, the heavens were formed by the division of the waters on the second day. It was far from clear what *could* have been going on in the long gap, if there was as yet no sun, moon, and stars. But, for Goodwin, the most telling point was the indignity of having men of science shuffling uneasily as they tried not to offend misplaced sensibilities:

The spectacle of able and . . . conscientious writers engaged in attempting the impossible is painful and humiliating. They evidently do not breathe freely over their work, but shuffle and stumble over their difficulties in a piteous manner; nor are they themselves again until they return to the pure and open fields of science.[3]

His conclusion, and the only way forward, was to acknowledge that the Mosaic cosmogony was "not an authentic utterance of Divine knowledge, but a human utterance, which it had pleased Providence to use in a special way for the education of mankind."

The convergence of scientific and historical criticism was even more conspicuous in the contribution of the Reverend Baden Powell, who argued that, whereas there had once been a time when the gospel miracles had constituted proof of Christ's divinity, they had now become an embarrassment and a liability. His argument deferred to the advancing tide of scientific naturalism, which he illustrated with the most topical example. Darwin's "masterly volume" would soon "bring about an entire revolution of opinion in favour of the grand principle of the self-evolving powers of nature."

Powell's reference to a pending revolution raises a point of the utmost importance. Darwin's revolution created a furor for many reasons, but prominent among them was that it was part of a much wider revolution in attitudes toward the Bible. It impinged on a sensitive issue at a sensitive time, compounding an intellectual crisis that was already proving divisive. *Essays and reviews* was condemned by the bishops of the Anglican Church, the archbishop of Canterbury issuing an encyclical against it. The efforts of the essayists to educate the public into a more truly historical understanding of Scripture were widely perceived as damaging to biblical authority. One sentence from the essay of Benjamin Jowett defined the new orientation, which, to more conservative minds, pointed toward an alarming conclusion: "Scripture has one meaning, the meaning which it had to the mind of the prophet or evangelist who first uttered or wrote, to the hearers or readers who first received it." Such was the storm generated by the ensuing debate that legal action was taken against two of the essayists, although the sentence (that they should be deprived of their livings for a year) was subsequently quashed by the Judicial Committee of the Privy Council. No fewer than 10,906 clergymen, however, signed a declaration protesting at the spread of this historically based liberalism, which had sprouted from within their own ranks.

If the convergence of scientific and historical criticism had the effect of intensifying alarm among those who found it difficult to adjust to the new perspectives, it had quite different consequences for the intelligentsia who had already readjusted. Clerics who were prepared to go some way with the higher critics were less likely to have a predisposition against evolutionary theory, although they might still take exception to the Darwinian mechanism. The opening contribution to *Essays and reviews* came from Frederick Temple, then headmaster of Rugby School, and already a firm believer in the proposition that a tolerant mentality was a sign of spiritual maturity. His thesis was that, in sensibility as in doctrine, there had been a historical development within Christianity, to which the physical sciences had contributed by revealing more things in heaven and earth than were dreamed of in patristic theology. To fear the result of scientific investigation was nothing short of "high treason against the faith." Indeed, the scientists' faith in the universality of physical law gave analogical support for the moralist's faith in the universality of moral law.

It was the same Frederick Temple who, on 1 July 1860, preached a sermon in the University Church when the British Association for the Advancement of Science met in Oxford. This was the meeting during which T. H. Huxley and Bishop Wilberforce had their celebrated encounter over their simian or episcopal ancestry. Deliberately distancing himself from the bishop's satire, Temple gave a long rein to scientific naturalism. Too often, he observed, theology had shown a tendency to take off wherever science had left off. In common with Richard Owen and Baden Powell, he insisted that the finger of God was to be discerned in the laws of nature – not in arbitrary limits placed on the scope of the sciences. Temple did not mention Darwin by name, but his congregation could well conclude, as one member certainly did, that "he espoused Darwin's ideas fully!"

CHAPTER VIII

Evolutionary Theory and Religious Belief

The Challenge of Darwinism

Charles Darwin concluded his *Origin of species* (1859) with the proclamation that there was grandeur in his view of life. From a simple beginning, in which living powers had been "breathed into" a few forms or even one, the most beautiful and wonderful organisms had evolved. Because he used that Old Testament metaphor, and because he also referred to "laws impressed upon matter by the Creator," it was possible to read into his conclusion a set of meanings and values associated with a biblical religion. His private correspondence suggests that this had not been his intention. He confided to the botanist J. D. Hooker that he had long regretted having truckled to public opinion by using the biblical term of creation, by which he had really meant "appeared by some wholly unknown process." It was not that he had deliberately concealed an underlying atheism. Rather, in retrospect, he saw that he had invited the attribution of a particular meaning to his science with which he was uncomfortable – especially when writing to sterner naturalists than himself.

Darwin's unease raises a point of general significance. Debates that have so often been interpreted in terms of the "conflict between science and religion" turn out, on closer inspection, to be debates in which rival claims are made for the "correct" meaning to be attached to scientific theories. Many of those who, in the eighteenth and early nineteenth centuries, had constructed their distinctive accounts of the earth's history felt themselves to be defending a particular set of social and religious values that they perceived to be under

275

threat from the cosmologies of their opponents. During the period through which Darwin worked on his evolutionary theory there had, however, been signs of a shift in sensibility, with scientists themselves wishing to exclude cosmological debate from the practice of science. Charles Lyell, for example, had argued that geology would only become a science when it disentangled itself from biblical precepts, narrowed its scope to the reconstruction of the past in terms of forces known in the present, and deliberately excluded speculation about origins, purposes, and ultimate meanings. Darwin, too, came to share this view. There were questions that lay beyond the purview of current science. To admit them would be to reintroduce metaphysical and theological issues alien to the quest for positive scientific knowledge. In the letter to Hooker in which he regretted his truckling, he stated that it was "mere rubbish, thinking at present of the origin of life; one might as well think of the origin of matter." It was one of his principal grievances that his account of the origin of species was so frequently judged by theological rather than scientifically informed criteria.

Yet Darwin could hardly have expected that theological considerations would be kept out of public debate. He had, after all, tried to persuade his public that, in the context of speciation, there was much that *could* be said about origins. Moreover, in presenting his case, he frequently stressed the superiority of his own theory over against that of "separate creation." And, in so doing, he often presumed to know that a rational Creator would not have distributed His creatures around the globe in the patterns that he himself had discovered. Although the antithesis he drew between evolution and separate creation was principally a device to point up the strengths of his theory, it could easily be read as a deliberate attack on the Christian faith. Two quite different meanings could therefore be attached to Darwin's *Origin* — that it was consistent with a biblical religion (as long as one did not take Genesis literally) and, conversely, that it undermined it. Because the prospect of evolutionary progress could become the basis of alternative religious creeds, the religious response to the Darwinian challenge is remarkable for its diversity.

To appreciate the force of the challenge it is useful to see how Darwin summarized his case for the power of his theory. In the last chapter of the *Origin* he claimed that, granted the truth of just three propositions, he could explain the emergence and gradual perfection of new species. The first was that gradations in the perfection of any organ or instinct could have existed, each good of its kind. The second: All organs and instincts are variable, however slightly. Third, there is a struggle for existence leading to the preservation of each profitable deviation of structure or instinct. Proceeding from these axioms, he explained why he believed his account had the edge over "separate creation." In the first place, he showed how it could resolve a problem in classification. This was where to draw the line between species (generally considered to have been separately created) and their varieties (generally considered to have been produced by secondary laws). In Darwin's theory of descent with modification, species were "only strongly marked and permanent varieties." There was no ultimate dividing line. Thus Darwin saw himself as a liberator: Taxonomists need no longer be plagued by the old question whether their specimens constituted *real* species.

His theory could also explain a recurrent pattern in geographical distribution. In each region where many species of a genus had been produced, these same species had usually produced many varieties. This was understandable, if all species first existed as varieties. He also showed how repeated divergence from common ancestors could give a historical explanation for the pyramidal, hierarchical arrangement of groups within groups that featured in most taxonomic schemes. That all forms of life could be placed in groups subordinate to groups, all within a few great classes, he deemed to be "utterly inexplicable on the theory of creation." On a theory of transmutation one could understand why species that had supplanted their predecessors should always have borne so great a resemblance to them.

Arguing for the superiority of his own explanatory program, Darwin also turned his attention to those creatures that had developed ways of surviving at variance with what one would expect on the assumption that they had been specially created. A favorite example was an upland goose, which, though equipped with webbed feet, deigned not to swim. There were birds resembling woodpeckers, which survived by picking insects from the ground; thrushes, which dived to feed on subaquatic prey. Such examples were an embarrassment to the old notion of preadaptation to preadapted conditions. They were, however, compatible with the view that each species was striving to increase in number, that natural selection favored the exploitation of every available niche.

A theory of evolution by natural selection had the further advantage that it could accommodate examples of imperfect adaptation. Darwin appealed to the sensibilities of an audience he believed would share his view that the sting of a bee was hardly a perfect defense if the bee died from using it. He underlined, too, his abhorrence of the young ichneumonidae feeding within the living bodies of caterpillars. Such a grizzly contrivance he found difficult to square with belief in a beneficent God; but it posed no threat to the concept of natural selection. Then there were vestigial and rudimentary organs. For modern writers on evolution these still supply some of the most telling evidence in favor of Darwin's theory. Of no use to their current possessors they are an emblem of an evolutionary past, a legacy from the day when they were once of service to an ancestor. Yet another phenomenon he considered inexplicable on the theory of creation: the occasional appearance of stripes on the shoulder and legs of several species of horse. This was simply explained if they had descended from a striped progenitor.

Continuing his tour de force, Darwin showed how repeated evolutionary divergence meshed with the fossil record. The more ancient the fossil, the more often it displayed characteristics intermediate between later related forms. For his coup de grace he returned full circle to the geographical distribution of species, to the evidence that had occasioned his original divergence from Lyell. The resemblances among species in a particular location were simply due to their sharing the same ancestors. The affinity between island species and the inhabitants of the nearest mainland was comprehensible if migration had preceded modification. He had earlier devised experiments to see whether seeds might still germinate after accidental transportation by swift ocean currents. That some islands could boast only one species, even though they were perfectly suited to others, was coupled with the now familiar refrain: "Such facts as the presence of peculiar species of bats, and the absence of all other mammals on oceanic islands, are utterly inexplicable on the theory of independent acts of creation."

Open-minded religious commentators would soon emphasize that a Christian doctrine of creation need not entail the view that every species had been independently created. Their point was that Darwin's critique did not touch the central thrust of their doctrine, which was that everything ultimately owed its existence and preservation to a power transcending the natural order. The Darwinian challenge could not be so lightly dismissed, however, in a related issue: The basis of morality and spirituality if men and women had emerged from lower forms of life. All that Darwin said on the subject in his *Origin* was confined to one cryptic sentence: "Light will be thrown on the origin of man and his history." Nevertheless, from the very inception of his theory, he had been utilizing concepts of the mind that would not obstruct an evolutionary account of human development.

Darwin was not alone in developing the case for human evolution. As we saw in Chapter VI, the Edinburgh publisher Robert Chambers also made the human mind the product of natural laws. In his anonymous *Vestiges of the natural history of creation* (1844), he appealed to statistical regularities in human behavior and to the "popular-science" of phrenology in arguing his case. Later, during the 1860s, before Darwin had yet written his *Descent of man* (1871), there was a further and wide-ranging debate on the issue of human origins. Comparative anatomists who, like Richard Owen, argued for a distinctive lobe in the human brain found themselves thwarted by Darwin's disciple T. H. Huxley (1825–95), who proved irrepressible in his pursuit of structural analogies between humans and the humanlike apes. Even in the important matter of cranial capacity, Huxley argued, "men differ more widely from one another than they do from the apes." He was particularly enamored of the gorilla's foot, for it was closer to man's than it was to the orangutan's. Huxley's strategy was beguiling in its simplicity. If there were closer anatomical resemblances between humans and the higher apes than among the apes themselves, then the same process of evolution could account for both.

The issue of human origins was also of keen interest in the 1860s because of its association with the problem of race. The decade that, in Britain, saw the Anthropological Society of London making a special study of racial characteristics saw, in America, a civil war in which concepts of racial superiority were never far from the surface. Long-standing debates between polygenists (who favored multiple origins for distinct human races) and monogenists (who preferred a single origin for a common species) were invested with greater theoretical

content once the relevance of evolutionary theory was perceived. In one respect, Darwin's theory, no less than Genesis, implied one ultimate origin; in another, however, it could
be used to underwrite the notion that different races were
incipiently distinct species, the "fittest" of which had their
superiority demonstrated by the very fact of their power and
success. Via evolutionary thought, the terms in which the
polygenist–monogenist debate had earlier been staged were
transcended.

If comparative anatomy and anthropology fed into the
stream of argument, so did the science of archaeology. This
was because stone tools spoke more convincingly of human
prehistory than the limited and often ambiguous data of fossil
remains. In his *Prehistoric times* (1865), John Lubbock refined
the sequence of Stone, Bronze, and Iron ages by dividing the
first into two. A Paleolithic period, in which stone tools were
chipped and flaked, had been succeeded by a Neolithic, in
which the stones had been polished. It was also during the
1860s that the Frenchman Edouard Lartet tried to bring greater
precision to the dating of human antiquity by studying the
fossil mammals that corresponded in time with particular tools.

In 1871, when Darwin published his *Descent of man*, there
was therefore a growing literature on human evolution on
which he himself could draw. Social and political theorists
were busy integrating concepts of social and biological evolution in ways that were soon to justify political creeds of
every hue. Consequently, Darwin's treatment of human evolution was not as electrifying in its originality as his *Origin of
species* had been. For many religious commentators, however,
its message was no less shocking for having been anticipated.
For insight into the full extent of the challenge it is instructive to see how Darwin handled the question of religious beliefs. By venturing to explain how such beliefs had arisen and
evolved from a primitive animism, he was often perceived to
have explained them away. He nevertheless believed that they
had played a role in human evolution by reinforcing ethical
codes; for they had the power to inculcate a sense of re-
morse. The moral sense could itself be attributed to its utility
in the early evolution of human societies. The emergence of
conscience had much to do with a social instinct, which expressed itself in the need for approval from one's fellows.
Obedience to the wishes of a community would be strengthened by habit so that an act of theft, for example, would in-

duce feelings of dissatisfaction with oneself. Religious beliefs, he seemed to suggest at one point, had actually arisen from confusion over the origin of conscience. When he later composed his *Autobiography,* he was prepared to think of these beliefs as inculcated in the early years of life and taking on the character of an instinct, meaning that they were followed independently of reason. For many of his contemporaries who were clinging to moral absolutes as the lifeline of a faith that had already been badly shaken, the effect of such an analysis could be desolating. His own wife could not stomach the proposition that all morality had grown up by evolution, and excised from the *Autobiography* a sentence comparing a child's belief in God with a monkey's fear of a snake. There is perhaps no more piquant expression of the Darwinian challenge than that Emma Darwin should confide in her son Frank that "where this sentence comes in, it gives one a sort of shock."

Contrary to what one might expect, Darwin had not argued for the relativity of moral values. It is more accurate to say that both he and Huxley wrote as if their liberal values could be authenticated from a natural order in which individual freedom, a meritocratic society, and evolutionary success went hand in hand. The golden rule ("as ye would that men should do to you, do ye to them likewise") was for Darwin the highest, but also the *natural,* outcome of the development of social instincts. But the relativity of moral values *was* the inference that others would draw from his work. And for those who still wished to ground ethical principles in religious belief there was the added inconvenience that secular writers of the caliber of Leslie Stephen were congratulating Darwin for his contribution to the task of showing how morality could be made independent of theology. The task had been begun before — by utilitarian moralists earlier in the century, who had sought to ground concepts of "good" and "bad" actions in the quantification of human pleasure or pain that they produced. There was, however, a new authority and vividness in Darwin's message.

The Darwinian challenge impinged on many facets of popular Christian doctrine: the nature of biblical authority, the historicity of the creation narratives, the meaning of Adam's fall from grace and (connected with it) the meaning of Christ's redemptive mission; the nature and scope of God's activity in the world; the persuasive force of the argument from design; what it meant for humankind to be made in the image of God; and the ultimate grounds of moral values. Because it was easy to set up a contradiction on each point between ostensibly "scientific" and ostensibly "religious" points of view, Darwin's theory soon came to symbolize that conflict on which militant secularists as well as militant fundamentalists still like to dwell. In Europe and America there were to be violent (though never universal) attacks on Darwin's science from both Catholic and Protestant communities. Among less well-educated Protestants there has been a recurrent pattern of asserting and reinforcing religious identity through opposition to evolutionary naturalism. Before examining the more subtle responses, it is therefore important to see something of the process by which the theory gained the kind of prestige that demanded a response.

The Success of Darwinism

Addressing the British Association for the Advancement of Science in August 1874, the physicist John Tyndall poignantly recalled a meeting near Boston at which both he and the Harvard professor Louis Agassiz had been present. The maple, he said, had been in its autumn glory. The exquisite beauty of the scene outside had suffused the intellectual action. Earnestly, almost sadly, Agassiz had confessed that he had not been prepared to see Darwin's theory received as it had been by the best intellects of the day. Its success had been greater than he had thought possible. Reflecting later on the reception of the *Origin of species,* Huxley claimed that it had taken but twenty years for evolutionary theory to win the almost total assent of the scientific community. This was not to say that naturalists were of one accord on the mechanism of speciation. The scope of Darwin's principle of natural selection remained highly controversial. Huxley himself had advocated sudden "saltations" (mutations) rather than the complete gradualism of the Darwinian process. It is difficult to deny, however, that Darwin convinced a younger generation of naturalists that evolution *had* occurred, and by a mechanism in which natural selection had played a significant part.

If Agassiz was saddened by Darwin's success, it was not because he was a slave to religious orthodoxy. He had, however, been campaigning for an alternative view of nature in which a divine plan could be affirmed. Referring to those animals that Cuvier had placed in the category of radiates, Agassiz claimed to see an "intellectual link" between them. They were "variations of an idea," emphatically not the result of diverse influences operating on them. Promoting a form of philosophical idealism similar to that adopted by Richard Owen in Britain, Agassiz claimed as evidence for a mind behind creation the facility with which the human mind could grasp the basic ideas on which each major group of creatures had been structured. He was astute enough to see that such a philosophy was threatened by Darwin's emphasis on "chance" variations as the material on which natural selection worked. The success of Darwinism was therefore a threat both to popular creationism and to models of the history of life based on the unfolding of a preordained plan.

One of the difficulties in discussing the spread of Darwinism arises, however, from the fact that the concept of evolutionary change was not intrinsically incompatible with such idealist assumptions. Some religious thinkers were not unhappy to accept that organic evolution had occurred, as long as the "chance" elements could be deleted. Why should evolution not be God's method of creation? Liberal Christians anxious to differentiate their position from more conservative factions might even find such an idea attractive. Religious motives could therefore play a role in assisting as well as resisting the spread of evolutionary theory. Among the less acute proponents of theistic evolution there was often a failure to perceive just how damaging the Darwinian mechanism was. Among the more enlightened, the strategy was simply to minimize the scope of natural selection, allowing the process to be controlled by such forces as the direct response of organisms to environmental change, which were more readily compatible with mental agency in a material world. In Britain, the duke of Argyll, the Roman Catholic St. George Mivart (1827–1900), and the Unitarian William Carpenter (1813–85), without denying that evolution had occurred, each claimed that Darwin's mechanism of natural selection was inadequate. Mivart's thesis, for example, was that species of widely differing ancestry nevertheless displayed marked similarity of structure. Why the close resemblance between ostriches and flying birds if, as his rival Huxley was contending, the latter were derived from dinosaurs, the former from pterosaurs? For Mivart it was convergence toward common structures rather than divergence from a single progenitor that had to be explained. No amount of fortuitous variation could establish such convergent trends. There had to be some preordained impulse from within. Mivart's strategy of opting for multiple origins of similar structures had the effect of replacing Darwin's image of a single branching tree with what Mivart himself described as a "grove of trees . . . greatly differing in age and size."

The fact that Mivart, one of Darwin's severest critics, could still argue for organic evolution shows that there is no simple way of gauging the success of Darwin's work. He could be successful on one level (in making the "fact" of evolution seem incontrovertible) even while failing on another (as the plausibility of his mechanism was called into question). The distinctions here are important because it is still sometimes assumed, quite erroneously, that it is only necessary to find defects in the Darwinian mechanism to disprove the historical reality of evolution itself. As Darwin was the first to recognize, it would never be difficult to pick holes in his theory. Indeed one of the reasons for his success was the openness with which he conceded difficulties, coupled with the skill with which he removed them. To admit that some had once seemed fatal to his theory was a most effective rhetorical device, for he went on to show how even the most daunting difficulty could be overcome, once the theory was properly applied.

The rarity of transitional forms in the fossil record was one of the more obvious problems. Even today their relative paucity has allowed models of "punctuated equilibrium" to vie with models of gradual, continuous change. Yet Darwin could appeal to the inevitable incompleteness of the fossil record and to the smaller population size of intermediates (precisely because they were intermediates between more successful species) to nullify the negative evidence. Not only the rarity but the viability of certain transition states was a problem. Surely the conversion of a land-based carnivore into one with aquatic habits would involve intermediates that could not have coped with the new environment? Yet the problem was not insuperable since Darwin could enlist known amphibians, such as the Mustela vison of North America, which could straddle both environments. During the summer this otterlike creature dived for fish; during the winter it left the freezing water to prey on land. The more difficult case was the conversion of an insectivorous quadruped into a flying bat. What advantage was there in an incipient wing, too incipient to work? Claiming again that he had once had no answer, Darwin pointed to the squirrel family, which showed the finest gradation from species with tails only slightly flattened, to others where the skin on their flanks was fuller, to the so-called flying squirrels in which a broad expanse of skin connected rear limbs to the base of the tail. Each structure had been useful in a particular habitat, even if it allowed the merest glide. The incipient stages of a bat's wing might similarly have conferred advantage by permitting gliding to occur. The ability to fly would be a later development.

Once naturalists learned to look at the world through Darwin's spectacles, such difficulties (which still loom large in creationist literature) began to appear inconsequential. There were, however, numerous scientific difficulties to consider. How, for example, could a single advantageous variant be passed on to succeeding generations and proliferate if it had first to mate with a "normal" member of the population? If one subscribed, as Darwin himself did, to a scheme of blending inheritance, would one not have to concede that the favorable variant would soon be swamped? From a quite different source there came the objection leveled by William Thomson (Lord Kelvin, 1824–1907) that estimates of the earth's cooling did not leave sufficient time for Darwin's gradual process to have taken place. In 1862 Thomson had allowed a maximum of 400 million years, but by 1868 it had shrunk to 100 million, and by 1876 to a mere 50 million. His technique was to apply Fourier's analysis of heat conduction to establish a temperature gradient from the earth's molten core to its surface. Knowing the thermal conductivities of rocks near Edinburgh and Greenwich, and assuming they were typical values for the earth's interior, he was able to calculate the rate of heat loss and so determine the duration of the process. Because he did take exception to natural selection, and because he complained that the design argument had recently suffered neglect, Thomson has been portrayed as a creationist. In his 1871 address to the British Association, however, he specifically warned against invoking an "abnormal act of Creative Power" if a solution to the origin and diffusion of life could be found "consistent with the ordinary course of nature." He was not averse to the idea that seed-bearing meteoric stones from another world had begun the process, all subsequent creatures having "proceeded by orderly evolution from some such origin." Darwin's mechanism may have been temporarily imperiled, but not the basic principles of an evolutionary naturalism.

There were philosophical objections, too, that Darwin's theory successfully withstood. What confidence could one place in a hypothesis that was not directly verifiable? While acknowledging the problem, the Darwinians gave the obvious reply that the time-scale for the formation of a new species meant that one would never be able to observe a transmutation taking place. It was in the nature of a historical science that economy and coherence of explanation would have to take precedence over direct verification. Such a reply, however, exposed the hypotheticodeductive structure of Darwin's science. For some observers this also marked an unacceptable departure from the hallowed norms of an inductive method. The difficulty here was one that the scientific community had partly brought upon itself. So often in the past it had pretended to be following the Baconian method in which preconceived hypotheses were outlawed. It was a pretense that had helped in projecting images of scientific objectivity and of a community united by a common methodology. So facile an image had already been challenged. The Cambridge philosopher William Whewell had underlined the differences between the sciences in the fundamental ideas and methods on which they depended. But the earlier propaganda had been remarkably effective. The popular magazine *Punch* carried the following objection to Darwin's *Descent of man:*

> "Hypotheses non fingo,"
> Sir Isaac Newton said.
> And that was true, by Jingo!
> As proof demonstrated.
> But Mr. Darwin's speculation
> Is of another sort.
> 'Tis one which demonstration
> In no wise doth support.[1]

Darwin knew there was much more to his science than inductivist slogans would allow. Without the testing of hunches and hypotheses, science would be reduced to a trivial pursuit. As he explained to his American correspondent Asa Gray:

I assume that species arise like our domestic varieties with *much* extinction; and then test this hypothesis by comparison with as many general and pretty well-established propositions as I can find made out – in geographical distribution, geological history, affinities etc. . . . And it seems to me that *supposing* that such hypothesis were to explain such general propositions, we ought, in accordance with the common way of following all sciences, to admit it till some better hypothesis can be found out.[2]

During the second half of the nineteenth century there was a greater willingness among scientists to celebrate in public the virtues of a hypotheticodeductive method. It was one way of underlining the value of scientific expertise, for the construction of technical hypotheses (unlike the classification of data) was a task beyond the uninitiated. The publication of Darwin's theory roughly coincided with this shift and may have even been partly responsible for it. Only partly, because there were developments in the physical sciences, notably the wave theory of light and the kinetic theory of gases, in which the power of a hypotheticodeductive structure was demonstrated. Nevertheless, an important aspect of Darwin's success was his advertisement of the fact that scientific creativity should not be bound by narrow and artificial definitions of legitimate scientific method.

The success of Darwin's theory among his fellow naturalists owed much to the fact that it did offer new insights into biological phenomena. H. W. Bates, who had traveled with Alfred Russel Wallace to Brazil in 1848, was quick to see that the phenomenon of mimicry, formerly ascribed to design, was elegantly explained by natural selection. Why, for example, among *Leptalis* butterflies were there some that, in their markings, resembled species of the *Heliconidae* family known to be distasteful to birds? A beneficial disguise, which would once have have elicited thoughts of design, was now explicable without them. Variants of *Leptalis* species having marks approximating to those among the *Heliconidae* would have had an advantage over those in which they were not so marked, for the simple reason that birds would tend to avoid them, having already learned to eschew the pattern. In due course what had once been a rudimentary copy would eventuate in a remarkable imitation – but merely through further variation and the selective predation of birds. This early application of Darwinian principles was contained in a paper that Bates read before the Linnean Society of London in November 1861, within two years of the publication of Darwin's *Origin*.

Figure VIII. 1. Woodcuts showing the markings on *Methona psidii* (*Heliconidae*) and *Leptalis orise* (*Pieridae*).

Those scientists such as T. H. Huxley who were already disposed to reject every trace of the supernatural from scientific explanation were the very ones most likely to suppress the fact that much of their data in support of Darwin (including the striking examples of mimicry) had been collected by earlier naturalists on the understanding that they were helping to demonstrate the unity of creation. Even if natural theology had dug its own grave, it had not been scientifically sterile. The great appeal of the Darwinian principles, however, was that they opened doors onto new lines of inquiry that would have been closed by the assumption of separate creation. In Germany, Darwin's popularizer Ernst Haeckel (1834–1919) boldly took upon himself the task of constructing evolutionary phylogenies – filling in the exact lines of descent as he conceived them. Darwin himself had been reluctant to speculate about the genealogies of modern species, given the fragmentary nature of the fossil record. Huxley, too, had expressed the reservation that many creatures could have come into existence long before they had left an imprint. Haeckel's enthusiasm, however, soon won Huxley round even though they disagreed on details. Haeckel looked to the Asian apes for the closest human relatives. It was this group, which also included the gibbon and orangutan, from which hypothetical ape-men, *Pithecanthropi,* had emerged. From these, according to Haeckel, there had been independent lines of evolution leading to the various human races. There was much that was rash in his speculations, but he conveyed to his readers the excitement of a science that was going somewhere, rather than one stultified by an obsolete theology. His hypothetical link with the Asian apes apparently encouraged Eugène Dubois to go to the East Indies in the hope of finding a missing link — a venture rewarded with the skull cap and thigh bone of what became popularly ascribed to "Java man."

The fact that evolutionary theories could stimulate new scientific research was not the only reason for Darwin's success. Huxley considered one of the greatest merits of Darwin's theory that, in his opinion, it was incompatible with Roman Catholicism. Haeckel, too, relentlessly pursued his thesis that a monistic religion based on evolution must supplant the matter–spirit and nature–supernature dualities of the Christian tradition. To understand the rapid dissemination of Darwin's theory it is therefore necessary to see how it could be used as a resource, not only for new types of secular theology, but also for social and political theories of almost every kind.

Darwinism as a Resource: The Varieties of Social Darwinism

That Darwin's theory could lend itself to a multiplicity of causes was partly due to tensions in his exposition that could be resolved in different directions. Although his account of human kinship with the animals carried a humbling and egalitarian message, the emphasis on development tended to restore a hierarchy with European man at the apex. Whereas the theory placed value on diversity and deviation from the norm, it also accorded value to compliance in the sense that an organism had to conform to the demands of its environment. There were both optimistic and pessimistic streaks running through the exposition — optimistic in that natural selection invariably worked for the good of the species, pessimistic in that nature was riven with struggle and strife. Again,

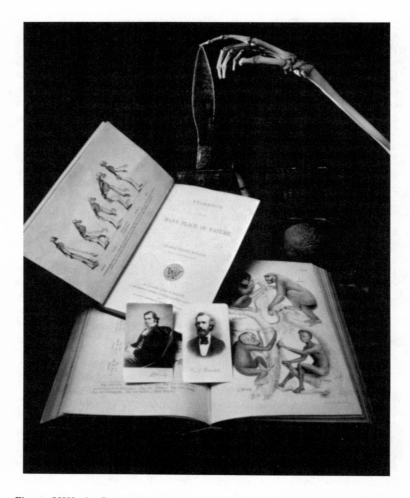

Figure VIII. 2. Composite photograph of two of Charles Darwin's most formidable disciples, Thomas Henry Huxley and Ernst Haeckel. Reproduced by courtesy of John Reader/Science Photo Library, London.

as we have already noted, although the theory favored a monogenist account of human origins, in that all races were ultimately derived from a common ancestor, it could be pressed in the direction of polygenism by treating geographical races

Figure VIII. 3. Evolutionary cartoon from *Harper's Bazaar* of 16 September 1876. The caption reads: Scientific monkey, "Cut if off short, Tim; I can't afford to await developments before I can take my proper position in Society." Reproduced by courtesy of the Science Photo Library, London.

as incipient, distinct species. Even the characteristic image of the branching tree was equivocal. Although bifurcation told against a preordained plan, the image nevertheless sanctioned belief in irreversible, upward growth.

Of the many forms of social Darwinism the best known are those that exploited the new science to infuse extra authority into Herbert Spencer's notion of the "survival of the fittest." In the accounts of human society that Spencer had offered during the 1850s, his principal concern had been progress within a species rather than transmutation of one species into another. Insofar as evolutionary biology had given him a license to speculate, it was the mechanism of Lamarck, and especially the transmission of acquired characteristics, that had proved attractive. But Darwin's theory also proved attractive to social evolutionists because it appeared to give access to their most intransigent problem: the emergence of the human race from a state of nature. And if humans had risen through nature's own powers of selection, what argument could there be for interfering with a process that produced such beneficent results? It was certainly not difficult to arrive at the conclusion that socialist conceptions of society were fundamentally irrational. One had to adapt to nature's laws if the human race were to prosper. Philosophies of individualism rather than collectivism appeared to be sanctioned by Darwinism, a laissez-faire capitalism offering a "natural" guarantee of economic prosperity. Or so it seemed to many capitalist entrepreneurs, to the many representatives of American big business (Andrew Carnegie among them) who gathered in New York to honor Spencer during his visit in the autumn of 1882. From Spencer and Darwin it was possible to justify the notion that wealth is a sign of worth. In his book suitably titled *The Gospel of wealth* (1890), Carnegie insisted that it was to the law of competition that civilization was indebted for its material development and improvement. A much-quoted American exponent of social Darwinism, William Graham Sumner, observed:

If we do not like the survival of the fittest, we have only one possible alternative, and that is the survival of the unfittest. The former is the law of civilization; the latter is the law of anti-civilization.[3]

If socialists cherished a plan for nourishing the unfittest and yet advancing in civilization, it was a plan "no man will ever find."

In both Britain and America, Darwinism was rapidly enlisted to serve interests that would now be generally regarded as conservative, racist, and even sexist. In its earliest forms, however, social Darwinism in Britain showed a liberal face. There were elements in Darwin's synthesis that pointed toward social reform, especially the removal of aristocratic privilege. Status conferred by birth rather than achievement was wide open to the criticism that it artificially protected idle and unproductive members of society. Landed property could discourage economic initiative, while the privileges enjoyed by firstborn males among the gentry had incited comment from Darwin himself: "Primogeniture is dreadfully opposed by selection; suppose the first-born bull was necessarily made by each farmer the begetter of the stock!" If one believed with Walter Bagehot and Leslie Stephen that social evolution occurred through the selection of advantageous social variations, there was a definite political conclusion to be drawn. There had to be a degree of intellectual freedom to allow new variations their chance. Accordingly the best political institutions had to be liberal. Conservatism could spell ossification and atrophy.

So many were the varieties of social Darwinism that the value of the expression has recently been called into question. There were, for example, aspects of Darwin's theory that gave succor to socialists – and socialists of several persuasions. Because the concept of "fitness" in Darwin's theory was defined in relation to environmental conditions, it was open to socialist commentators to stress the advantages that might accrue if social structures were changed. They could accuse Spencer of having misunderstood Darwin, for it was ultimately the conditions of life that had produced different instincts and habits. The appeal of Darwin's theory to Marx and Engels has often been noted. The attraction consisted largely in the fact that Darwin had abandoned traditional conceptions of teleology and transformed the history of life into one of the interplay of unconscious forces, independent of a directing will – a concept of history that resonated with Marx's own emphasis on the apersonal social and economic forces that had shaped human history. The potential for a fully secular account of human nature was also a notable attraction. But, as for the inference that poverty was an inevitable consequence of a struggle for existence, Engels replied that it was a piece of arrant circularity to project Malthusian principles onto nature and then, by reversing the projection from nature to history, to claim that they had the status of eternal laws.

It was not only secular-minded socialists who felt there was leverage in Darwin's theory. The Christian socialist Charles Kingsley responded so positively that he was cited by Darwin himself as a clergyman who had claimed that his understanding of divine activity had been enhanced by the new theory. Instead of a God who created as if by magic, Kingsley embraced a God so wise that He could make all things make themselves. And if there could be metamorphosis in the organic world, why not the social? Organic transformation had religious meaning for Kingsley because it helped to restore a sense of the wonderful in nature; but it also had social meaning because it offered a metaphor for the hope of social improvement, for the creation of a society in which orphan boys would not be trapped as chimney sweeps. The burden of a Darwinian socialism was expressed years later by Keir Hardie when he recalled Darwin's statement that those communities that included the greatest number of the most sympathetic members would flourish best − a statement that, in Hardie's view, conceded the whole case for which the socialist contended.

The notion that those societies that encouraged human sympathy were best fitted to survive had appealed to Darwin's fellow evolutionist, Alfred Russel Wallace, whose socialism was neither Christian nor (since he was drawn into spiritualism) fully secular. In 1870 Wallace argued that at a certain stage in human evolution, mental and moral qualities had assumed so great an importance that natural selection acting on purely physical characteristics had ceased to be the overriding mechanism. He caused Darwin anxiety by suggesting that certain mental qualities − a pronounced aesthetic sense, mathematical skill, or musical appreciation − were beyond the power of natural selection to explain. It was simply not clear what advantage these particular attributes conferred, especially considering their absence in large numbers of the human race. As both spiritualist and socialist, Wallace showed an independence of mind that had earlier manifested itself in his development of evolutionary theory. He had originally accepted the view that other races were disappearing from the "inevitable effects of an unequal mental and physical struggle" with superior whites; but, by 1869, he had rejected that preconception. By then he was prepared to say that it was "among people in a very low state of civilization" where some approach to a perfect social state could be seen. In their case there was "none of those wide distinctions, of education and ignorance, or wealth and poverty, master and servant, which are the product of our civilization." In Wallace there was support for the nationalization of land and for the economic emancipation of women. The latter reform he actually justified in evolutionary terms, thereby giving rise to a form of social Wallaceism. His point was that women were currently prevented, by their social and economic disadvantages, from fully exercising their selective role in the choice of mate. Although he sometimes felt that Darwin attached too great an importance to sexual selection in the mechanics of evolution, Wallace was nevertheless convinced that female emancipation could only benefit posterity. How difficult it was for exponents of biological evolution not to draw social and political conclusions!

And how difficult it was for social and political theorists to renounce Darwinism as an ideological resource. In the burgeoning imperial culture of Britain, colonial expansion could easily be rationalized with the notion that the most efficient peoples must ultimately prevail. By projecting a set of individual qualities onto an entire nation, by turning colonial conquest into natural law, apologists for the partition of Africa, for example, were able to exonerate themselves from charges of cruelty and had reason to be grateful to Darwin for their cue. A particularly vivid example is provided by F. C. Selous whose *Sunshine and storm in Rhodesia* (1896) sought to exonerate the British South Africa Company of responsibility for the Ndebele rebellion of 1896. It might seem a hard and cruel fate, Selous observed, that if the black would not conform to white man's laws, he had to go, or die in resisting them; but it was

a destiny which the broadest philanthropy cannot avert, while the British colonist is but the irresponsible atom employed in carrying out a preordained law – the law which has ruled upon this planet ever since . . . organic life was first evolved upon the earth – the inexorable law which Darwin has aptly termed the "Survival of the Fittest."[4]

The use of Darwin to justify the whole gamut of social and political creeds was a remarkably pervasive and enduring phenomenon. It may be tempting to say that Darwinian metaphors carried weight only because the theory in which they were embedded had already gained the highest scientific prestige. But there is the alternative view that the theory gained its prestige, certainly beyond the confines of a scientific elite, by the very process of its incorporation into statements of social policy. Because it was so readily incorporated into secular visions of an evolving society, it is perhaps not surprising that some religious thinkers would not give it credence. Reporting and lamenting the fact that "official Spanish science is *evolutionist,* that is, enemy of independent creation and therefore of the doctrine of Genesis," a schoolteacher from Valencia, Manuel Polo y Peyrolon, devoted himself to proving that the spread of Darwinism had been due to the forces of materialism, not to the competence of Darwin's science. By its refusal to consider the scientific fertility of Darwin's work, such a view expressed a false disjunction. That it was even remotely plausible, however, suggests that in discussing the relations between Darwinism and religious belief, it is simply not possible to abstract the twin entities of "science" and "religion" to see how they fit together. As in other instances we have examined, though perhaps most transparently in this case, scientific and religious beliefs were so enmeshed in broader social and political debates that attempts to extricate them and relate them one to the other can be extremely artificial. This can perhaps best be seen in the fact that the history of Darwinism has a different complexion in different European countries.

The Relevance of Political Context: Darwinism in France and Germany

The reception accorded to Darwin's theory did vary from country to country, according to prevailing scientific ethos and to the distribution of political power between ecclesiastical and secular forces. In France, Darwinian evolution made virtually no inroads among the scientific elite of the Second Empire, but soon began to gain ground under the anticlerical regime of the Third Republic. In Spain, the Revolution of 1868 suddenly allowed the ideas of Darwin, Spencer, and Haeckel to flourish among dissident intellectuals – to such a degree that the Restoration of 1874, though removing conspicuous Darwinians from their university posts, had difficulty in containing the threat. By contrast, Catholic authorities in Italy were almost totally ineffective in opposing what rapidly became a consensus in favor of Darwin among Italian scientists, one of whom, Filippo De Filippi, a practicing Catholic and minister of education, expounded the relevance of Darwinism to the study of man several years before Darwin's *Descent* appeared. The peculiarities of the Italian situation have been ascribed to the fact that Italian Catholicism had already been reduced to a veneer over a basic unbelief, to a consequent dislike among the public for clerical attacks on modernity, and to the absence of an alternative scientific paradigm (comparable with that of Cuvier in France or Agassiz in America) that might have lent *brio* to the opposition. In a predominantly Protestant Germany, various forms of scientific materialism had already been sufficiently publicized to ensure that Darwin's thesis did not create a sense of shock comparable with that often experienced in Britain and America.

The comparison between France and Germany is particularly instructive. France, with its oscillation between empire and the forces of secular republicanism, shows perhaps more clearly than elsewhere the extent to which the fortunes of Darwinism could depend upon changes of political power. When Darwin's *Origin* made its debut, a somewhat conservative scientific element in Paris had just been protecting itself against materialist accounts of evolution, claims for spontaneous generation, and polygenist models of the human race. Moreover, the aggressive preface that Clemence Royer stuck to her French translation of the *Origin* (1862) could not have been better calculated to alienate Roman Catholic opinion. Her readers were presented with a stark choice between the "rational revelation" of scientific progress and the obsolete revelation of the Christian religion. Between the two, no synthesis was possible. Her message might appeal to those who felt excluded by the scientific and religious establishments, but it did little to win a sympathetic hearing for the Darwinian thesis, which suffered the further disadvantage that, if read through the spectacles of Cuvier, it would be resisted as erroneous; if read through the spectacles of Lamarck or Geoffroy St. Hilaire, it could be resisted as unoriginal.

During the 1870s the situation palpably changed, as the political establishment of the Third Republic adopted a scientistic ideology that contained strong Darwinian motifs. The University of Paris was purged of scholars with clerical leanings and the minister of education appointed in 1879, Jules Ferry, epitomized the change when he contrasted "modern scientific education" and the "old literary education" of the Church. One consequence of the change in sensibility was that scientists with Catholic sympathies were kept under careful surveillance, especially if they used their university podium to criticize secular values. Another consequence was that conservative Catholic scientists could find themselves isolated from current trends in their discipline — though, as the contributions of Paul Sabatier (1854–1941) to organic chemistry have indicated, this was not always to the detriment of their science.

With the shift to a large republican majority in the French Assembly, circumstances proved auspicious for theories of evolution. In 1889 the Darwinian Alfred Giard defined an intellectual mood with which Darwinism was often allied: "Among the most happy public expressions of opinion toward the end of this century must be counted the tendency of science to replace gradually the role hitherto enjoyed by religion." Eight years earlier, the alliance between Darwinism and anticlericalism had been voiced by the atheist Charles Contejean, who claimed that, irrespective of the strength of the proofs, he would have to be an evolutionist as the only way of dispensing with miracles. Not that all republican positivists were atheists, however much their Catholic opponents might pretend otherwise. One minister of education, Léon Bourgois, argued that since a positivist philosophy limited itself to affirming or denying only that which was open to experimental inquiry, it actually left room for religious beliefs, which simply lay outside that category. But the hostility of Catholic commentators to the pretensions of scientific naturalism can be readily understood in the light of such overtly secular societies as the Paris school of anthropology, one of whose members argued in 1878 that the behavior of Christ suggested he had been a victim of meningitis.

A common charge against Catholic intellectuals was that it was impossible to be both a believer and a savant, that the demands of faith would tend to close or deflect certain lines of inquiry. In response to such allegations, and in the knowledge that the Catholic Church would continue to lose its grip unless it came to terms with scientific culture, attempts were made by Catholic scholars to achieve a rapprochement. A new journal, the *Revue des Questions Scientifiques,* first published by the Scientific Society of Brussels in 1877, provided a forum. A recurrent theme among contributors was the relationship between scientific advance and moral freedom. Extending the domain of physical laws did not impugn that freedom because it was exercised in the very pursuit of scientific knowledge.

But how amenable were these Catholic philosophers to Darwinian evolution? The gravity of the problems posed by natural selection may be appreciated from the fact that, until the late 1890s, the editorial policy of the *Revue* was to publish antievolutionary articles. During the 1890s, however, a change of heart has been discerned among French Catholics. It has been illustrated by attitudes expressed at five international congresses held by Catholic scientists between 1888 and 1900. At the first there could still be a proposal calling for a declaration that it was the duty of Catholics to oppose evolution. Yet the proposal met with a cool reception, several speakers urging that scientific issues should be freely discussed. There was already a growing awareness that a distinction had to be drawn between the evolutionary hypothesis as a scientific instrument and its exploitation by materialists. No consensus emerged, however, on the status of the hypothesis. By 1894, when the third congress was held at Brussels, the ambience was distinctly more favorable. The anthropological section succeeded in passing a resolution supporting those who "under the supreme magistrature of the teaching church, devote themselves to research the role that evolution can have had" in bringing the physical world to its present state. The language was circumspect but the change in outlook was sufficient to worry representatives of the old guard. When the fifth congress met in Munich, in 1900, Darwin's theory was explicitly adopted by one contributor who freely speculated on the probable links between humans and animals. With the assurance that Darwin had favored special creation for the first living things, scientific views that had seemed "heretical" a few years earlier were incorporated into respectable Catholic science.

Not that one can generalize about the attitudes of the Catholic clergy. Among certain members there was clandestine support for evolutionary concepts – clandestine because they were always liable to be denounced to an unsympathetic papacy preoccupied with other modernist threats. Darwinian principles were certainly uncongenial to many, not only on religious grounds but also because they constituted a political threat. The Church looked to mutual aid and a conciliatory temper in dealing with the social problems of poverty and class. The fear was that Darwinism might encourage the taste for a self-centered ruthlessness. Catholic authorities in France were not always averse to the idea that organic evolution might be responsible for the human body, but there was invariably the qualification that it could not account for the soul. Even in the twentieth century, Teilhard de Chardin's vision of an evolutionary "ascent toward consciousness," as depicted in *The phenomenon of man,* was refused publication during his lifetime. A significant landmark came with the encyclical *Humani generis* of 1951, which explicitly invited examination of the case for the evolution of the human body. Catholic scientists were nevertheless instructed that they must still be ready to submit to the judgment of their Church.

If French Catholics had difficulty in dissociating Darwin's theory from monism and materialism, this was largely due to the avidity with which that association had been fostered in Germany. It was here that Haeckel had turned Darwin's science into a popular movement with its own world-view – a substitute religion with its own catechism of nature worship. Of the major European countries, Germany had seen the greatest surge in mass literacy, creating the conditions for Darwinism to engage a wider public. An expanding market for popular science created opportunities that the churches seemingly overlooked but which were seized by the advocates of scientific rationalism. Prominent among them were Friedrich Ratzel, Carl Vogt, Ludwig Büchner, Arnold Dodel, Edward Aveling, and Wilhelm Bölsche – each of whom added their volumes to those of Haeckel and, in their different ways, peddled the notion that Christianity was defunct, evolution the victor. Scientific progress had not only rendered special creation obsolete. It had made it inconceivable. So argued Aveling in his 1887 volume on Darwin's theory. Any act of creation violated the conservation of energy. In the same rationalist spirit, Vogt had used the principle that "from nothing can nothing come" to deride the creation of matter as "palpable nonsense." Such was the stark antithesis between scientific rationalism and conventional religion that eventually took hold among a receptive proportion of the German working classes, for whom popular Darwinism appears to have been a more seductive philosophy than Marxism. The effect may even have been to prevent a full understanding of what Marx had meant by socialist liberation. Where human development was seen as part of an evolving, ever-improving universe, the will to revolutionary action could easily dissolve.

Why did popular Darwinism enjoy such prosperity in Germany? Part of the explanation may be that two quite different intellectual movements had helped to prepare the way. The *Naturphilosophie* of the early nineteenth century had not been concerned with any physical mechanism for the derivation of one species from another, but it had fostered a vision of nature in which living forms could be arranged in a progressive sequence with humanity as summation and microcosm of the universe. For those so inclined, it was not too great a step to translate the progressive sequence into a history of material change. Haeckel himself incorporated elements of *Naturphilosophie* into his vocabulary. Second, the materialism of the 1850s, as proclaimed by Vogt, Büchner, and the Dutchman Jacob Moleschott, paved the way for a positive assessment of Darwin both in its rejection of a dualistic account of man and its deliberate assault on the God of the theologians. Although the materialists claimed their philosophy was a deduction from science, much of their inspiration came from Ludwig Feuerbach (1804–72), then enjoying notoriety for his argument that images of God came from essentially human projections, which the churches too readily objectified. Feuerbach, and the materialists indebted to him, regarded sensation as the only source of knowledge. They were intolerant of the idealists' claim that the very possibility of coherent experience already presupposed such principles as those of causality and the uniformity of nature, which were as much imposed on the world by the mind as inferred from experience. This suspicion of idealist philosophies was common to Darwin and the German materialists, who quickly saw the attractions of an evolutionary naturalism. During 1860 Büchner was already criticizing Agassiz's position on the fixity of species and calling out for more research on the topic. When Darwin's *Origin* appeared in German translation, Büchner welcomed it as the answer to a prayer. To a German audience, already exposed to the view that the relation of mind to brain was as that of urine to kidneys, the news that humans were somehow derived from apes came as no great surprise.

That so many of the German Darwinists had already abandoned a traditional Christianity may also help to explain the swift assimilation. As we saw in the preceding chapter, Germany was the home of the most radical forms of biblical criticism in which naturalistic assumptions were made of essentially the same kind as those embodied in Darwin's science. The peculiar timbre of Darwinism in Germany has also been related to the 1848 revolution and the consequences of its failure. Defiant liberals, disappointed to see their political hopes dashed, found in scientific materialism an ideological weapon with which to continue the fight. Büchner, Vogt, and Moleschott had lost their university posts (at Tübingen, Giessen, and Heidelberg respectively), accused – in the case of Vogt – of overt revolutionary activity or of corrupting the minds of students with doctrines that undermined morality. Not surprisingly, Vogt became even more venomous in his attack on a system that had excluded him, his philosophical position becoming even more deterministic. During the 1850s Feuerbach had written of the natural sciences as a revolutionary force, having the particular power to dispel belief in the miraculous. Darwin's theory could be construed as the perfect embodiment of that point. Even Haeckel, whose popular works did not shoot a precise political line, nevertheless pitted Darwin against social institutions such as church and school, which he considered the bastions of reaction.

During the early 1870s German Darwinists were able to capitalize on a more pervasive anticlerical, and particularly anti-Catholic, mood that had set in under Bismarck, provoked in part by the declaration of papal infallibility in 1870. There was, nevertheless, a backlash to contend with when Haeckel's determination to have Darwinism taught in schools met with stiff resistance, even from a fellow scientist, Rudolph Virchow, who objected that evolutionary theory did not yet constitute certain knowledge and was therefore inappropriate as part of the curriculum. Teachers who did introduce their pupils to Darwin's theory were liable to harassment, as in the case of Hermann Müller, who took several newspapers to court in order to clear himself of the charge that he was an enemy of Christianity. Part of the appeal of popular Darwinist literature undoubtedly stemmed from the fact that it did have the taste of forbidden fruit. It flourished for fifty or sixty years until, a victim of its own success, it was no longer sustained through engagement with an effective opposition.

Comparative studies of the reception of Darwinism in different European cultures indicate that the popularization of evolutionary science was rarely, if ever, a straightforward process in which the science of an elite simply diffused downward to a mass audience. Darwin's science was actively seized. It was vulgarized in the promotion of particular political goals and these, in turn, often reflected local circumstances. Provoked by the refusal of Roman Catholic authorities in Ireland to heed a request from their own laity that physical science be made part of a Catholic university curriculum, John Tyndall took advantage of his presidential address to the Belfast meeting of the British Association (1874) to launch an offensive. This was the address in which he referred to his meeting with a disillusioned Agassiz. Men of science, he declared, shall wrest from theology the entire domain of cosmological theory. Darwin had shown how it could be done.

Despite the variability of context, one message was put out loud and clear by leaders of secular movements in every European country. Taken to its logical conclusion, Darwin's theory was the apotheosis of a scientific naturalism that simply could not be squared with a historic Christianity. In Germany, Haeckel was adamant that there was no middle ground. It was either Darwinian evolution or miracles. Tyndall took the same line. Detecting a note of prevarication in Darwin's treatment of the first living forms, he stressed that:

We need clearness and thoroughness here. Two courses and two only are possible. Either let us open our doors freely to the conception of creative acts, or abandoning them, let us radically change our notions of matter.[5]

This insistence on a straight choice between incompatible paradigms needs further examination for at least two reasons. First, the power of such rhetoric did leave an indelible impression upon popular opinion. Second, there emphatically were elements in the Darwinian challenge that, to many observers, would make any attempt at compromise appear incongruous.

"Two Courses and Two Only": The "Religion" of Evolutionary Naturalism

The proposition that Darwin's description of natural selection and Christian images of divine activity were fundamentally incompatible was not the invention of secularists alone. In his book *What is Darwinism?* (1874), the Princeton theologian Charles Hodge reached the same conclusion. His was no diatribe against Darwin. He saw no reason in Scripture to reject evolution out of hand. Nor was there any intrinsic objection to the idea of theistic evolution, in which the development of new species was under divine control. In fairness to Darwin, he also acknowledged that neither the theory nor its author were atheistic in the sense that an original Creator was denied. In the last analysis, however, he could not see how a process in which natural selection worked on random variations could be said to be anything other than *effectively* atheistic, since the doctrine of an active providence working to specific designs was evacuated.

Christians who agreed with Hodge could only have their suspicions confirmed by the eagerness with which Darwin's science was welded into a scientistic world-view. Back in 1838 Darwin had recognized that, in one vital respect at least, his theorizing coalesced with the positivist doctrines of Auguste Comte (1798–1857). In his schematic history, Comte had argued that human societies had passed through three stages, the theological, the metaphysical, and, finally, the positive scientific stage when what passed for knowledge had to be expressed in terms of natural laws. Final causes had no place in that final stage. In his metaphysical notebook, Darwin had written: "M. Le Comte argues against all contrivance . . . it is what my views tend to." Darwin did not disallow the possibility that the laws of nature might have been designed, but at the heart of his theory was the recognition that traditional conceptions of divine contrivance were old bottles for heady new wine.

The assertion of incompatibility between a Christian and a Darwinian world-view often hinged on the antithesis between chance and design. The challenge of the Darwinian position was succinctly expressed by the twentieth-century evolutionist G. G. Simpson when he declared man to be the product of a process that had never had him in mind. By "chance" variations, Darwin did not mean occurrences that were uncaused. He sometimes used the word merely to cover his ignorance of the cause. At other times he meant by "chance" the intersection of otherwise independent causal chains. The crux of the matter, however, as for neo-Darwinians in the twentieth century, was that variations, however they were produced, could not be said to be a response to the adaptational needs of the organism. In that respect the process seemed to lack direction.

There was a deeper level, too, at which the very enterprise of a Christian natural theology was put in jeopardy. In his more agnostic moods, Darwin found himself asking whether he should place any confidence in his own convictions, such convictions as that the universe as a whole could not be the result of chance. Moreover, he resorted to his theory to justify his agnosticism. If the human mind were only that bit more refined than the mind of a dog, what guarantee was there that it could resolve metaphysical problems? An evolutionary ancestry for humankind raised formidable problems because it wrecked the belief, defensible on creationist assumptions, that the human mind had been designed for its quest. For Darwin the very enterprise of natural theology had become incurably anthropocentric, reflecting man's arrogance in believing himself the product of special creation.

It should not be difficult to see why intelligent people have often taken the view that Darwin's theory, properly understood, and Christian conceptions of an active providence are not merely incompatible but belong to two mutually exclusive worlds of thought. This does not mean that it was impossible to build bridges between them. Theologians could adapt as they had to the scientific innovations of the past. But from the standpoint of twentieth-century humanism, Darwin's theory was the focus, if not the cause, of a transition into a modern world in which humanity could no longer delude itself that there was a caring providence, that pain and suffering had an ultimate rationale, or that there was any destiny other than engineering the future course of evolution.

This was, however, a destiny of a kind. And if social improvement, even human perfectibility, was grounded in a law of nature, then there was a basis for a secular religion pursued with all the fervor of the sacred. The vocabulary of its exponents does indeed suggest that scientific naturalism could take on the mantle of a religion in which human values were corroborated, if not positively derivable, from the facts of biology. T. H. Huxley would preach what he called "lay sermons" to a public whose consciousness of the value of science he sought to raise. His campaign to gain greater social prestige for the scientific professional, having as its corollary the exclusion of the clerical amateur, can easily be parodied as the bid to create a "church scientific." For Herbert Spencer there was a power behind evolution, an "Unknowable Power" that nevertheless made for righteousness. In 1884 he declared that it was a power that "stands towards our general conception of things, in substantially the same relation as does the Creative Power asserted by Theology." For Darwin's cousin Francis Galton, as for Huxley, it was vital that any prerogatives claimed by the clergy to control the machinery of education should be denied. The pursuit of science, he wrote, is "uncongenial to the priestly character." As a member of the scientific priesthood, Galton had his alternative religion, which he called practical Darwinism. Its creed was a eugenics program, which was also a response to the fears of middle-class intellectuals in Britain that their social values were at risk from the higher rates of reproduction of the poorer classes. The gist of Galton's case was that genius ran in families (like his own), that eminence could be a criterion for constructing a statistical distribution of intelligence, and that the dysgenic should be discouraged from multiplying. Nature needed a little help in the selecting. In Britain, at the turn of the century, variants of Galton's program could hold considerable appeal for intellectuals not deterred by the degree of social administration that was implied. To the Fabian socialist Sydney Webb it provided an attractive recipe for social progress.

The ease with which Darwin's science could be inflated into a naturalistic world-view, and thence into a rival religion, can be seen most clearly in the writings of Haeckel, who cherished a vision in which Christian churches would not be so much empty as taken over by like-minded monists who would refurbish them with symbols of nature and science. The altar would give way to Urania, the Greek muse of astronomy, while the walls would be decked with exotic flowers, trees, and aquaria. In due course Haeckel would be elected antipope by the apostles of free thought. His view of nature, as one historian has observed, "resembled a giant work of art, almost yearning for the creator he kept begrudging it." The possibilities of Evolution as an alternative religion were similarly perceived by a later popularizer, Wilhelm Bölsche, who spoke of the scientific movement as having effected a "Second Reformation." There had been an Old and a New Testament; now there was a third, the testament of science, which transcended both.

Just as traditional religions had their dogmas, so did the religion of evolutionary naturalism. As he reflected on evolutionary theory in 1889, the French Catholic physician Pierre Jousset protested in these terms:

Anti-Christian science has perhaps never been more dangerous than at this moment. Rich in positive knowledge, it has become proud and authoritarian. The intolerance it blames on the Catholic Church has become its supreme law. It imposes its theories as dogmas, its hypotheses as incontestable truths; the dreams of imagination become articles of faith — science is infallible![6]

DANGERS OF DOGMATISM.

Brown (a mild Agnostic, in reply to Smith, a rabid Evolutionist, who has been asserting the doctrines of his school with unnecessary violence). "ALMOST THOU PERSUADEST ME TO BE A CHRISTIAN !"

Figure VIII. 4. Cartoon from *Punch*, 5 June 1880. How an aggressive proevolutionary stance might give rise to divisions within unbelief. Reproduced by permission of the Syndics of Cambridge University Library.

Such sentiments were sometimes expressed by scientists themselves, resentful of the doctrinaire mentality of their colleagues. One such was the Lucasian Professor of Mathematics at Cambridge, George Stokes (1819–1903), who reported in 1883 that Darwin's theory had been accepted by many eminent biologists with an alacrity that was puzzling to one accustomed to the rigor of the physical sciences. As an evangelical Christian who believed that Genesis spoke of successive creations rather than evolution, Stokes found it difficult to sympathize with the Darwinians. He has a place in the history of Victorian religious thought as one of a group of evangelicals committed to making their Christianity more credible by replacing the doctrine of eternal torment with what they considered a more biblical view – that those who had not experienced redemption would simply perish after their earthly life. There were limits, however, to his reforming zeal. That the Darwinian image of nature had been so pervasive puzzled and saddened him as it had Agassiz. It was indeed as if some other explanation for the theory's success was required than could be vouchsafed by its scientific credentials alone. The clue was that its advocates sometimes spoke as if evolution "must" be true, thereby indicating what Stokes described as an "*animus* in the direction of endeavoring to dispense with a Creator."

The *animus* to which Stokes referred probably manifested itself most keenly in the bitter dispute between Huxley and his former pupil Mivart. The career of Mivart was that of mediator and it illustrates the difficulties encountered by one who took that role. He wanted to show that it was perfectly possible for a good Catholic to be an evolutionist. He claimed to have found in the scholastic philosopher Suarez (1584–1617) a precedent for admitting the possibility of creation through secondary causes. And, as we have seen, he took the sting out of Darwin's science by knocking natural selection and by replacing a divergent with a convergent scenario. Huxley, who liked to claim that "extinguished theologians lie about the cradle of every science as the strangled snakes beside that of Hercules," decided it was time to extinguish one more. Plunging into Suarez himself, he found what he had expected and wanted – that Mivart had stretched his meaning to create space for theistic evolution. Huxley's *animus* was undisguised. Evolution occupied a position of "complete and irreconcilable antagonism to that vigorous and consistent enemy of the highest intellectual, moral, and social life of mankind – the Catholic Church." Two alternative courses and two only were open. It was impossible for Mivart to be "both a true son of the Church and a loyal soldier of science."

Excommunicated by the church scientific, Mivart's intellectual pilgrimage eventually resulted in excommunication by the Church whose true son he had tried to be. It is a complex story because initially at least his efforts at mediation were rewarded, not resisted by the papacy. The degree of doctor of philosophy was conferred upon him by Pius IX. In time, however, Mivart's theologizing increasingly bore the print of a scientific rationalism. As with Galileo, some two hundred fifty years earlier, he ventured to suggest that men of science might have a truer perception than ecclesiastical authorities of the true meaning of Scripture. He began to see parallels between clerical attacks on evolution and seventeenth-century attempts to suppress a heliocentric cosmology. When the reaction set in against attempts to modernize the Catholic faith, the parallel became an intensely personal issue. Accused by critics of having misrepresented Catholic principles, he eventually fell foul of the encyclical of 1893, which now prohibited even the evolution of the human body. After years of effort as a peacemaker, he was himself forced to the conclusion that Catholic doctrine and science were "fatally at variance."

Repression of his evolutionary science was not the only source of his disillusionment. In common with the evangelical Stokes, he protested at the doctrine of eternal torment. Surely a merciful God, Mivart argued, would leave some hope, some means of redemption, some respite, for those who found themselves in hell? It was for such liberal views on eternity that he was actually denounced. And he in turn was to denounce his Church. As he observed the persecution of Alfred Dreyfus in France (1899), he became increasingly disturbed by the involvement of the clergy in what would be construed as an anti-Semitic campaign. When the pope himself remained silent, what, Mivart asked, was left of the *moral* authority that his Church arrogated to itself?

Mivart's misfortunes show that when evolution was the issue, the peacemakers were not always blessed. And the same was true for many Protestant mediators. Bridges could be built, but powerful interests on either side were apt to detonate them. Vulnerable though many of them were, some do bear serious consideration because that sense of polarity we have been exploring was certainly not felt by every Christian thinker faced with the Darwinian universe. Claims were made, often repeated since, that evolutionary theory could actually enrich the Christian faith, making certain religious beliefs more realistic.

Evolution within Religion

To impose order on theological reactions to Darwin is a difficult task. There was a marked diversity of response even among members of the same religious denominations. As we saw in the previous chapter, it was possible for two Anglican dignitaries, Samuel Wilberforce and Frederick Temple, to strike quite different attitudes when Darwin's theory invaded Oxford during the British Association meeting of 1860. In America, Charles Hodge (1797–1878) may have concluded that Darwinism was effectively atheistic, but his Presbyterian colleague at Princeton James McCosh (1811–94) wrote reams reconciling evolution with a skillful providence. Nor does it help to suppose that theologians had to be either for Darwin or against. The Calvinist McCosh, like the Catholic Mivart, presented a view of evolution in which the "chance" was taken out of "chance variation" and a more Lamarckian mechanism proposed. A further reason for the complexity lies in the fact that scholars, perhaps more in Continental Europe than in Britain or America, could regard the conclusions of natural science as more or less irrelevant to their theological concerns. In Germany, the school associated with Albrecht Ritschl (1822–89) at Göttingen made religious experience the cornerstone of its theology: Human feelings and human relationships were its subject matter. How humans had come to be humans was neither here nor there. Similarly, those who felt that the essence of the religious life was a sense of contact with a reality beyond the senses were unlikely to be impressed by claims that Darwin had undermined religion. Because apes and humans had ancestors in common, it did not follow that humans were nothing but apes. Precisely because they had evolved to higher planes of consciousness, they had moral capacities and spiritual needs, which membership in religious communions could help to fulfill. This sense that the concerns of theology transcended those of science had been voiced many times in the past, but it was now powerfully expressed by the German theologian Ernst Troeltsch (1865–1923):

No physics and no biology, no psychology and no theory of evolution can take from us our belief in the living, creative purpose of God, and . . . no anti-teleology, no brutality and no fortuitousness of nature . . . can take from us our belief in redemption as the destination of the whole world.[7]

To an outsider this could look like willful blindness to scientific advance, but from within a communion bound by the profession of faith it was an honest affirmation of priorities.

A theology in which the whole world was destined for redemption was one that some scholars found perfectly compatible with a theory of evolution. Indeed, they drew on Darwin and Spencer to enrich their Christian vision. The moves they made deserve attention because they were often accompanied by the claim that insights from evolutionary science could actually resolve outstanding theological problems. This incorporation of evolution within religion added further complexity to the post-Darwinian debates, not least because the strategies adopted by different apologists were often inconsistent among themselves.

In general terms, concepts of theistic evolution could be both plausible and appealing. The parallel that Spencer had drawn between his "Unknown Power" and the "Creative Power" of the theologians almost invited the conflation of the two. Why posit an unknown power behind the evolutionary process when the God of Abraham, Isaac, and Jacob would do just as well? This was very much the view of Henry Drummond, whose fusion of Christianity and science we encountered at the beginning of Chapter I. Both Christianity and evolution could be said to have the same object: the creation of more perfect beings. Then again, it was possible to treat the Genesis story as "myth" without having to renounce the view that it contained authoritative insight into both the dependence of the world on a Creator and the alienation of man from his Maker. With the privilege of hindsight, it was even possible to say that the making of man from the dust of the earth was a poetic way of expressing the truth that man was a product of evolution rather than creation ex nihilo. More significantly, however, evolutionary motifs were brought to bear on a range of theological problems: the nature of biblical authority in the light of the "higher criticism," the meaning of "original sin," the meaning of God's immanence in the world, and the problem of suffering.

A general evolutionary framework helped in formulating replies to radical biblical critics. By admitting the evolution of religious sensibilities and coupling it with the concept of "progressive revelation," it was possible to ditch obsolete notions of biblical inspiration but at the same time to retain a privileged role for the Bible as a historical record of spiritual ascent – toward a more refined understanding of God. The belief that Scripture revealed the evolution of human spirituality was not, of course, derived from biological theory. Even in England, the notion of progressive revelation had been developed early in the nineteenth century. The Oxford mathematician and philosopher Baden Powell made use of it to emphasize the culture-dependence of Old Testament cosmology. Liberal theologians nevertheless saw themselves as completing Darwin's work by documenting, through Scripture, the spiritual evolution of humankind.

As a solution to the problem of scriptural authority, this had both conservative and radical possibilities. Among conservatives a strong sense of divine initiative was retained by insisting that God had revealed His truth gradually, as His people were able to understand it. More radical possibilities emerged with the question whether the religious experience of the gospel writers could be said to be normative for subsequent generations. A new line of approach was to regard the Scriptures as the raw material for a psychological analysis of the religious consciousness as it had passed through its several phases – without construing the notion of revelation in terms other than the mind's own intuitions of an unseen reality. This line, taken by Troeltsch, promised fresh insights in the field of comparative religion. Troeltsch could argue that each of the world's religions had its own distinctive development, but underlying them all was a common impulse. This was neither primitive animism nor the craving for happiness. Nor was it the speculative impulse to explain the world. It was rather a mystical and intuitive contact with a supersensory being. A history of religions showed that what Troeltsch called "the presence of God in the human soul" had assumed varied forms of expression. One could nevertheless detect a sequence and progression in ethical and spiritual terms. Early in his career, Troeltsch had looked for a common essence in all religion as a way of asserting the superiority of Protestant Christianity. Later, however, he shifted to the more pragmatic view that Buddhism was as "true" for Buddhists as Christianity for Christians.

Linked to the problem of biblical authority was that of original sin. The difficulty was that Genesis referred not merely to the creation of humankind, but also to the Fall of Adam, which had repercussions throughout the Bible – notably in the New Testament interpretation of Christ's mission as the "second Adam" atoning for human sin. On one level Darwin had made the problem worse. If man had risen, not fallen, what would be left of the scheme of redemption? How could Christ be the second Adam if there had never been a first? If human beings had been created by natural evolutionary processes, would this not place upon the divine author of those processes (rather than humankind) the responsibility for their sinful state? Conservative evangelicals have been particularly sensitive to such difficulties. Among nineteenth-century liberals, however, evolutionary theory offered the hope of a realistic redescription of the traditional doctrine. Spencer's contention that sin was merely a vestige of an animal past, which the future progress of society would eliminate, may have seemed tendentious, but to redescribe sin in terms of vestigial animality was not to deny that it existed. Of human self-centeredness and self-assertion Huxley wrote that it had been the condition of victory in the struggle for existence. But as an inheritance it was real enough, nothing less than "the reality at the bottom of the doctrine of original sin." It has even been said that for Darwin himself, savage man was fallen man, in the sense that he was appalled by behavior more wretched than anything he had witnessed in the animal world. In 1897 Lyman Abbott, who has been described as America's outstanding representative of theological liberalism, published his book *The theology of an evolutionist.* It is clear that he saw several advantages in a theology informed by Darwinism, one of which was that the traditional notion of sin could be reformulated in modern terms: New meaning could be given to the old adage that immoral acts were lapses into animality.

A theology informed by Darwinism could also throw light on the question of God's involvement in the world. Despite the challenge to traditional forms of the design argument, Christian writers sometimes rejoiced that they now had a more sophisticated understanding of how God could be both transcendent over and immanent in His creation. The laws of nature that allowed the evolutionary process to occur were the work of a transcendent Being; yet the emergence of novelty pointed to the continuous participation of that Being in what was an unmistakably creative process. Apologists who developed this point often observed that a mechanistic universe, into which an external God occasionally intruded with miraculous acts of creation, had the defect of emphasizing

His transcendence at the expense of His immanence. Evolutionary perspectives allowed a more balanced view. That, at least, is how Charles Kinglsey came to see it. If God were truly Father of mankind, then His relationship to the world could not be one of mechanic to machine. Creation could not mean, as it often had, so many acts of a supermagician. An evolving universe provided tangible symbols of a God continually involved. A theology in which the doctrine of the Incarnation came to the fore, in which the presence of God in the world was stressed, could be quite hospitable toward evolutionary science. The British theologian Aubrey Moore had this to say in his contribution to the theological essays *Lux mundi* (1889):

The one absolutely impossible conception of God, in the present day, is that which represents him as an occasional visitor. Science has pushed the deist's God further and further away, and at the moment when it seemed as if He would be thrust out all together, Darwinism appeared, and, under the disguise of a foe, did the work of a friend.[8]

Either God was everywhere present in nature, Moore continued, or He was nowhere.

The idea of God as a sympathetic participant in the evolutionary process was developed in different ways. Attempts to specify precisely how He was involved usually ran into trouble, as when Asa Gray suggested to Darwin that, in ignorance of the source of variation, it would be wise to suppose that God was responsible. Such a god-of-the-gaps was always liable to become a casualty of further scientific advance. But where there was evidence of convergent trends in the process, the belief in a divine participant was not obviously vacuous. The British physiologist William Carpenter claimed to have found such evidence in his comparative study of ancient and modern forms of *Foraminifera,* a group of simple marine organisms. Constructing a hypothetical pedigree for the group, he showed how the simplest spiral shell had gradually developed into a complex cyclical form through an apparently regular progression. His conclusion was that such evolution had taken place along a definite course, every stage having been one of progress, and each a preparation for the next. The insufficiency of natural selection was underlined by stressing that all the members of the series were still extant. If each step had to correspond to an increase in fitness, the earlier members would have been exterminated. Such forms of reasoning, leading to a theological conclusion, became increasingly rare among biologists in the late nineteenth century, but they remained attractive to religious apologists.

Even the elements of chance in the Darwinian mechanism were turned to theological advantage. If the human race were the product of so many contingencies, then either it was a more monumental fluke than atheists had ever dreamed, or, as McCosh wryly put it, the prevalence of accident could not be accidental. Darwin did not so much destroy the concept of design as sharpen the choice between chance and design. The point that struck McCosh was that the evolution of man had required "adjustment upon adjustment of all the elements and all the powers of nature towards the accomplishment of an evidently contemplated end." The argument was worthless to those who had renounced belief in that "evidently contemplated end." But to those who had not, a sense of divine involvement could become more, not less, real. The sheer improbability of the emergence of man deeply impressed Alfred Russel Wallace. Contingency had piled upon contingency with each critical stage in evolutionary divergence. In a book written late in life, *Man's place in the universe* (1903), he turned the argument against physicists and astronomers who were scouring the heavens for planets having a physical environment comparable with that of the earth and on which intelligent life might be presumed to have evolved. Properly understood, Wallace argued, the theory of evolution told against such a possibility – certainly against the emergence of intelligence akin to human. However close the physico-chemical environment to that on earth, it was inconceivable that the evolutionary process on other worlds could have followed the same nuanced path as on earth. One minor deviation at an early stage and the whole process would take an entirely different course.

The argument that the process had been so chancy that it could not have been left to chance was not the only way of preserving providence. The American philosopher of science C. S. Peirce (1839–1914) was prepared to accept that there *is* an element of indeterminacy, spontaneity, or absolute chance in nature. Far from contradicting a personal Creator, such a possibility was in keeping with a basic feature of personality. The execution of predetermined ends was a purely mechanical process, leaving no room for development or growth. If there was a personality behind the universe and not merely a predetermined program, diversification, spontaneity, and potential for growth were precisely what one would expect. The recognition that Darwinian evolution was "evolution by the operation of chance" meant for Peirce that it was inseparable from the idea of a personal Creator. Darwin himself would not have gone that far, but there were times when he had considered the possibility that a universe in which the laws were designed, but the details left to chance, might relieve the Creator of direct responsibility for the more macabre features of creation.

Any serious discussion of divine activity in the world had to address itself at some stage to the problem of suffering. It was another of the problems that, at one level, Darwin's theory acccentuated. The presence of so *much* pain and suffering in the world Darwin considered to be one of the strongest arguments against belief in a beneficent God. But, he added, it accorded well with his theory of natural selection. It was what one would expect from a process driven by a relentless struggle for existence. Here was a genuine difficulty for the Christian evolutionist. Would a merciful God have instigated such a tortuous, wasteful, and bloodstained scheme? No one wrestled with the problem more strenuously than Darwin's friend George Romanes (1848–94), one of the pioneers in the study of animal behavior. Romanes kept vacillating between theism and skepticism. In his *Thoughts on religion,* published posthumously in 1895, he drew a poignant contrast between the personality of the deity as inferred from evolutionary biology and the qualities of love, mercy, and justice as proclaimed in the highest forms of religion. The two sets of qualities, he wrote with a sense of despair, were almost exactly opposites. One can see why it has been said that the theologians' problem had become Darwin's solution.

But if this was so, then there was arguably a sense in which Darwin's solution had also to be the theologians' solution. There was another level at which Darwin's science could come to the aid of the apologist rather than embarrass him. This point was grasped by Asa Gray (1810–88), who did so much to make evolution respectable in America. However much waste and suffering there may have been in nature, Gray argued, it was more intelligible when analyzed from a broader Darwinian teleology than from a creationist platform:

Darwinian teleology has the special advantage of accounting for the imperfections and failures as well as the successes. It not only accounts for them, but turns them to practical account. It explains the seeming waste as being part and parcel of a great economical process. Without the competing multitude, no struggle for life; and, without this, no natural selection and the survival of the fittest, no continuous adaptation to changing surroundings, no diversifications, and improvement, leading from lower up to higher and nobler forms. So the most puzzling things of all to the old-school teleologists are the *principia* of the Darwinian.[9]

Precisely because waste, pain, and suffering were endemic to the creative process they could be given a new rationale. Nor was Gray alone in arguing that Darwinism could liberate natural theology from common objections. In England, Frederick Temple conceded that the separate creationism underpinning Paley's argument from adaptation to design was vulnerable to Hume's objection that one might as well infer that several designers had been at work. So long, however, as Darwinian evolution could be construed as a unified process in which potential was actualized in higher organic forms, the inference, he suggested, had to be to a single designer. Natural theology had not yet lost its power to corroborate a preexisting faith.

The Question of Truth

Attempts to show that Christianity could be enriched by evolutionary theory did not define the only way forward for scholars anxious to create space both for science and for their religious heritage. In America the pressure to create that space was nowhere more keenly felt than among the circle of Boston intellectuals with whom the young William James became acquainted in the 1860s. The impact of Darwin had been nothing less than shattering. James himself recorded that day after day he had awaked with a feeling of horrible dread. It seemed that the foundations of morality had collapsed, the freedom of the will fallen victim to scientific determinism. If the sense that one could control one's thoughts was pure illusion, what release was possible? By April 1870 James had snapped out of his despair. "My first act of free will," he told himself, "shall be to believe in free will." But that still left unanswered the burning question of the relationship between scientific and religious truth. The attempt to accommodate both was a powerful stimulus and it was in part responsible for the creation of a new philosophical system — the pragmatism with which James is associated.

The fundamental tenet of pragmatism was set down by C. S. Peirce: Beliefs were not to be understood as mental entities but as habits of action. They could be analyzed by examining behavior. Different kinds of belief could then be distinguished according to the different modes of action to which they gave rise. Extending the scope of Peirce's work, James looked to the use that belief in a particular concept might have in benefiting a person's psyche. He tended to assume that behavior inspired by belief in a deity was beneficial to society. Whereas advocates of scientific naturalism made adjustment to the known laws of nature the touchstone of what was morally acceptable, James argued that the supreme good lay in adjustment to that "force-greater-than-ourselves" which was assumed in the religious confessions. In his celebrated study *The varieties of religious experience* (1902), he insisted that the essence of religion did not consist in theological reasoning but in the promise of richer and more satisfying lives consequent upon the assurance that this unseen force was aligned on one's side when fighting moral battles. A belief that was luminous, reasonable, and morally supportive could be "verified" if it did have consequences for one's life. Belief in God, if accompanied by action on the basis of that belief, went a very long way toward making the belief "true." To explain how religious beliefs had arisen was emphatically not to explain them away. They had consequences by which they could be corroborated, as did scientific hypotheses in their contexts.

In the conclusion to his study of religious experience, James explicitly discussed the relationship between scientific and religious beliefs. He acknowledged that anthropologists often regarded the world's religions as anachronistic "survivals" from a prescientific age. It was not difficult to understand such an attitude, for religion did perpetuate traditions of primeval thought. To get the spiritual powers on one's side had, for enormous tracts of time, been *the* great object in dealings with the natural world. He acknowledged, too, that the days had gone when it was true for *science* that the heavens declared the glory of God. Darwinism symbolized a new "temper of the scientific imagination," unable to find in the drifting of cosmic atoms anything but a kind of aimless weather. The old books on natural theology now seemed grotesque, boasting a God who conformed the largest things of nature to the paltriest of human wishes. But for James all this was gain rather than loss. In a popular essay of 1895, addressed to the Harvard Young Men's Christian Association, he argued that to jettison the God of physico-theology was the very act of rebellion required in order to achieve "healthy ultimate relations with the universe." The need for rebellion arose from the fact that the God of religious experience was not an intellectual inference, but a God with whom one entered into union or harmonious relations as one's true end in life.

However successful the natural sciences might be in articulating general laws of nature, the deepest realities experienced by human beings would always be those pertaining to their inward life. For James, there had to be room for both science and religion because each satisfied different human needs, the former a craving for ideal mathematical harmonies, the latter a craving for something more spiritual and eternal than the world of nature itself. And if human needs outrun the visible universe, why, he asked, may not that be a sign that an invisible universe is there? In keeping with his pragmatism, he would suggest that "Does God exist?" "How does he exist?" and "What is he?" were so many irrelevant questions. The real question was whether those who lived in a state of faith actually experienced an enrichment of life, a greater endurance and moral courage. He had no doubt that they did.

In studying the experience of those who subscribed to an invisible universe, James was deeply influenced by psychical research and by a science of psychology that purported to establish different levels of human consciousness. That there were such levels he did not consider a threat to religion. It was rather an indication that individual consciousness might graduate into a wider realm of consciousness, which religion ascribed to higher spiritual powers. From his study of conversion and other mystical experiences, including prayer, he concluded that invasions from the subconscious played a striking part in the religious life. The concept of a deeper, subliminal self allowed him to elucidate one of the more general characteristics of religious experience — the awareness in the believer that the better part of oneself was continuous with a "more" of the same quality operating in the universe beyond. James proposed that "whatever it may be on its *farther* side, this 'more' with which in religious experience we feel ourselves connected is on its *hither* side the subconscious continuation of our conscious life." In this way the theologian's contention that the religious man is moved by an external power could be vindicated, for it was one of the peculiarities of invasions from the subconscious to take on objective appearances and to suggest to the subject an external control.

Of one thing James was truly convinced. A religious interpretation of life could not possibly be vanquished by the forces of scientific rationalism. In an address to the Yale Philosophical Club in 1891 he contrasted a "strenuous mood," which arose when religious belief conferred objectivity on moral imperatives, with an apathetic mood when religious belief was absent. His conclusion may not have been welcome to every member of his audience, but he seemed to be saying that in a Darwinian universe it was the religious who were best fitted to survive:

Every sort of energy and endurance, of courage and capacity for handling life's evils, is set free in those who have religious faith. For this reason the strenuous type of character will on the battle-field of human history always outwear the easy-going type, and religion will drive irreligion to the wall.[10]

How different from the common view that nineteenth-century scientific naturalism had driven religion to the wall.

Science and Religion in the Twentieth Century

Introduction

The principal aim of this book has been to reveal something of the complexity of the relationship between science and religion as they have interacted in the past. Popular generalizations about that relationship, whether couched in terms of war or peace, simply do not stand up to serious investigation. There is no such thing as *the* relationship between science and religion. It is what different individuals and communities have made of it in a plethora of different contexts. Not only has the problematic interface between them shifted over time, but there is also a high degree of artificiality in abstracting the science and the religion of earlier centuries to see how they were related. As we saw in the opening chapters, part of what was meant by natural philosophy in the seventeenth century involved a discussion of God's relationship to nature. Religious beliefs could operate *within* science, providing presupposition and sanction as well as regulating the discussion of method. They also informed attitudes toward new conceptions of nature, influencing the process of theory selection. In the eighteenth and nineteenth centuries, despite vigorous attempts to separate scientific and religious discourse, the meaning attributed to scientific innovations continued to be reflected in the often conflicting social, religious, and political uses to which they were put. Part of the meaning of Darwin's theory for T. H. Huxley lay in its potency as an antidote to what he saw as the poison of Roman Catholicism.

At first sight, this complicated story would seem to have reached a much simplified conclusion in the twentieth cen-

321

tury. With the emergence of more pragmatic conceptions of truth, such as we have just observed in William James, it has been much easier to insist on the mutual irrelevance of scientific and religious discourse – each reflecting distinct practices and preoccupations. Scholars who have followed Wittgenstein in analyzing the functions of language have recognized the several levels on which it may operate: The world may be described in many different ways, without any one having to be reduced to another. On such a view it is possible, in principle, for science and religion to coexist without mutual interference.

In this concluding postscript we examine a small sample of twentieth-century developments that are widely perceived to have a bearing on how the relations between science and religion should be discussed. The coverage is bound to be superficial; but, however disparate they may seem, the topics selected share a common and recurring theme – whether a philosophy of scientific reductionism is sufficient to meet human needs. We begin with a brief inspection of Freud's critique of religion, which had the effect, certainly for Freud himself, of disqualifying religious belief from functioning as a complement to the conclusions of science. But we also consider developments in particle physics; for, within science itself, a need was felt for complementary descriptions of subatomic states, thereby reopening the debate concerning complementarity between scientific and religious discourse. For many commentators, another consequence of the revolution in physics has been a rekindling of interest in holistic conceptions of reality, so it is appropriate here to try to explain what this has meant and how it has impinged on theological discussion. Finally, we consider a question that has been implicit in much of our discussion, but which has recently acquired a new urgency: the relevance of human values in shaping priorities for scientific research and in directing the application of scientific knowledge.

Freud on Religion

At first sight, the intensification of both individualism and secularism in Western cultures has had the effect of reinforcing boundary fences – with religious language games often dismissed as not merely different but defunct. The twentieth century has witnessed that long process of cultural readjustment in the West that has sometimes been described as de-conversion — a learning to live with an absent God. In earlier societies, a sense of moral purpose was usually associated with commitment to some communal organization. Religious belief could fulfill a therapeutic role through the transformation of the individual, sometimes through an intense experience of conversion, more commonly through membership of a church or communion with its own mechanisms for forgiveness and release. One of the distinctive features of modern Western societies has been the emergence of an individualism suspicious of all institutions, religious or otherwise, that pretend to be normative.

A tolerant individualism would still leave room for religious practices. The pragmatic philosophy of William James was to prove a consoling doctrine for many twentieth-century Americans, since any religious belief was to be permitted if it met a cry for help. But the twentieth century has also witnessed a less tolerant individualism, with religious beliefs disparaged for their alleged irrationality. The patron saint of that more secular mentality was not James but Sigmund Freud (1856–1939). He, perhaps more than anyone, provided an analysis of religion that would allow twentieth-century thinkers to rationalize their unbelief. In Freud's celebrated opinion, humanity had to be liberated not *by* religious belief but *from* it. Any value that religions pretended to confer on human life had to be discounted.

Freud frequently observed that those who still believed in a caring God focused their illusion on the figure of an exalted father. He found it painful to reflect that the majority of mortals would never rise above so infantile a view of reality. In the spirit of much Enlightenment philosophy, he placed the highest value on the quest for truth enshrined in the sciences. But the quest for a still higher meaning and value in life was a symptom of neurosis. He apparently liked the American advertisement that read: "Why live, when you can be buried for ten dollars?"

Since Freud's vocabulary has become common currency, it is hard to recapture the challenge that he issued to prevailing models of self-understanding. Transferring the Oedipus complex from his own soul to the soul of primitive nations, he argued that at the origin of all civilization had been a primordial crime — the murder of a father by his sons, who desired sexual relations with their mother. Every facet of civilization, its art and science as well as its religion, was ultimately a consequence of sexual repression. The blow to human self-esteem, capping those administered by Copernicus and Darwin, was the realization that man was no longer master of himself. He was victim of the conflicts between different levels of his mind, the product of childhood repressions over which he had no conscious control.

Freud's treatment of guilt had also challenged a Hebraic–Christian understanding. On at least two levels there was an intriguing similarity. His insistence on a primordial crime in every civilization bore so close a resemblance to the doctrine of original sin that Freud could describe it as the dogma on which he built his faith. Nor did he deny the guilt feelings of the pious. The difference came over their interpretation. Within the Christian traditions, guilt was not merely a feeling but a state in which all humanity was bound by the failure, experienced by every individual, to live in perfect obedience to divine commands. The only release could come through divine forgiveness. It was right that men and women should have a deep sense of guilt because they *were* unworthy of God's love. It was a potent force in bringing them to repentance. By contrast, the object of Freud's psychoanalysis was to help patients understand their guilt. If it could be shown to have arisen naturally, it could be deprived of its potency.

Describing himself as a "completely Godless Jew," Freud developed a strong aversion to Christianity, and to Roman Catholicism in particular. This is likely to have been rooted, in part, in the persecution faced by a Jewish family in a predominantly Christian culture. But it also reflected the conviction that to turn the other cheek was too weak a response to the kind of insult his father had experienced when his hat had been knocked off in the street. The inadequacy of his father's submissive response apparently made a deep and lasting impression on him.

The vigor of Freud's critique of religion owed much to the reductionist rhetoric he employed. A large part of the mythological component of modern religions was "nothing but" a psychological projection. God was "nothing other than" an exalted father. The events of human history were "no more than" a reflection of the conflicts between the "ego," the "id," and the "superego." Freud's critics would cavil at this reductionism, especially if they took a more sympathetic view of the therapeutic power of religious commitment. When, however, they declared that, on the basis of Freud's own theories, his sentiments toward religion could be dismissed as *mere* wish-fulfillment, they were guilty of the same derogatory device.

With the analyst as surrogate priest, it is tempting to bring out the parallels between the psychoanalytical movement and previous religious structures. Of the early meetings of the Vienna Psychoanalytical Society, Max Graf could say that there was "an atmosphere of the foundation of a religion in that room." Wilhelm Stekel described himself as "the apostle of Freud who was my Christ." The correspondences are indeed striking. What had once been the state of the soul became the state of the mind. If the soul had once transcended the conscious mind, so did the unconscious as Freud described it. Where the Christian's principal concern had once been his own salvation, it was replaced in a secular society with the goal of survival. What had once been treated as sin would be treated as sickness. The honesty that the sinner brought to the confessional had its counterpart in the honesty without which the services of the analyst would be pointless. If religion, in the judgment of Karl Marx, had once been an opiate for the masses, faith in psychoanalysis would eventually be described as the new opiate of the bourgeoisie.

Such parallels are instructive and suggestive. It is, however, necessary to add the qualification that the subsequent history of psychoanalysis has seen many movements and therapeutic practices, which, deviating from Freud, have re-mapped the unconscious mind to make room for what one critic of Freud, Roberto Assagioli, calls the higher unconscious. Evidence for this stratum of the unconscious is said to consist in those irruptions into the conscious mind of material with seemingly little or no connection with previous experiences. Its existence is that which allegedly makes possible the attainment of higher states of awareness and spiritual consciousness, associated with the highest forms of human creativity. The object of such systems of "psychosynthesis" is not to defend religious institutions. But the symbols of traditional religions (such as the "inner Christ" in certain forms of Christian piety) are valued in the treatment of individuals for whom they have meaning. Inevitably, such eclectic techniques are seen by hard-line Freudians as dilutions of Freud's original insights.

Freudian psychology has not, of course, supplied the only resources by which to rationalize unbelief. Vienna also saw the rise of a philosophical movement – logical positivism – by the canons of which religious language was hopelessly defunct. Compared with scientific propositions, which were empirically verifiable, propositions about God were not strictly testable. They could be dismissed as meaningless, even nonsensical. Although Freudian explanations for human behavior were to be attacked for their unfalsifiability, Freud had captured the spirit of a positivist outlook when he had written that his science was no illusion, but "an illusion it would be to suppose that what science cannot give us we can get elsewhere."

This scientistic attitude has had a good run for its money in twentieth-century societies. It has permeated popular cultures and has fed on technological success. But it is impossible to be an informed citizen in the late twentieth century and to imagine that such neat dichotomies between science and nonscience can be sustained. Ironically, it is partly because of Freud's provision of what by some standards could be described as a "nonscientific" interpretative discourse that such dichotomies have collapsed. Our complicated story appeared to have a simple resolution, but appearances can be deceptive. Despite the pressures to insulate scientific and religious vocabularies, there have been profound changes in our understanding of science itself that have created space for renewed dialogue between scientist and theologian.

Toward a New Understanding of Science: The Revolution in Physics

Positivist accounts of scientific theory have been successfully challenged by research in the history, philosophy, and sociology of science. It is no longer possible to regard scientific theories as independent deductive systems, each proposition of which acquires its meaning by infusion, as it were, from the verifiable facts with which it ostensibly connects. Theoretical constructs, appearing in different branches of science, have been shown to be interdependent, and also underdetermined by the data they purport to explain. It has been necessary to adjust to the idea that the concepts of theoretical science are linked together in complex networks, with some components of the network more open to modification than others. In fact new parallels have been drawn between scientific and religious beliefs – in the sense that, in both, one often finds a protected core of received wisdom surrounded by belts of more negotiable doctrine. This does not, of course, entitle religious concepts to intrude into scientific conceptualization. But the realization that theories are underdetermined by their supporting data explains one phenomenon that has been stressed throughout this book — that aesthetic and religious beliefs *have* played a selective role in the past, giving priority to one theoretical model rather than another.

In creating the space for a less scientistic view of science, the development of subatomic physics has played a central role. In a famous dialogue with Einstein, the Danish physicist Niels Bohr (1885–1962) argued in 1935 that recent developments in quantum mechanics demanded a complete renunciation of the classical ideal of causality and a radical revision of attitudes toward the problem of physical reality. A British astronomer and Quaker, Arthur Stanley Eddington (1882–1944), even made the extraordinary remark that religion first became possible for a reasonable scientific man about the year 1927.

A central assumption of classical physics had been the possibility of achieving a description of the world that was essentially independent of the means by which it was investigated. The physicist was a creator in the sense that he constructed the mechanical models and the mathematical equations with which to describe a world external to himself. But he was quintessentially a spectator, detached from the world that he probed with ever more sophisticated instruments. This did not mean that nineteenth-century physicists eagerly identified their models with physical reality. They were usually sophisticated enough to present them as tentative, exploratory devices. But the idea that nature is picturable was widely assumed, especially by British physicists who shared the opinion of William Thomson (Lord Kelvin) that a mechanical model was essential to the understanding of every physical phenomenon. As with the kinetic theory of gases, the mathematical formalism might express statistical regularities to which large groups of molecules conformed. But there was scarcely any doubt that the behavior of individual molecules was rigorously determined by the forces to which they were subjected. And if the character of physical explanation was deterministic, it was often reductionist too. The macroscopic phenomena associated with chemical and biological systems were often considered reducible in principle to the laws governing microscopic particles. It was these ideals of picturability, determinism, and reductionism that were challenged by a new physics.

The essential dilemma was publicized by Einstein. In its propagation, light behaves like a wave. But in its interaction with matter it resembles a particle. During the 1920s, it became clear that electrons as well as photons could exhibit both wave and particle characteristics. Their behavior was not schizophrenic. One always knew which aspect of their personality would show up under which experimental circumstances. But there was still a dilemma. Did it make sense to ask what photons and electrons were in themselves? And what kind of limitation was implied by a negative reply? In 1927 Werner Heisenberg (1901–76) answered that question with the enunciation of what became known as his uncertainty principle.

Heisenberg's argument was that certain pairs of variables were linked in such a way that attempts to measure them at the subatomic level were subject to constraint. One could determine the position of an electron, but in so doing one lost the opportunity to determine its momentum. To measure momentum with great accuracy one had to be prepared to accept a very inaccurate measurement of position. The limitation can be illustrated with a thought experiment. Suppose one tries to measure the velocity of an electron by bouncing a low-frequency wave off it. By listening to the echo, it would be possible to determine whether the electron is approaching or receding. But the trouble with a low-frequency signal is that the large wavelength limits the precision with which the electron can be located. Why not shorten the wave length of the outgoing signal? True, the position could then be more precisely determined. But the shorter the wavelength of the signal, the more it disturbs the electron's momentum — the very quantity one is also trying to measure. At the quantum level, the very act of measurement can interfere with what it is one is trying to measure. Heisenberg later recalled lengthy discussions with Bohr that had ended almost in despair. He had repeatedly asked himself whether nature could possibly be so absurd.

To minimize the absurdity, Bohr focused attention on the measuring instruments — pointing out that they defined the conditions under which different manifestations of a subatomic entity were produced. Instead of seeing the apparatus as the means of disclosing a hidden reality, it had to be seen as a constitutive part of the reality under investigation. Whether an electron behaved as a particle or a wave was determined by the experimental setup. To amplify his argument, Bohr introduced a concept of "complementarity," soon to become important in theological discussion. Information obtained under one set of experimental conditions was said to be *complementary* to that obtained under another when the information related to the same object and when the two sets of conditions were mutually exclusive. Bohr could then argue for the retention of classical terms for the description of atomic phenomena. It was simply that information obtained under different experimental conditions could not be comprehended within a single picture.

Just how revolutionary such proposals could seem can be gauged by the reaction of Einstein, who would not renounce his belief in the reality of an external world controlled by causal mechanisms that it was the object of science to disclose. Underlying Einstein's objection was the conviction that, if only more were known about the laws governing subatomic entities, a causal account of their behavior could be given, which need not be compromised by the very act of observation. In a celebrated paper of 1935, Einstein asked whether quantum mechanical descriptions of physical reality could be considered complete. The paradox that so disturbed him was this. Suppose two subatomic particles, with a large momentum, interact at a known position. Later, an observer might track one, far from the point of interaction. He could then choose whether to determine its position or momentum. By Heisenberg's principle, he could not simultaneously know both. If he chose to measure its momentum, he could, at the same instant, know the momentum (but not the position) of the particle with which it had interacted. If he chose to determine its position, he could also fix the position (but not the momentum) of the partner. The perplexing question was how the final state of that second particle could apparently be influenced by a measurement performed on the first, when all physical interaction had ceased. Both the position and the momentum of the second could be predicted with certainty, depending on which measurement was made on the first. They could therefore be considered elements of physical reality. But, by Einstein's criterion, a theory could only be complete if every element of that reality had a counterpart in the theory. And a simultaneous specification of both the position and momentum of the second particle was precisely that which was precluded by quantum theory. Therefore the quantum theory must be incomplete.

It was a vigorous attempt to defend a relationship between physical theory and physical reality that, until the twentieth century, had been almost axiomatic. But Bohr had a rejoinder. The error consisted in treating the second particle as an isolated object when it should be considered as a part of a *system* incorporating both the first and whichever measuring instrument had been used on it. Quantum mechanics required a holistic approach, as Bohr later explained: "The feature of wholeness typical of proper quantum phenomena finds its logical expression in the circumstance that any well-defined subdivision would demand a change in the experimental arrangement incompatible with the definition of the phenomena under investigation." That license for a holistic philosophy was to reinvigorate religious critiques of reductionism.

In the new physics, it still made sense to say that one state of a physical system gave rise to the next. But because a quantum mechanical description could not be given without reference to probabilities, it seemed reasonable to conclude that each state was only one of several possibilities permitted by its antecedent. Heisenberg accordingly argued that the new physics had restored the concept of potentiality. It had shattered the famous illusion of Laplace, who had argued that from the known laws of mechanics, and from a full knowledge of the current state of the universe, every future state could in principle be predicted. Insofar as theologians had been chilled by that vision, they could warm to new possibilities.

Several features of the new physics have been exploited in theological discussion. First, the recognition that physicists dealt with models of an elusive reality, and that no one model could give an exhaustive account of subatomic phenomena, allowed a little humility to enter the dialogues between scientists and theologians. In *The physicist's conception of nature* (1958), Heisenberg spoke of a modesty that science had largely lost during the nineteenth century but which modern physics had helped to restore. In Bohr's view, too, the limitations of a mechanical conception of nature had been decisively exposed. As long as theologians were willing to reformulate their doctrines in terms of models rather than absolutes, there really were new opportunities for truce.

The indeterminacy of the new physics was a second feature to create extra space for dialogue. Was it possible to give new meaning to concepts of human freedom and divine activity? Was it possible, for example, to correlate the freedom of the will with the indeterminacies inherent in any physical description of the brain? Sophisticated attempts have been made to show how such a correlation might be possible but they have not enjoyed a wide following. Most have been open to the objection that the occurrence of chance or random events in the microstructure of the brain would detonate responsibility for one's actions rather than secure it. Nevertheless, the physicist James Jeans (1877–1946) was not alone in expressing a feeling of liberation, as if classical physics had imprisoned the human spirit.

A third feature of the new physics to attract theological attention was the metaphor of complementarity. Because Bohr applied it in many contexts other than physics, theologians had few inhibitions in adopting it for their purposes. If complementary modes of description were required within science itself, why not allow them between science and religion? Why should two accounts, say of human origins, or of human nature, not complement each other rather than be seen as mutually exclusive? The suggestion was hardly new. Even within a mechanistic framework, seventeenth-century natural philosophers had seen no contradiction in giving both a mechanical and a providentialist explanation for one and the same event. Nor was the new version of the argument free from difficulty. After all, the whole point of the complementarity principle within physics was to stress that the two descriptions were mutually exclusive if specified simultaneously. Skeptics would also point out that within physics each of the descriptions (if it included the measuring apparatus) was complete within itself. To switch from complementarity at the same level to complementarity between different levels (especially if one involved the transcendent) was not a straightforward move.

It has to be said that religious apologists have not been as circumspect as they might have been in drawing lessons from the new physics. The mathematical formalism of quantum mechanics is open to different philosophical interpretations, making it rash to extrapolate from one alone. The old trap of the god-of-the-gaps has had its victims among those who tried to correlate God's activity with the actualization of one, rather than another, set of physical possibilities inherent in particular states of the subatomic world. Other apologists have fallen victim to a form of circularity, which it is important to expose.

The problem of circularity had often arisen in the past. In Newton's mechanics, the universality of the law of gravitation had been justified by reference to God's unity and omnipresence. And then the next generation of natural theologians had proved the unity of the Godhead from the unversality of gravitation! In the case of the new mechanics the circularity was rarely so blatant, though it was surely there in Sir James Jeans's remark that the universe had begun to resemble a great thought more than a great machine. The critical point is that a specific cultural output might be obtained from the new physics, but only because there had already been a cultural input.

The example of Bohr is instructive because the meaning he attached to quantum physics, via his principle of complementarity, was sustained by considerations drawn from other contexts. Physiology, psychology, even philosophical theology provided insights that he felt would support his interpretation. Physiology was the subject in which Bohr's father occupied a chair at the University of Copenhagen. He was known to take a distinctive view on the question whether living organisms could be completely described in mechanical terms. Denying that mechanistic explanation rendered teleological explanation superfluous, Christian Bohr had argued for their complementarity. For an exhaustive account of animal behavior, both were required. The frequency with which Niels Bohr stressed the parallel with quantum models suggests that his father's stance had made a deep impression upon him.

By Bohr's own admission, another source of insight had been the psychology of William James, whose conception of a stream of consciousness had caught his attention for the limitations it implied for introspective analysis. To interrupt the stream, by looking, as it were, at one's own thoughts before they reached their terminus, was to destroy them. Subject and object were intimately bound together. The role of the subject in disturbing the object was central in Bohr's interpretation of quantum mechanics, in which the image of a detached observer was renounced. His concept of complementarity was partly sustained by the belief that it could illuminate the paradoxes of psychology as well as those of physics. Accounts of human behavior that presupposed the freewill of the agent were "complementary" to those that gave a deterministic explanation.

A cultural input from philosophical theology came from the nineteenth-century Danish philosopher Søren Kierkegaard (1813–55), whose existentialist outlook had attracted Bohr as a young man. Reacting against those trends in Enlightenment philosophy that deprived men and women of their individuality by making them the object of scientific study, Kierkegaard had stressed the primacy of human choice and decision. And a leap of faith was always required in choosing to accept one set of values rather than another. Human life was a series of either/or choices. At each crossroads, the one turning precluded the other. Attempts to achieve an overarching coherence of thought were doomed to failure since, in the practice of life, one had to choose between incompatible courses of action. A long way from subatomic physics – and yet the parallel is there. The physicist had to make a choice between mutually exclusive descriptions. Moreover, Kierkegaard's stress on the individual had its parallel in Bohr's emphasis on observer-dependent description. The point is not that Bohr simply translated Kierkegaard's theology into physical terms. Rather, he had assimilated a view of the human subject with which his interpretation of quantum physics resonated. With so rich an input into the Copenhagen interpretation of quantum mechanics, it is not surprising that religious apologists should obtain an output. But the danger of circularity is transparent.

One extrapolation from the new physics deserves special attention because, circular or not, it has proved influential. This is to justify critiques of scientific reductionism in the name of science itself. No longer, it is argued, does one have the license to reduce the behavior of complex systems to the laws governing the behavior of their parts. An objective description of elementary particles (and thence of subelementary components such as the "quarks") seems to be the very thing that modern physics denies us. To speak of the ultimate building blocks of matter seems inappropriate when there is a sense in which it is we, with our machines, who bring their paradoxical manifestations into being. On the interpretation

of quantum mechanics that regards the atom as a system vi-
brant with possibilities, reduction to a single description of
the state of its components proves to be elusive. The separate
identity of elementary particles is lost in the atom, the laws
describing the system not derivable from those describing its
components. The holistic implications to which Bohr drew
attention were expressed by Heisenberg in the statement that
"even in science the object of research is no longer nature
itself, but man's investigation of nature." The corollary seems
to be that nature should be envisaged in organic terms, as
systems in symbiotic relationship with the systems that en-
fold them. At the same time, the scientific investigator acquires a new self-image, for his search for truth modifies the
very truth he seeks.

The World as a System: The Rise of Holistic Conceptions

The belief that reductionist accounts of natural phenomena
must always be complemented by holistic perspectives has
gained ground recently, largely through an awareness among
the public of ecological interdependencies. A society that can
worry about the damage it inflicts on the earth's ozone layer
is one in which it may seem less mystical than it once did to
speak of our planet as an organism, or of the universe as an
unbroken whole. And where it is acknowledged that reductionist accounts of human consciousess are not exhaustive –
in that they miss what consciousness feels like, as it were,
from the inside – there may also be greater tolerance of the
view that scientific descriptions of the natural world do not
exhaust the totality of that unbroken whole. The extrapolation to a divine consciousness will always be a controversial
step, but it has been repeatedly taken by American "process"
theologians who have followed A. N. Whitehead and Charles
Hartshorne in refining the concept of a deity who both par-
ticipates in, and is enriched by, the processes of the natural
order.

In Western cultures, where belief in any form of life after
death has largely receded, process philosophy has offered an
alternative rendering of immortality by stressing two respects
in which actions of moral worth may have permanent value.
They have immediate consequences, which, through the creation of further opportunities, leave their mark on a world

that will henceforth be richer than it would otherwise have been. Less prosaically, they have lasting value through their enrichment of a God who, in one aspect of His nature, is not already perfect but unsurpassable except by Himself. In Hartshorne's paraphrase, the good deed is enjoyed by Him forever, and this immortality in God is the creature's only value in the long run. To crave for more is to succumb to that same egocentrism that, according to Darwin, had fostered belief in man's special creation.

The general reader is more likely to encounter the case for a holistic metaphysics in works such as Fritjof Capra's *The turning point* (1982) in which the logic of a systems analysis is employed to show how scientific and mystical views of consciousness might be unified. Capra's analysis has excited interest for at least two reasons. One is his claim that the conceptual shift in modern physics has profound social implications. To reflect the harmonious interrelatedness observed in nature, a cultural revolution is required – underpinned by a radically different social and economic structure. The other is his claim in *The Tao of physics* (1975) that modern science and oriental mysticism offer parallel insights into man's relationship with nature.

A congruence between modern science and Eastern mysticism can easily be overstated. Long before Capra pronounced, Bohr had warned of the danger. He, as with Capra later, saw the parallels. He acknowledged that the development of atomic physics had forced an attitude toward the problem of explanation that reminded him of ancient wisdom: "When searching for harmony in life one must never forget that in the drama of existence we are ourselves both actors and spectators." He would mention Buddha and the sage Lao Tzu in the same context. When he was awarded the Danish Order of the Elephant in 1947, he designed a coat of arms that featured the symbol for Yin and Yang: Contraries were complementary. The parallels were engaging. But he was adamant that to use them for illustrative purposes was not to imply "acceptance in atomic physics of any mysticism foreign to the true spirit of science."

In Capra's presentation it is science (and especially ecology) that confirms the wisdom of a holistic vision. He then argues that to achieve harmony, indeed to survive, it is imperative that science and its application should be regulated by values that have been largely eclipsed in the West. Holistic philosophies have undoubtedly gained ground because they do provide a theoretical base for the rejection of cultural and political norms in favor of alternative technologies, alternative medicine, and alternative economic priorities. In Capra's view, nuclear power, gas-guzzling cars, petroleum-subsidized agriculture, even computerized diagnostic tools, are antiecological, inflationary, and unhealthy. Ecological sensitivity requires the development of softer technologies, which operate with renewable resources – hence Capra's emphasis on solar energy, wind generators, organic farming, regional and local food production, and the recycling of waste products. The message is explicitly political, for it is suggested that highly centralized national governments are able neither to act locally nor to think globally. Britain in the early 1990s may well make an interesting test case. For, at the time of writing, it is a sensitive political issue whether market forces could possibly come to the rescue – if both producers and consumers of pollutants were taxed. It is nevertheless difficult to see how such global problems as the disposal of hazardous waste could be solved by such a mechanism.

The direction and application of scientific research clearly can be different under different value systems. And since human values are often organically linked with religious beliefs, the latter can still be presented as relevant to the orientation of science and technology. Although Capra stresses the insights of the Eastern mystics, the theme of stewardship in the Christian tradition is one that also finds application in ecological discussion. It may, therefore, be useful to take a parting glance at the issue of science and ethical values.

Science and Human Values

The twentieth century has seen a profound shift in our understanding of the mutual relevance of science and moral values. As late as the 1930s and 1940s, it was commonly asserted that because the sciences aimed at achieving consensus on an objective description of the world, they had no relevance to the subjective questions of personal and corporate morality. Although this view placed restrictions on the scope of the sciences, it was attractive because it protected the freedom of scientists to pursue their research without fear of external controls. It was also attractive to moralists and theologians in that it gave them a similar professional autonomy. It was a convenient view for those scientists who wished to absolve themselves of responsibility for any inhumane application of their research. It was equally convenient for the theologian who wished to say that science and religion could not possibly conflict because they had two quite separate provinces. And there was the additional convenience that a clear line of demarcation could be defended with a respectable philosophical argument – that from scientific statements that purport to say what *is* the case, no conclusion can be drawn about what *ought* to be the case, without smuggling in external value judgments.

Had they been more familiar with the history of science, the proponents of that neat division of labor might have found it difficult to sustain. As soon as one asks why science should be pursued at all, questions of value immediately arise. For Francis Bacon in the seventeenth century, an empirical study of nature was to be valued because it promised an increase in the power of the state, the relief of human suffering, and the restoration of dominion over nature that humanity had lost at the Fall. It was to be valued because it encouraged humility, in contrast to the arrogance of the academic philosophers. The ideal of sharing knowledge for the public good had also informed Bacon's critique of the selfish pretensions of the magician.

That the practice of science would encourage human virtue was a theme taken up by philosophers of the Enlightenment. In eighteenth-century France science was often valued because of the threat it posed to popular religious superstition. At the same time, it was invested with other values that almost made of it a substitute religion. In his eulogies of the great figures of the Paris Academy, Fontenelle constructed an image of the scientific character replete with virtue. Seriousness of purpose, dedication, the absence of self-interest, unswerving allegiance to truth, fortitude, tranquillity, conscientiousness, even righteousness were the qualities he assigned to the scientific prophet. Moral values were not so much deduced from religion as inscribed in the very practice of science. Physics, wrote Fontenelle, "becomes a kind of theology when it is pursued correctly."

During the nineteenth century, many links were forged between evolutionary science and human values. It was not simply that Darwin's theory challenged traditional religious authority. As we saw in the preceding chapter, there were many attempts to justify social and political creeds in terms of evolutionary naturalism. Among scientific humanists in the twentieth century, the same tendency has continued. In *Evolution in action* (1953), Julian Huxley defined a progressive culture as one that contained the seeds of its own further transformation. But how could one say that such further transformation could be good? Huxley's ultimate reference point was the "realization of inherent capacities by the individual and of new possibilities by the race." But why place value on the individual life? Huxley's claim was that evolution required nothing less, since human personalities were its highest products. The primacy of human personality, he declared, had been a *postulate* both of Christianity and liberal democracy, but it was a *fact* of evolution. Values were being naturalized through science.

To affirm a clear separation between science and values may represent a philosophical ideal, but it flies in the face of a history of science sensitive to the manifold ways in which scientific knowledge has been value-laden. The shift that has occurred in the twentieth century has been prompted by the realization that there is no simple equation linking scientific advance to social benefit. The twentieth century has had its crisis of faith, with the loss of that confidence that in the Enlightenment had been placed in science as the key to solving all human problems. Even before World War I shattered illusions of human progress, doubts had been expressed about the negative consequences of a technological society, in which unemployment might be increased and the environment rendered more hazardous. It is not merely that atomic bombs and industrial pollution have turned utopian fantasies into nightmares. Contrary to the triumphalist images of the nineteenth century, science has generated its own penumbra of ignorance, surrounding, for example, what may be considered a safe level of exposure to nuclear radiation. It has also become a platitude to say that it has generated ethical problems that it is itself powerless to solve. A public alerted to experiments on human embryos and to the prospects for genetic engineering can no longer share the illusion that scientific and ethical imperatives belong to two quite different spheres. How research is oriented and how it is applied have become matters of unprecedented public concern.

The interface between science and moral accountability is one along which religious values do still find expression. They are sometimes expressed through outright opposition, as when the Roman Catholic community in Britain recently made clear its aversion to in vitro fertilization. A technique, which many would approve on humanitarian grounds if it permitted an otherwise infertile couple to have children, is excluded on the basis of a sacred bond between the physical expression of sexual love and the begetting of a child. To fracture the process through medical intervention is seen as a distortion, not justified by the end. Experiments on spare human embryos have been similarly opposed on a priori grounds. The objection usually is that the embryo is potentially a person and therefore has rights of protection. To experiment upon it, to kill it, or even to deny it a uterine environment necessary for life, is then said to be immoral. That Protestant as well as Catholic sensibilities can foster resistance on such issues is clear from the character of antiabortion constituencies in America.

Religious values do not always find expression in such black and white terms. They may define the limits of acceptable inquiry and so merely reinforce conclusions based on pragmatic criteria. Thus the principle that the degree of protection should be graded according to growth can be shown to have a long history in the Christian West and a meaningful application in fetal research were experimentation would be precluded at the point where the neural development of the organism would have made it a sentient being. It has been observed that the words used by modern embryologists to describe the capacities of a thirty- to forty-day fetus (such as sentience and awareness) were exactly those used by Dante to describe the stage at which it was ready to receive its "soul."

Have scientific advances called for the revision of basic human values? In a recent discussion of this question, Loren Graham has observed that at times of bewildering social and technological change, many cling more closely to familiar values in the hope of finding guidance and stability. Scientific innovation may facilitate a change of emphasis but hardly forces it. And even then it is likely to be derivative rather than primary values that are changed. One of his examples concerns the hypothetical case of a Roman Catholic couple who might be more likely to agree to an abortion if an amniocentesis showed the wife to be carrying a child with Down's syndrome. If the couple had previously rejected abortion on principle, the new technique might be thought responsible for a change in values. But Graham's point is that the two primary principles involved, that one should not kill and that parents bear a heavy responsibility for the welfare of their children, were not themselves changed by the scientific knowledge. It is rather that new information illuminated a conflict between the two principles that may not have been so clearly perceived before.

This example might almost serve as a model for the subtlety of the relationship between scientific advance and secularization. As we have seen many times in this book, scientific innovations have facilitated the growth of secular attitudes, but they have by no means forced them. Even if one were obliged to speak in terms of causal connection, the most that could be said is that, where science has been a determinant, it has been one among many. Almost all the shifts, for which the term *secularization* is conventionally employed, have to be related to social, economic, and political changes, of which scientific activity could hardly be the determinant. The processes whereby the clergy in the West gradually lost their monopoly over learning, and the power to persecute the unorthodox, require elaborate historical explanation, as does the loss of their power in influencing the policies of governments. The decline in the wealth, power, and status of the dominant Christian churches was not the result of a simple linear process; nor was the concomitant increase in the autonomy of the laity. Mechanistic science and melioristic technology have both played their part in shrinking the jurisdiction of providence, but only in conjunction with metaphysical assumptions that are not strictly derivable from science. The replacement of spiritual by material values would seem to owe more to the security of modern medicine, to the seduction of urban comforts and economic prosperity than to any scientific imperative. And insofar as the social functions once performed by religion have been taken over by secular groups and institutions, it is to their social and political origins one must look for insights into the redistribution of power.

The pluralism of most Western societies now makes it difficult for any one religious group to succeed in establishing its own values as those by which publicly sponsored science should be regulated. It has been pointed out that, in America, the influence of scientists in protecting their own freedom of research grew, during the 1970s, with the growing importance of administrative decision making. Groups opposing fetal research, for example, tended to voice their dissent through the media or the courts, rather than through established administrative channels, which proved more receptive to submissions from scientific organizations. There are in fact many reasons for supposing that the relationship between science and secularization has been more slippery than was once commonly assumed. The very fact of a resurgence of religious fundamentalism in highly technological societies is causing eminent social scientists to think again. With the rise of Islamic consciousness, the question asked by Muslim scholars has not been How might Islam be damaged by the results of scientific research? but How might Islamic values be invoked to purge Western science and technology of its materialistic ethos?

In the early 1980s, two seminars were held, in Stockholm and Granada, at which Western and Muslim scholars assessed the possible relevance of Islamic values to the orientation of scientific research. Although no consensus was achieved, Muslim contributors converged on a series of imperatives that they felt were rooted in their religious heritage. A stress on the unity of God required a stress on the unity of nature of which humanity was an integral part. Nature was not to be seen as "out there" and exploitable, but as a cosmic web of which we are part. Any damage inflicted on the environment rebounds on ourselves. By contrast with Western secular philosophy, humanity is accountable to God for the abuse of natural resources. Moreover we suffer the penalties now, not in some future destiny. Metaphors of conquest and domination, which have often informed scientific attitudes toward nature in the West, have to give way to those of trusteeship and stewardship. The pursuit of knowledge has to be seen as a contemplative exercise – an expression of the obligation to worship. More specifically, those forms of science and technology that promote alienation, dehumanization, and the concentration of wealth in fewer hands are to be resisted. The orientation has to be toward a science for the people in which the means as much as the end must always be subject to moral scrutiny.

It was recognized that these were lofty ideals, in the pursuit of which Muslim countries were plainly negligent. Oil-rich states had imported Western technologies with minimal concern for the environment or for traditional mores. It was as if the policy makers were as detached from their religious roots as those in the West. Tower blocks and expressways had shredded the extended family and other socially cohesive customs. The rapidity with which foreign technologies were implanted had prevented a proper evaluation of their draw-backs. Reappraisal had become an urgent necessity and, in the process, the West might have something to learn from Islam.

The unease with which that last suggestion would normally be greeted in the West may help to focus attention on the difficulties faced by the leaders of any religious movement who might claim prerogatives for controlling scientific research and its practical application. One difficulty concerns the applicability of a unique set of ethical priorities to the problem under discussion. It might be convenient if a set of values derived from a specific religion could provide the solution to each moral problem erupting from scientific research. But on so many issues there is no neat fit. The high value that Islam places on a conventional family structure, with sexual relations confined to marriage, may provide clear guidance on certain issues. In the treatment of infertility, for example, one might expect tolerance toward artificial insemination by the husband, but not toward the use of anonymous donors. But, in the example raised earlier, of malformation revealed by amniocentesis, the issue at once becomes cloudy. If abortion were undertaken in the name of mercy killing it would be contrary to Islamic law, but if it were undertaken in the name of preserving the mother's health, it might be allowed. A similar cloudiness arises in other contexts. One Muslim scholar notes that painful experiments on animals seem intuitively wrong on the basis of Islamic values, and yet he also acknowledges that, without such experiments in the past, significant medical advances would not have occurred.

A further difficulty arises for the religious protagonist in that the absence of a clear correspondence between problem and solution can generate embarrassing internecine divisions. This problem is not, of course, confined to Islam. In July 1985, the General Synod of the Church of England welcomed a suggestion in the Warnock Commission's report that there should be a national licensing authority to regulate fetal research and infertility services. On the specific issue whether the use of third-party sperm and ova should be condemned as contrary to Christian standards, the Synod was, however, deeply divided. A motion for condemnation was narrowly defeated by 195 votes to 183. A working-party report prepared for the Synod's Board for Social Responsibility was also divided on most of the issues, except for its rejection of surrogate motherhood. For the outside observer, such divisiveness can encourage the suspicion that values ostensibly derived from religious belief are no more able to illuminate complex ethical problems than those with no pretense to religious authentication.

The greatest difficulty for the religious apologist would seem to be how he might best allay the fear that the pursuit of science under his own prescription would not be abused in deference to fanatical extremes. In America, the defender of liberal religious values has the burden of differentiating his own position from that growing body of creationists whose attacks on the theory of evolution constitute a threat not merely to the orientation of research but to the cognitive content of academic science. Hitherto, federal courts in America have been successful in overturning such state laws as those passed in Arkansas and Louisiana, which had prescribed the teaching of "creation" as well as evolution in their schools. Despite their legal defeat, the proponents of a so-called scientific creationism have been increasingly militant in their efforts to censor textbooks at a local level and to ensure that their children are not indoctrinated, as they see it, with an evolutionary account of human origins. The militancy of those seeking "equal time" for creationist accounts has been such that the American Association for the Advancement of Science condemned what, in 1982, it perceived as a "real and present threat to the integrity of education and the teaching of science."

The resurgence of creationist literature is undoubtedly one expression, among many, of a cultural reaction against the libertarian values of the 1960s when, as one commentator puts it, "recreational sex, open marriage, abortion-on-demand, gay pride, women's liberation, and the fission of nuclear families" went "hand-in-hand with the rise of the Pill and the decline of the work ethic, with ecological consciousness, ethnic consciousness, drug-induced consciousness, and a profusion of narcissistic therapies and fads." For the new creationists a "secular humanism" rooted in evolutionary theory was to blame for this corruption of traditional religious values. To attack evolutionary science was to attack both a symbol and allegedly a cause of all that was going awry in

American society. As early as June 1963, in an address before the American Scientific Affiliation, which allegedly
marked the separation of the modern creationist movement
from the evangelical mainstream, Henry M. Morris insisted
that the choice between creation and evolution had to be made
on the basis of moral and spiritual considerations.

The populist appeal of a fundamentalist religion claiming
for itself scientific credentials has also been explained by the
certainties and securities it offers in an increasingly uncertain
and turbulent world. Sociologists describing that loss of security often point to the nuclear-arms race, the sophistication
of Soviet space research, the economic ascendancy of Japan,
the stability of Castro's Cuba, humiliation in Vietnam, and
impotence in Tehran – all contributing to a disconcerting
awareness of new balances of power. That the appeal of "scientific creationism" is essentially populist is suggested by data
that indicate the generally lower educational profile of its adherents and by the claims of creationist leaders that they speak
for the mass of Christian lay people. In that gap between the
lay person and the professional biologist lies a further reason
for the growth of the movement – a growing suspicion of
experts and frustration at having to take their word. It has
been observed that there is an almost religious mystique about
the inner gnoses of a specialized science, which can be forbidding to those indoctrinated with simpler verities. If we
add to this the political conviction that one should have greater
control over the education of one's children, the strength of
the movement, especially in those areas with indigenous bib-
licist traditions, becomes less surprising. It does, however,
become no less alarming to the community of professional
biologists who rightly insist that the plausibility of scientific
theories cannot be settled by popular votes. For the sociologist it is a commonplace that the ideas of intellectuals are
easily resisted or ignored when they fail to resonate with significant sections of public opinion. For professional scientists
this can be an uncomfortable notion, because they have so
little control over those social and political forces that sway
popular perception.

Discussion of the interface along which science and human
values meet has been given further impetus with the advent
of new research programs taking their cue from Edward O.
Wilson's *Sociobiology* (1975). If human values and moral claims
can be *explained* by factual assertions about our evolutionary

past, is the rug not pulled from under the feet of those who like to see them as authenticated by cultural and religious tradition? If the origin of ethical principles is to be found in developing instincts (and their genetic correlates) that once conferred evolutionary advantage on altruistic behavior, can the moral scruples shared by a society be anything more, as one commentator puts it, than a "collective illusion foisted upon us by our genes"? Wilson's claim was that science would soon be in a position to investigate the origin and the *meaning* of human values. To provide a biological explanation for religious behavior was the ultimate challenge. But as long as there was a mechanism by which altruistic and conformist tendencies could be encoded within human genes, religious beliefs and practices, as exemplars of conformity, might be comprehended.

Many of Wilson's critics have caricatured his position. While describing himself as a "scientific materialist," he refuses to be labeled a determinist or reductionist. The argument is not that genes determine behavior but that, through the mediation of biological processes, they constrain it. Indeed the reality of choice is central to the argument, for it is the cultural choices of the past that have been subsequently reinforced by the genetic recoding that may accompany them. By emphasizing "coevolution," the two-way interaction between cultural and genetic evolution, Wilson and his associates avoid glaring forms of reductionism. From his perspective, the real reductionist is the social scientist who, treating the human mind as a *tabula rasa,* simply discounts the biological constraints.

The territorial ambitions of sociobiology nevertheless continue to cause disquiet. Its critics detect a reluctance to admit that, at the level where human choices are experienced, they are principally constrained by political institutions, economic limitations, and social conventions. Religious critics resent the notion that the values for which they stand derive their ultimate meaning from considerations of biological utility. Wilson's protestations of political neutrality notwithstanding, left-wing critics have reacted sharply to the exploitation of his program by those on the political right who have claimed a new "natural" justification for the nuclear family and conventional roles for the sexes. There has even been the suggestion that, ill-fitted as they are to be bedfellows, sociobiology and scientific creationism are twin manifestations of the same cultural reaction in American society.

Irrespective of political preferences, there would be strange consequences if the meaning of human values could be fully explained in terms of evolutionary science. For it would then be reasonable to suppose that the same science could tell us how they might need to be modified. The problem, as Loren Graham has pointed out, is that the aims of the modification also embody values that are, presumably, susceptible to scientific research and further modification. The specter of an infinite regress hangs over the enterprise. It may even be a specter with a sense of humor. For it has been asked what good could come from the efforts of a sociobiologist to convince us that a belief necessary for the common good really is a collective illusion foisted upon us by our genes. The question is: How have the genes that sustained the illusion suddenly arranged that some other good might now be served by its removal? Might not human sociobiology, by its own precepts, turn out to be maladaptive?

According to Wilson, three values must be paramount in the articulation of any new moral code. One is the long-term survival of the human gene pool. The second is the diversity of that pool. The third: universal human rights. The justification for the first two can be given in straightforward biological terms. A serious problem, however, arises with the third. A man may feel a deep sense of obligation toward his brother, even to the point of sacrificing his own interests to those of his kin. The sociobiologist will even explain the origins of that moral sense. In the case of two brothers, at least half the genes of the one will be identical to those of the other. It is possible to envisage altruistic behavior that would deprive one of the brothers of the opportunity to be as active in procreation as the other in whose interests he acted. But if the act of altruism more than doubled the latter's representation in the next generation, then the altruist would actually have gained in genetic terms. Because many of the genes shared by the brothers would be those that encoded the tendency toward altruistic behavior, that tendency would recur and be further reinforced in subsequent generations.

The problem concerns the step from altruism within a family to the proclamation of universal human rights. The latter, as a principle, has arrived relatively late in human civilization. Its constant violation also raises doubts about its justification or explanation merely in terms of biological adaptation. That it has been violated in the name of so many tribal gods remains one of the principal reasons why the secular moralist refuses to countenance a theistic premise in the presentation of a humanitarian ethic. There has, however, never been a simpler way of getting from the brotherhood of brothers to the brotherhood of man than via an affirmation of the fatherhood of God. Many would say that Freud has cost the twentieth century the luxury of that premise. But whether belief in the supreme worth of every human life, and the action such an ideal requires, can be sustained without reference to the transcendent, is a question unlikely to be laid to rest.

Bibliographic Essay

In constructing a guide to further reading, I have chosen to follow the order in which themes have been introduced in the main body of the text. By so doing, I hope that two objectives will be met: to provide an indication of sources to which I have been particularly indebted and to suggest avenues by which the issues raised may be pursued in greater depth. It should, therefore, be appreciated that what follows makes no pretense toward comprehensiveness. Topics that have not been covered in preceding chapters, such as the ramifications of the physical sciences in the nineteenth century, are also underrepresented in this essay. Access to a literature on the scientific developments not considered in the text can, however, be achieved via such works of reference as the *Dictionary of scientific biography,* ed. C. C. Gillispie, 16 vols. (New York, 1970–80), and the *Companion to the history of modern science,* ed. R. C. Olby, G. N. Cantor, J. R. R. Christie, and M. J. S. Hodge (London, 1990).

Introduction

There is a prolific literature having "the relations between science and religion" as its organizing theme. Much is suspect because of thinly veiled apologetic intentions; much is vitiated by an insensitivity to the richness of past debates that historical analysis alone can remedy. Among recent studies, it is possible to recommend the essays contained in David C. Lindberg and Ronald L. Numbers, eds., *God and nature: historical essays on the encounter between Christianity and science* (Berkeley, 1986), which is also valuable for its bibliographic

348

guidance. One of the most encyclopedic treatments remains Ian G. Barbour, *Issues in science and religion* (Englewood Cliffs, N.J., 1966). Although colored by a distinctive Protestant neo-orthodoxy, the historical interpretation in John Dillenberger, *Protestant theology and natural science* (London, 1961), has also retained much of its value.

Several introductory texts, designed to broaden horizons on the interaction between science and other beliefs, were published by the Open University as constituents of two courses: *Science and belief from Copernicus to Darwin* (Milton Keynes, 1974); and *Science and belief from Darwin to Einstein* (Milton Keynes, 1981). Although these courses are no longer extant, the associated texts, geared to the analysis of primary sources, may still provide a convenient induction. From the first (1974) these include: *The conflict thesis and cosmology* (Units 1–3); *Towards a mechanistic philosophy* (Units 4–5); *Scientific progress and religious dissent* (Units 6–8); *New interactions between theology and natural science* (Units 9–11); *The crisis of evolution* (Units 12–14); *The new outlook for science* (Units 15–16). From the second (1981): *Beliefs in science: an introduction* (Unit 1); *Science and metaphysics in Victorian Britain* (Units 2–3); *Time, chance and thermodynamics* (Units 4–5); *Modern physics and the problem of knowledge* (Units 6–9); *The mystery of life* (Units 10–11); *Problems in the biological and human sciences* (Units 12–14); *The future of science and belief* (Unit 15). The accompanying anthologies were compiled by D. C. Goodman, ed., *Science and religious belief 1600–1900: a selection of primary sources* (Dorchester, 1973); C. A. Russell, ed., *Science and religious belief: a selection of recent historical studies* (London, 1973); Noel Coley and Vance M.D. Hall, eds., *Darwin to Einstein: primary sources on science and belief* (Harlow, 1980); Colin Chant and John Fauvel, eds., *Darwin to Einstein: historical studies on science and belief* (New York, 1980).

For the parallel between the rise of Christianity and the rise of modern science, see Herbert Butterfield, *The origins of modern science*, 2d ed. (London, 1957), pp. 175–90. That the relationship between the two forces has been essentially one of conflict was most famously argued in J. W. Draper's *History of the conflict between religion and science* (London, 1875) and A. D. White's *A history of the warfare of science with theology in Christendom* (New York, 1896). A succinct review of current alternatives is provided by Ian G. Barbour, "Ways of relating science and theology," in *Physics, philosophy and the-*

ology: a common quest for understanding, ed. Robert John Russell, William R. Stoeger, and George V. Coyne (Vatican City and Notre Dame, Ind., 1988), pp. 21–48. The customary grounds for asserting that science and theology should have nothing to do with each other are explored and criticized by William H. Austin, *The relevance of natural science to theology* (London, 1976). The thesis that seventeenth-century science was, in certain respects, a derivative of medieval theology was stated in A. N. Whitehead, *Science and the modern world* (New York, 1925), pp. 9–25. The claim that religious values could work to the advantage of the sciences has often been made with reference to the stimulus allegedly supplied by the Protestant Reformation. Exemplars of this historiography include R. Hooykaas, *Religion and the rise of modern science* (Grand Rapids, Mich., 1972, and Edinburgh, 1973), and, more recently, Colin A. Russell, *Cross-currents: interactions between science and faith* (Leicester, 1985). An earlier version of the argument, that of Robert K. Merton, *Science, technology and society in seventeenth-century England* (New York 1970), which had first appeared in *Osiris* 4 (1938), part 2, 360–632, has generated a continuing scholarly debate, reference to which will be found under Chapter III below.

The specific case of Franklin and electricity is discussed in Stanley L. Jaki, *The relevance of physics* (Chicago, 1966). As an antidote to the popular mythology that the sciences deal only with "facts," there is a classic statement of the view that facts are theory-laden in Norwood Russell Hanson, *Patterns of discovery* (Cambridge, 1958). On the driving force of scientific *theory,* see also the discussion by Barry Barnes, *Scientific knowledge and sociological theory* (London, 1974). The openness of scientific practice (and theory evaluation) to sociological analysis is also argued by Michael Mulkay, *Science and the sociology of knowledge* (London, 1979). The classic critique of the notion that science proceeds by the linear accumulation of undisputed "fact" was formulated by T. S. Kuhn in his *The structure of scientific revolutions* (Chicago, 1962). The myth that there is a single, privileged "scientific method" is most thoroughly exposed in John A. Schuster and Richard R. Yeo, eds., *The politics and rhetoric of scientific method* (Dordrecht, 1986).

The application of sociological techniques to the study of religion has a longer history. As an introduction, two works by Peter Berger may be recommended: *The sacred canopy* (New

York, 1967) and *The social reality of religion* (London, 1969). For an alternative approach, see Bryan S. Turner, *Religion and social theory: a materialist perspective* (London, 1983). An anthology of classic texts in the academic study of religion will be found in Roland Robertson, ed., *Sociology of religion* (London, 1969). An acquaintance with other religions than one's own can be gained from Ninian Smart, *The world's religions* (Cambridge, 1989). The inclusion of theological discussion within the boundaries of "natural philosophy" can be seen most clearly in the *General Scholium* from the second edition of Newton's *Principia* (1713) and in Queries 28 and 31 from the later editions of his *Opticks*. These are available in H. S. Thayer, ed., *Newton's philosophy of nature* (New York, 1953), which also contains other primary sources in which Newton spoke of God's relation to nature. Thomas Burnet's *The sacred theory of the earth* (1691) is accessible in a reprint edition, with an introduction by Basil Willey (London, 1965).

For a first introduction to the political dimensions of the Galileo affair, see Olaf Pederson, "Galileo and the Council of Trent," *Journal of the History of Astronomy* 14 (1983), 1–29. Two studies reveal in a striking manner the artificiality of abstracting the "science" and the "religion" of an earlier period from their political contexts: Steven Shapin, "Of gods and kings: natural philosophy and politics in the Leibniz–Clarke disputes," *Isis* 72 (1981), 187–215; and L. S. Jacyna, "Immanence or transcendence: theories of life and organization in Britain, 1790–1835," *Isis* 74 (1983), 311–29.

The conventional view that the sciences have been primary agents of secularization is challenged by Mary Douglas, "The effects of modernization on religious change," *Daedalus* (Winter 1982), issued as vol. 111, no. 1, of the *Proceedings of the American Academy of Arts and Sciences,* 1–19. Certain difficulties with the concept and meaning of "secularization" are discussed by Michael Hill, *A sociology of religion* (London, 1973), pp. 228–51.

I. Interaction between Science and Religion: Some Preliminary Considerations

The fusion of science and religion in the writings of Henry Drummond is placed in context by James R. Moore, "Evangelicals and evolution," *Scottish Journal of Theology* 38 (1985), 383–417. Conventional contrasts between science and religion are drawn by Anthony Giddens, *New rules of sociological method* (London, 1976), pp. 138–44. The riposte of Leibniz to Newton on the theological connotations of "force" can be found in H. G. Alexander, ed., *The Leibniz–Clarke correspondence* (Manchester, 1956), pp. 90–94. My example of a "holy war" within science is taken from Herbert C. Brown, *Boranes in organic chemistry* (Ithaca, N.Y., 1972), pp. 140–1. On science as a form of worship, see H. Fisch, "The scientist as priest: a note on Robert Boyle's natural theology," *Isis* 44 (1953), 252–65.

There is an extensive literature on the doctrine of creation as a presupposition of the uniformity of nature and of modern science. The case was argued with philosophical sophistication by Michael Foster in three articles that appeared in *Mind* 43 (1934), 446–68; 44 (1935), 439–66; 45 (1936), 1–27. Strong but controversial claims for the uniqueness of Christian doctrine in making modern sicence possible will be found in Stanley L. Jaki, *Science and creation: from eternal cycles to an oscillating universe* (New York, 1974). That a particular tradition within Christian theology, in which the free will of a Creator is stressed, was conducive to a new experimental philosophy having as its object the discovery of natural *laws* has been argued by Francis Oakley, "Christian theology and Newtonian science: the rise of the concept of laws of nature," *Church History* 30 (1961), 433–57; by J. R. Milton, "The origin and development of the concept of the 'laws of nature,' " *European Journal of Sociology* 22 (1981), 173–95; and by Eugene M. Klaaren, *Religious origins of modern science* (Grand Rapids, Mich., 1977). The argument has not been confined to religious apologists. It occurs in the work of the Marxist historian E. Zilsel, "The genesis of the concept of the physical law," *Philosophical Review* 51 (1942), 245–79; and in attenuated form in Joseph Needham, *The grand titration: science and society in East and West* (London, 1969), pp. 299–330, in which explicit contrasts are drawn with Chinese culture. In a critique of Zilsel's thesis, it has been argued that, prior to Descartes, a concept of *laws* of nature had been formulated independently of the metaphor of divine legislation: see Jane E. Ruby, "The origins of scientific 'law,' " *Journal of the History of Ideas* 47 (1986), 341–59.

On the epistemological implications for William Whewell of a doctrine of creation, see Richard R. Yeo, "William Whewell, natural theology and the philosophy of science in mid nineteenth century Britain," *Annals of Science* 36 (1979), 493–516. Francis Bacon's sanction for a new form of practical knowledge is discussed by Charles Webster, *The great instauration: science, medicine and reform 1626–1660* (London, 1975); the eventual demise of the "two books" analogy by James R. Moore, "Geologists and interpreters of Genesis in the nineteenth century," in *God and nature: historical essays on the encounter between Christianity and science,* ed. David C. Lindberg and Ronald L. Numbers (Berkeley, 1986), pp. 322–50. On natural theology as a sanction for science, the case of Thomas Sprat is reviewed by Paul B. Wood, "Methodology and apologetics: Thomas Sprat's *History of the Royal Society,*" *British Journal for the History of Science* 13 (1980), 1–26; and that of Adam Sedgwick by John H. Brooke, "The natural theology of the geologists: some theological strata," in *Images of the Earth: essays in the history of the environmental sciences,* ed. L. J. Jordanova and Roy S. Porter, British Society for the History of Science monograph 1 (Chalfont St. Giles, 1979), pp. 39–64. For the historical links between doctrines of the millennium and the idea of progress, see E. L. Tuveson, *Millennium and utopia* (New York, 1964). On the links forged between natural theology and natural history in the works of John Ray, see Charles Raven, *John Ray, naturalist: his life and works,* 2d ed. (Cambridge, 1950); the connections in the case of Priestley's science are explored by John H. Brooke, " 'A sower went forth': Joseph Priestley and the ministry of reform," in *Motion toward perfection: the achievement of Joseph Priestley,* ed. A. Truman Schwartz and John G. McEvoy (Boston, 1990), pp. 21–56.

The penetration of religious concerns into discussions of scientific methodology has been discussed in the case of Paracelsus by Owen Hannaway, *The chemists and the word* (Baltimore, 1975); in that of Mersenne by Stanley L. Jaki, *Planets and planetarians* (Edinburgh, 1978), pp. 25–6; and in early nineteenth-century geology by Charles C. Gillispie, *Genesis and geology* (Cambridge, Mass., 1950).

An underdetermination of scientific theories by the experimental facts they purport to explain has been increasingly recognized by philosophers and sociologists of science. A seminal work in this respect is Mary B. Hesse, *The structure of scientific inference* (London, 1974). The inadequacy of the view that experimental results speak for themselves is cleverly exposed, with reference to Robert Boyle and his critics, by Steven Shapin and Simon Schaffer, *Leviathan and the air-pump: Hobbes, Boyle and the experimental life* (Princeton, 1985). Further entry into the specialist literature is provided by D. Gooding, T. Pinch, and S. Schaffer, *The uses of experiment* (Cambridge, 1989). For the problems associated with the replication of experiments, see H. M. Collins, *Changing order* (London, 1985). The theological connotations of belief in the simplicity of nature are discussed in the case of Michael Faraday by Geoffrey Cantor, "Reading the book of nature: the relation between Faraday's religion and his science," in *Faraday rediscovered,* ed. F. James and D. Gooding (London, 1985), pp. 69–81. For the selective role of religious belief in Kepler's resistance to the vision of Bruno, see Paolo Rossi, "Nobility of man and plurality of worlds," in *Science, medicine and society in the Renaissance,* 2 vols., ed. A. G. Debus (New York, 1972), 2:131–62; and A. Koyré, *From the closed world to the infinite universe* (New York, 1958), pp. 58–87. A specialist literature on the role of aesthetic criteria in theory selection and justification may be approached through James W. McAllister, "Truth and beauty in scientific reason," *Synthèse* 78 (1989), 25–51.

Typical of pleas from Christian writers to avoid a god-of-the-gaps is that of C. A. Coulson, *Science and Christian belief* (Oxford, 1955). The constitutive role played by the legend of Noah's ark in the emergent science of biogeography is analyzed by Janet Browne, *The secular ark: studies in the history of biogeography* (New Haven, 1983), pp. 1–31. That questions prompted by religious concerns could promote new lines of scientific inquiry is illustrated in the case of Bentley and Newton by Michael Hoskin, "Newton, Providence and the universe of stars," *Journal of the History of Astronomy* 8 (1977), 77–101. The contrary creed of ninteenth-century scientific naturalism is expounded by Frank M. Turner, *Between religion and science: the reaction to scientific naturalism in late Victorian England* (New Haven, 1974), pp. 8–37. For Durkheim's view that a fully secularized society would be a contradiction in terms, see Anthony Giddens, ed., *Emile Durkheim: selected writings* (Cambridge, 1972), pp. 239–49.

A forceful and tenacious critique of the works by Draper and White is presented in James R. Moore, *The post-Darwinian controversies: a study of the Protestant struggle to come to terms with Darwin in Great Britain and America, 1870–1900* (Cambridge, 1981), which is also valuable for its rich bibliography. My own critique, developed in this chapter, can also be supplemented by David C. Lindberg and Ronald L. Numbers, "Beyond war and peace: a reappraisal of the encounter between Christianity and science," *Church History* 55 (1986), 338–54; Ronald L. Numbers, "Science and religion," *Osiris,* 2d ser. 1 (1985), 59–80; and Colin A. Russell, "The conflict metaphor and its social origins," *Science and Christian Belief* 1 (1989), 3–26. T. H. Huxley's concession that complementarity between science and theism was still possible, even after the assimilation of Darwin's theory, was made in his "On the reception of the 'Origin of species,' " in *The life and letters of Charles Darwin,* 3 vols., ed. Francis Darwin (London, 1887), 2:179–204. The deficiencies of a conflict model for embracing the diversity of opinion among the early church fathers is stressed by David C. Lindberg, "Science and the early Christian Church," *Isis* 74 (1983), 509–30. On Servetus, see Allen G. Debus, *Man and nature in the Renaissance* (Cambridge, 1978), pp. 64–65; and for his heresies concerning infant baptism, the Trinity, and the nature of God's immanence, Roland H. Bainton, *Hunted heretic: the life and death of Michael Servetus 1511–1553* (Boston, 1960), pp. 182–201. That the burning of Bruno should not be pressed into the mold of religion versus science was argued by Frances Yates, *Giordano Bruno and the hermetic tradition* (Chicago, 1964), pp. 348–59. Lyell's problems at King's College London are reinterpreted by Martin J. S. Rudwick, "Charles Lyell F.R.S. (1797–1875) and his London lectures on geology, 1832–33," *Notes and Records of the Royal Society* 29 (1975), 231–63. Revisionist accounts of the Huxley–Wilberforce debate include J. R. Lucas, "Wilberforce and Huxley: a legendary encounter," *The Historical Journal* 22 (1979), 313–30; Sheridan Gilley, "The Huxley–Wilberforce debate: a reconstruction," in *Religion and humanism,* ed. Keith Robbins (Oxford, 1981), pp. 325–40. For a recent attempt to achieve a balanced view, see J. Vernon Jensen, "Return to the Wilberforce–Huxley debate," *British Journal for the History of Science* 21 (1988), 161–79.

On the problem of cultural chauvinism in the historiography of science, there is a succinct critique of Western myopia in R. Rashed, "Science as a western phenomenon," *Fundamenta Scientiae* 1 (1980), 7–21. Whether such questions as Why did the Chinese fail to develop an anlytical science of nature comparable with that of the West? already betray a cultural bias may be considered in the light of commentaries on the work of Joseph Needham by Robert S. Cohen, "The problem of 19(k)," *Journal of Chinese Philosophy* 1 (1973), 103–17; A. Koslow, "More on 19(k)," *Journal of Chinese Philosophy* 2 (1975), 181–96; Chung-Ying Cheng, "On Chinese science: a review essay," *Journal of Chinese Philosophy* 4 (1977), 395–407; and N. Sivin, "On the word 'Taoist' as a source of perplexity, with special reference to the relations of science and religion in traditional China," *History of Religions* 17 (1978), 303–28. The spectacle of contending claims for Protestant and Catholic contributions to modern science is provided by the dispute between R. Hooykaas and F. Russo reproduced in *The evolution of science,* ed. G. S. Métraux and F. Crouzet (New York, 1963), pp. 258–320. Internecine divisions have similarly affected the historiography of Islamic science. The metaphysics presupposed in such influential works of S. H. Nasr as *Science and civilisation in Islam* (Cambridge, Mass., 1968) and *An introduction to Islamic cosmological doctrines* (London, 1978) is criticized as partisan by Z. Sardar, "Where's where? Mapping out the future of Islamic science," *MAAS Journal of Islamic Science* 4 (1988), 35–63.

Social pressures for religious conformity, of the kind that religious apologists sometimes neglect, are identified in Howard Gruber and Paul Barrett, *Darwin on man* (London, 1974), from which the Hooker quotation is taken (p. 26). A more nuanced treatment of the consequent ambiguities in Darwin's theistic remarks is given by David Kohn, "Darwin's ambiguity: the secularization of biological meaning," *British Journal for the History of Science* 22 (1989), 215–39. The context in which Galileo had presented his version of the harmony between Scripture and science is outlined in Stillman Drake, *Discoveries and opinions of Galileo* (New York, 1957), pp. 145–216; and by William R. Shea, "Galileo and the Church," in Lindberg and Numbers, *God and nature,* pp. 114–35. On the eventual failure of an idealist metaphysics in the life sciences, see Peter J. Bowler, "Darwinism and the argument from design: suggestions for a re-evaluation," *Journal of the History of Biology* 10 (1977), 29–43; and David M. Knight, *Ordering the world* (London, 1981), pp. 112–17.

My critique of the apologetic use of revisionist history may be supplemented by reference to Rolf Gruner, "Science, nature and Christianity," *Journal of Theological Studies* 26 (1975), 55–81, where the potentially inhibitive effects of the doctrine of the Fall are given prominence. The social transformation in nineteenth-century Britain, through which the clergy lost their place as arbiters of scientific culture, is defined by Frank M. Turner, "The Victorian conflict between science and religion: a professional dimension," *Isis* 69 (1978), 356–76. The qualification that one did not have to be a professional scientist to be anticlerical, or an amateur to have religious sympathies, is made by Ruth Barton, " 'An influential set of chaps': the X-club and Royal Society politics 1864–85," *British Journal for the History of Science* 23 (1990), 53–81.

II. Science and Religion in the Scientific Revolution

A detailed guide to the innovations in each branch of the sciences during the period conventionally known as the scientific revolution is provided by A. Rupert Hall, *The revolution in science 1500–1700* (London, 1983). A classic text that still deserves attention is E. A. Burtt, *The metaphysical foundations of modern physical science* (London, 1949). Penetrating insights into the conceptual changes associated with Galileo and Newton are contained in the collected papers of Alexandre Koyré, *Metaphysics and measurement* (London, 1968) and *Newtonian studies* (London, 1965). Access to current issues of scholarly concern can be gained through David C. Lindberg and Robert S. Westman, eds., *Reappraisals of the scientific revolution* (Cambridge, 1990). The case for an unprecedented fusion (not separation) of science and theology in the seventeenth century has been made, with great erudition, by Amos Funkenstein, *Theology and the scientific imagination from the Middle Ages to the seventeenth century* (Princeton, 1986), which also constitutes a rich bibliographic resource.

The dispute between Wilkins and Ross on the province of biblical authority is discussed in Grant McColley, "The Ross–Wilkins controversy," *Annals of Science* 3 (1938), 153–89. On the effects of the printing press in underlining religious divisions and facilitating the transmission of secular forms of learning, see Elizabeth Eisenstein, *The printing press as an agent of change,* 2 vols. (Cambridge, 1979). For the wider significance of the establishment of seventeenth-century scientific academies, see Joseph Ben-David, *The scientist's role in society* (Englewood Cliffs, N.J., 1971), pp. 45–87; A. Rupert Hall, *From Galileo to Newton* (London, 1963), pp. 132–54;

and Michael Hunter, *Science and society in Restoration England* (Cambridge, 1981), pp. 32–58, from which the Oldenburg quotation is taken (p. 37). A differentiation between natural philosophy and theology on the basis of their respective sources of authority was argued by Pascal in the projected preface to his *Traité du vide* (1647).

My case against characterizations of seventeenth-century natural philosophy that imply a full separation from religion can be complemented by reference to Richard H. Popkin, "The religious background to seventeenth-century philosophy," *Journal of the History of Philosophy* 25 (1987), 35–50. On the relevance of Reformation theology for an understanding of the new mechanical philosophies of nature, see Gary B. Deason, "Reformation theology and the mechanistic conception of nature," in Lindberg and Numbers, *God and nature,* pp. 167–91. That Calvin's theology had permitted a degree of autonomy for astronomy was stressed by Edward Rosen, "Calvin's attitude towards Copernicus," *Journal of the History of Ideas* 21 (1960), 431–41; and earlier by R. Hooykaas, "Science and Reformation," *Journal of World History* 3 (1956), reproduced in Métraux and Crouzet, *The evolution of science,* pp. 258–90. Even among scholars who have chosen to speak of a separation of science and religion in the mind of Francis Bacon, there has been the acknowledgment that he still invested scientific activity with religious meaning; see, for example, Christopher Hill, *Intellectual origins of the English Revolution* (Oxford, 1965), pp. 85–130.

The earlier subordination to theology is illustrated in the case of Roger Bacon by David C. Lindberg, "On the applicability of mathematics to nature: Roger Bacon and his predecessors," *British Journal for the History of Science* 15 (1982), 3–25; and by N. W. Fisher and S. Unguru, "Experimental science and mathematics in Roger Bacon's thought," *Traditio* 27 (1971), 353–78. See also David C. Lindberg, *Roger Bacon's philosophy of nature* (Oxford, 1983). The stance of Thomas Aquinas toward Aristotle, both respectful and critical, is brought out by Vernon J. Bourke in his introduction to Thomas Aquinas, *Commentary on Aristotle's "Physics"* (London, 1963), pp. xxii–xxiv; by E. J. Dijksterhuis, *The mechanisation of the world picture* (Oxford, 1961), pp. 126–35; and by Frederick Copleston, *Thomas Aquinas* (1955; reprint, London, 1976), pp. 63–83. There are additional insights into the role of Aquinas in Stephen Gaukroger, *Explanatory struc-*

tures (Hassocks, 1978), pp. 134–42. The strategy of Oresme in raising the question of the earth's motion is clearly explained by Edward Grant, *Physical science in the middle ages* (Cambridge, 1977), pp. 66–9. How a stress on divine omnipotence could be an antidote to Aristotelian dogmatism is shown by Steven J. Dick, *Plurality of worlds: the extraterrestrial life debate from Democritus to Kant* (Cambridge, 1982), pp. 23–43.

A useful historical introduction to the relationship between magic and science is given by John Henry, "Magic and science in the sixteenth and seventeenth centuries" in Olby et al., *Companion to the history of modern science,* pp. 583–96, and a fuller analysis by Charles Webster, *From Paracelsus to Newton: magic and the making of modern science* (Cambridge, 1982). The flavor of the magical philosophies of the Renaissance can also be tasted in D. P. Walker, *Spiritual and demonic magic from Ficino to Campanella* (London, 1958); Wayne Shumaker, *The occult sciences in the Renaissance* (Berkeley, 1972); and Peter J. French, *John Dee: the world of an Elizabethan magus* (London, 1972). The contention that magic appealed to Renaissance intellectuals as a higher form of religion can be found in Nicholas H. Clulee, *John Dee's natural philosophy: between science and religion* (London, 1989). One of Frances Yates's more balanced statements on the role of the hermetic corpus in inspiring visions of power over nature occurs in her essay, "The hermetic tradition in Renaissance science," in *Art, science and history in the Renaissance,* ed. Charles S. Singleton (Baltimore, 1968), pp. 255–74. Even this must be tempered with such critical perspectives as those offered in Robert S. Westman and J. E. McGuire, *Hermeticism and the Scientific Revolution* (Los Angeles, 1977), and Owen Hannaway, "Laboratory design and the aim of science," *Isis* 77 (1986), 585–610. Further perspectives on the historical relevance of hermeticism can be obtained from Allen G. Debus and Ingrid Merkel, eds., *Hermeticism and the Renaissance* (London, 1988). For a characterization in which natural magic and natural science are seen as antithetical, see the editor's introduction in Brian Vickers, ed., *Occult and scientific mentalities in the Renaissance* (Cambridge, 1984) – an interpretation, however, that contrasts with that offered by several contributors to that volume, including R. S. Westfall, "Newton and alchemy," pp. 315–36.

For an in-depth study of Paracelsus one still turns to Walter Pagel, *Paracelsus: an introduction to philosophical medicine in the era of the Renaissance* (Basel, 1958). This may be supplemented by Allen G. Debus, *The chemical philosophy: Paracelsian science and medicine in the sixteenth and seventeenth centuries,* 2 vols. (New York, 1977). My discussion in the text is also indebted to Hannaway, *The chemists and the word.*

Protestant critics of magical practices are discussed by Walker, *Spiritual and demonic magic from Ficino to Campanella,* pp. 152–66. Francis Bacon is presented as both assimilator and critic of hermetic motifs in Paolo Rossi, *Francis Bacon: from magic to science* (Chicago, 1968). Protestant uses of the Bible to challenge Catholic demonologies and exaggerated claims for the power of witches are identified by Keith Thomas, *Religion and the decline of magic* (Harmondsworth, 1973), pp. 681–98.

The presentation of scientific innovation in theological terms is explored in the case of van Helmont by Walter Pagel, "The debt of science and medicine to a devout belief in God," *Journal of the Transactions of the Victoria Institute* 74 (1942), 99–115. See also A. Browne, "Van Helmont's attack on Aristotle," *Annals of Science* 36 (1979), 575–91, and Klaaren, *Religious origins of modern science,* pp. 61–9. The integral role of a metaphysical theology in Bruno's vision of an infinite universe is affirmed by Robert S. Westman, "Magical reform and astronomical reform: the Yates thesis reconsidered," in Westman and McGuire, *Hermeticism and the Scientific Revolution,* pp. 1–91. On Descartes's philosophy as a response to skepticism, see Richard H. Popkin, *The history of scepticism from Erasmus to Spinoza,* 2d ed. (Berkeley, 1979). Descartes's theistically grounded natural philosophy and the changing perceptions that it engendered are discussed by Alan Gabbey, "Philosophia Cartesiana Triumphata; Henry More 1636–1671", in *Problems in Cartesianism: studies in the history of ideas,* ed. Thomas M. Lennon (Montreal, 1982), pp. 171–250. Boyle's critique of Descartes was contained in "A disquisition about the final causes of natural things" (1688), reproduced in Goodman, *Science and religious belief 1600–1900,* pp. 103–18. An emphasis on the sovereignty of God in Boyle's version of the mechanical philosophy may be read in his "Free enquiry into the vulgarly received notion of nature" (1686), reproduced in *Selected philosophical papers of Robert Boyle,* ed. M. A. Stewart (Manchester, 1979), pp. 176–91. For further discussion of Boyle, see Shapin and Schaffer, *Leviathan and the air-pump,* pp. 201–24.

Further information on Galileo's handling of scriptural objections to his Copernicanism can be gained from Jerome J. Langford, *Galileo, science and the Church* (New York, 1966), pp. 50–78. The difficulty experienced by Halley in effecting a separation of science from religion, even by the end of the seventeenth century, is examined by Simon Schaffer, "Halley's atheism and the end of the world," *Notes and Records of the Royal Society* 32 (1977), 17–40.

III. The Parallel between Scientific and Religious Reform

Much of the literature on the relations between Protestantism, capitalism, and the expansion of the sciences has had the work of Max Weber as its inspiration. An accessible introduction to Weber's thesis is given by Gordon Marshall, *In search of the spirit of capitalism* (London, 1982). In the affirmation of parallels between scientific, political, and religious reforms in seventeenth-century England, different emphases can be detected in Christopher Hill, *Intellectual origins of the English revolution* (Oxford, 1965); Hooykaas, *Religion and the rise of modern science;* Merton, *Science, technology and society in seventeenth-century England;* Theodore K. Rabb, "Puritanism and the rise of experimental science in England," in *The rise of science in relation to society,* ed. Leonard M. Marsak (New York, 1964), pp. 54–67; Webster, *The great instauration: science, medicine and reform 1626–1660.*

For an introduction to the other major theme of this chapter, see Thomas S. Kuhn, *The Copernican revolution* (Cambridge, Mass., 1957). This should be supplemented with Robert S. Westman, ed., *The Copernican achievement* (Berkeley, 1975). Of the reception of Copernican astronomy in different national contexts there are useful accounts in J. Dobrzycki, ed., *The reception of Copernicus's heliocentric theory* (Dordrecht, 1972). The challenge of the Copernican system, and more especially that of Giordano Bruno, to a geocentric cosmology was analyzed by Arthur O. Lovejoy, *The great chain of being* (1936; reprint, New York, 1960), pp. 99–143 – a discussion that, though dated, still constitutes a useful corrective to popular misunderstanding. The same can be said of another classic comparison between old and new cosmographies: C. S. Lewis, *The discarded image: an introduction to medieval and Renaissance literature* (Cambridge, 1964). The instructive dispute between Wilkins and Ross is brought to life by McColley, "The Ross–Wilkins controversy," pp. 153–89. Kepler's aversion to the loss of cosmic identity implied by Bruno is discussed by Paolo Rossi, "Nobility of man and plurality of worlds," in *Science, medicine and society in the Renaissance*, 2 vols., ed. Allen G. Debus (New York, 1972), 2:131–62. The conviction of Nicholas Cusa that human uniqueness does not depend on cosmic location is discussed by Dorothy Koenigsberger, *Renaissance man and creative thinking: a history of concepts of harmony 1400–1700* (Hassocks, 1979). The impetus given by the Copernican innovation to speculation concerning a plurality of worlds is discussed by Steven J. Dick, *Plurality of worlds: the extraterrestrial life debate from Democritus to Kant* (Cambridge, 1982), pp. 61–105. A comprehensive treatment of this recurrent theme in the modern period is provided by Michael J. Crowe, *The extraterrestrial life debate 1750–1900: the idea of a plurality of worlds from Kant to Lowell* (Cambridge, 1986).

The crucial distinction between an astronomical innovation as a mathematical convenience and as a representation of physical reality was emphasized by Pierre Duhem, *To save the phenomena,* trans. Edmund Doland and Chaninah Maschler (Chicago, 1969), and is elaborated in the context of the Wittenberg school by Robert S. Westman, "The Melanchthon circle, Rheticus and the Wittenberg interpretation of the Copernican theory," *Isis* 66 (1975), 165–93. The importance of the same distinction in the evolving relationship between Galileo and the Roman Catholic authorities is made clear in Stillman Drake, *Galileo at work: his scientific biography* (Chicago, 1978). Useful data on the adoption of the Copernican system in the sixteenth century, and the special case of de Zuñiga, are presented by Robert S. Westman, "The Copernicans and the Churches," in Lindberg and Numbers, *God and nature,* pp. 76–113. The same author suggests reasons why it was *court* astronomers who did most to promulgate the Copernican system in "The astronomer's role in the sixteenth century: a preliminary study," *History of Science* 18 (1980), 105–74. The Medici court, as a liberating source of patronage for Galileo, is discussed by Mario Biagioli, "Galileo the emblem maker," *Isis* 81 (1990), 230–58. The attractions of adopting one but not all the Copernican motions is illustrated in Paul H. Kocher, *Science and religion in Elizabethan England* (New York, 1969), pp. 189–200. The reasons for Tycho's unwillingness to adopt a heliostatic model are explored by Kristian P. Moesgaard in Dobrzycki, *The reception of Copernicus's heliocentric theory,* pp. 31–55. For an illuminating account of the Tychonic system, the position taken by Ursus, and Kepler's intervention, see Nicholas Jardine, *The birth of history and philosophy of science: Kepler's A defence of Tycho against Ursus, with essays on its provenance and significance* (Cambridge, 1984). The metaphysics that underpinned Kepler's quest for physical causes is expounded by Gerald Holton, "Johannes Kepler's universe: its physics and metaphysics," in *Thematic origins of scientific thought* (Cambridge, Mass., 1973), pp. 69–90. Kepler's enduring preoccupation with a geometrical harmony of the heavens is the subject of Judith V. Field, *Kepler's geometrical cosmology* (London, 1988). The integration of his cosmography with a Trinitarian theology is considered by Richard S. Westfall, "The rise of science and the decline of orthodox Christianity: a study of Kepler, Descartes and Newton," in Lindberg and Numbers, *God and nature,* pp. 218–37.

The complexity of Bacon's response to the current state of astronomy is reviewed by Antonio Perez-Ramos, "Francis Bacon and astronomical enquiry," *British Journal for the History of Science* 23 (1990), 197–205. For a general survey of religion in the periods of the Reformation and Counter-Reformation, see John Bossy, *Christianity in the West 1400–1700* (Oxford 1985). The Lutheran movement is discussed by R. W. Scribner, *The German Reformation* (London, 1986). The Calvinist Reformation is the subject of Menna Prestwich, ed., *International Calvinism 1541–1715* (Oxford, 1985). See also Michael Mullett, *Calvin* (London, 1989). Perspectives on the Catholic recovery are provided by Jean Delumeau, *Catholicism between Luther and Voltaire: a new view of the Counter-Reformation* (London, 1977). Osiander's role in the public presentation of Copernicus's book is sympathetically reconsidered by Bruce Wrightsman, "Andreas Osiander's contribution to the Copernican achievement," in Westman, *The Copernican achievement,* pp. 213–43. On the shift from the exegetical principles of the Reformers to more legalistic attitudes toward biblical authority, see P. Lehmann, "The Reformers' use of the Bible," *Theology Today* 3 (1946),

328–44, and John Dillenberger, *Protestant thought and natural science* (London, 1961), pp. 50–74. The documentary basis for Luther's alleged (and legendary) disparagement of Copernicus is called into question by Wilhelm Norlind, "Copernicus and Luther," *Isis* 44 (1953), 273–6. For Calvin's attitude, see Robert White, "Calvin and Copernicus: the problem reconsidered," *Calvin Theological Journal* 15 (1980), 233–43.

The problems created for Galileo by decisions taken at the Council of Trent are highlighted by Olaf Pederson, "Galileo and the Council of Trent," *Journal of the History of Astronomy* 14 (1983), 1–29. The role of Galileo's friend Sarpi in the Venetian revolt against Rome is outlined by Paul F. Grendler, *The Roman Inquisition and the Venetian press, 1540–1605* (Princeton, 1977). That Galileo misrepresented Copernicus's own relations with Rome is shown by Edward Rosen, "Galileo's mis-statements about Copernicus," *Isis* 49 (1958), 319–30. Some of the problems confronting Urban VIII are identified by E. A. Gosselin and L. S. Lerner, "Galileo and the long shade of Bruno," *Archives Internationales d'Histoire des Sciences* 25 (1975), 223–46. That Galileo had a genuine wish to uphold the credentials of Catholic science and had the best interests of his Church at heart is an interpretation explicit in the several studies of Galileo by Stillman Drake, including his brief biography, *Galileo* (Oxford, 1980). The extent of Galileo's debt to the Jesuits on questions of methodology is analyzed by William A. Wallace, *Galileo and his sources: the heritage of the Collegio Romano in Galileo's science* (Princeton, 1984). Galileo's protracted dispute with Grassi can be followed in Stillman Drake and C. D. O'Malley, *The controversy on the comets of 1618* (Philadelphia, 1960), while the issues in the earlier dispute with Scheiner are discussed by William R. Shea in *Galileo's intellectual revolution,* 2d ed. (New York, 1977), and in his "Galileo, Scheiner, and the interpretation of sunspots," *Isis* 61 (1970), 498–519. The role of certain Jesuits in Galileo's downfall is given special emphasis by Giorgio de Santillana, *The crime of Galileo* (Chicago, 1955). Access to crucial documents in the affair has been made easier by Maurice A. Finnocchiaro, ed., *The Galileo affair: a documentary history* (Berkeley, 1989).

The career of John Wilkins, which so contrasts with that of Galileo, is discussed in detail by Barbara J. Shapiro, *John Wilkins* (Berkeley, 1969). Wilkins also features prominently in her later study: *Probability and certainty in seventeenth-*

century England: a study of the relationships between natural science, religion, history, law and literature (Princeton, 1983). On the state of science at Oxford during Wilkins's early career, see Nicholas Tyacke, "Science and religion at Oxford before the Civil War," in *Puritans and revolutionaries: essays in seventeenth-century history presented to Christopher Hill,* ed. Donald H. Pennington and Keith Thomas (Oxford, 1978), pp. 73–93. That discrimination is required when assessing the oppressive effects of the Inquisition is argued by Grendler, *The Roman Inquisition and the Venetian press,* chap. 10. Important perspectives on Jesuit educational institutions and their role in promoting the physical sciences can be found in John L. Heilbron, *Elements of early modern physics* (Berkeley, 1982), pp. 93–106.

The state of the debate over Merton's thesis at the time his original essay [1938] was republished [1970] can be ascertained from the collection of essays edited by Charles Webster, *The intellectual revolution of the seventeenth century* (London, 1974). Among earlier critiques that of A. Rupert Hall, "Merton revisited," *History of Science* 2 (1963), 1–16, was severest. There had also appeared the iconoclastic view that the scientific movement in seventeenth-century England was the product of a hedonist rather than a puritan ethic: Lewis Feuer, *The scientific intellectual: the psychological and sociological origins of modern science* (New York, 1963). By contrast, the existence of structural parallels between reformed theology and the presuppositions of modern science was affirmed by Stephen Mason, "The Scientific Revolution and the Protestant Reformation," *Annals of Science* 9 (1953), 64–87 and 154–75. The relevance of a millenarian theology to the high value placed on the applied sciences by intellectual leaders of the puritan revolution was then shown by Webster, *The great instauration.* Webster has also provided a useful update and a sensitive reply to critics in "Puritanism, separatism, and science," in Lindberg and Numbers, *God and nature,* pp. 192–217. There are also useful insights in P. M. Rattansi, "The social interpretation of seventeenth-century science," in Peter Mathias, ed., *Science and society 1600–1900* (Cambridge, 1972), pp. 1–32.

The problems that arise in *testing* Merton's thesis and its derivatives are discussed by Arnold Thackray, "Natural knowledge in cultural context: the Manchester model," *American Historical Review* 79 (1974), 672–709, and by John H. Brooke, "Joseph Priestley (1733–1804) and William

Whewell (1794–1866): apologists and historians of science,"
in *Science, medicine and dissent: Joseph Priestley,* ed. R. G. W.
Anderson and Christopher Lawrence (London, 1987), pp. 11–
27. To add to the complications, it has been argued that historians have failed to understand Merton's original thesis: Gary
A. Abraham, "Misunderstanding the Merton thesis: a boundary dispute between history and sociology," *Isis* 74 (1983),
368–87. The notion that there might be distinctive mind-sets
characteristic of either Catholic or Protestant science is opposed by William B. Ashworth Jr., "Catholicism and early
modern science," in Lindberg and Numbers, *God and nature,*
pp. 136–66. The methodological objection that scholars favoring a puritan stimulus to science have examined only those
"puritans" who were interested in science is urged by John
Morgan, "Puritanism and science: a reinterpretation," *The
Historical Journal* 22 (1979), 535–60. The same author's *Godly
learning: puritan attitudes towards reason, learning and education, 1560–1640* (Cambridge, 1986) gives an authoritative
account of the limits and proper uses of reason as originally
conceived by puritan divines. A statistical test of Merton's
thesis based on the allegiance of early Fellows of the Royal
Society is provided by Lotte Mulligan, "Civil War politics,
religion and the Royal Society," in Webster, *The intellectual
revolution of the seventeenth century,* pp. 317–39. See also her
"Puritanism and English science: a critique of Webster," *Isis*
71 (1980), 456–69. The thesis that latitudinarian rather than
puritan religious attitudes were the more propitious for the
theoretical sciences is argued by Barbara J. Shapiro, "Latitu-
dinarianism and science in seventeenth century England," in
Webster, *The intellectual revolution of the seventeenth century,*
pp. 286–316. A possible way out of the impasse created by
the conflicting correlations is suggested by James R. Jacob
and Margaret C. Jacob, "The Anglican origins of modern science," *Isis* 71 (1980), 251–67. Additional, more recent,
commentaries on Merton's thesis have featured in special
thematic issues of two journals: *Isis* 79 (1988), 571–605, and
Science in Context 3 (1989).

IV. Divine Activity in a Mechanical Universe

The so-called mechanization of the world-picture has attracted discussion at many levels of sophistication. Introductory texts include John H. Brooke and David C. Goodman,

Towards a mechanistic philosophy (Milton Keynes: Open University, 1974); Brian Easlea, *Witch-hunting, magic and the new philosophy* (Hassocks, 1980); and for a feminist perspective: Carolyn Merchant, *The death of nature: women, ecology and the scientific revolution* (San Francisco, 1979). The consequences of a mechanistic program for each branch of the sciences are analyzed by Richard S. Westfall, *The construction of modern science* (New York, 1971). The reinterpretation in mechanical terms of the Aristotelian concept of *form* is discussed in more detail by E. J. Dijksterhuis, *The mechanization of the world-picture* (Oxford, 1961); Robert H. Kargon, *Atomism in England from Hariot to Newton* (Oxford, 1966); Marie Boas Hall, *Robert Boyle and seventeenth century chemistry* (Cambridge, 1958), and Norma E. Emerton, *The scientific reinterpretation of form* (Ithaca, N.Y., 1984). Although it may overstate both the novelty and gravity of the problems posed for doctrines of providence by the mechanical philosophies of nature, Richard S. Westfall's *Science and religion in seventeenth century England* (New Haven, 1958) remains a standard text. There is a useful complement in J. S. Spink, *French free thought from Gassendi to Voltaire* (London, 1960). Two of the more interesting philosophical problems associated with the assimilation of nature to machinery are discussed respectively by Laurens Laudan, "The clock metaphor and probabilism: the impact of Descartes on English methodological thought, 1650–65," *Annals of Science* 22 (1966), 73–104, and Maurice Mandelbaum, *Philosophy, science and sense-perception* (Baltimore, 1964), pp. 61–117.

By stressing the diversity of "mechanical philosophies" and their continuities with previous forms of naturalism, recent scholarship has undermined older attempts to depict a triumph of mechanism over the irrational. See, for example, Ron Millen, "The manifestation of occult qualities in the scientific revolution," in *Religion, science and worldview: essays in honor of Richard S. Westfall*, ed. Margaret J. Osler and Paul L. Farber (Cambridge, 1985), pp. 185–216; John Henry, "Newton, matter, and magic," in *Let Newton be!*, ed. John Fauvel, Raymond Flood, Michael Shortland, and Robin Wilson (Oxford, 1988), pp. 127–45; and Simon Schaffer, "Godly men and mechanical philosophers: souls and spirits in Restoration natural philosophy," *Science in Context* 1 (1987), 55–85. My reference in the text to Robert Fludd's conception of the weapon salve as an example of premechanical science is taken from Allen G. Debus, "Robert Fludd and the use of Gilbert's *De magnete* in the weapon-salve controversy," *Journal of the History of Medicine* 19 (1964), 389–417. It should be stressed, however, that when mechanical models were applied to magnetic attraction, the result was a proliferation of possibilities, not convergence toward a uniquely rational solution: see Stephen Pumfrey, "Mechanizing magnetism in Restoration England – the decline of magnetic philosophy," *Annals of Science* 44 (1987), 1–22.

The importance for the physical sciences of a method of abstraction leading to the construction of idealized mathematical models is emphasized by Alexandre Koyré, *Metaphysics and measurement* (London, 1968), pp. 1–43. That the mechanical philosophies reflected a new ideal of knowing-by-doing is one of the theses of Funkenstein, *Theology and the scientific imagination from the middle ages to the seventeenth century,* pp. 290–345. That mechanical metaphors may have conferred dignity on the use of apparatus and instruments hitherto associated with mathematicians is suggested by J. A. Bennett, "The mechanics' philosophy and the mechanical philosophy," *History of Science* 24 (1986), 1–28. On the diversity of particles employed by mechanical philosophers, and for the point that English mechanists did not always follow Descartes in divesting matter of all inherent powers, see John Henry, "Occult qualities and the experimental philosophy: active principles in pre-Newtonian matter theory," *History of Science* 24 (1986), 335–81. That a mechanical philosophy did not remove the prospect of metallic transmutation is argued in the case of Boyle by Thomas S. Kuhn, "Robert Boyle and structural chemistry in the seventeenth century," *Isis* 43 (1952), 12–36. That there was no direct correlation between espousal of a mechanical philosophy and skepticism toward the possibility of witchcraft is made clear in T. H. Jobe, "The Devil in Restoration science: the Glanvill–Webster witchcraft debate," *Isis* 72 (1981), 343–56.

Access to a further literature on the theological ramifications of seventeenth-century atomism can be gained from John Henry, "Atomism and eschatology: Catholicism and natural philosophy in the Interregnum," *British Journal for the History of Science* 15 (1982), 211–39. For the English Catholic Sir Kenelm Digby an impregnable order in the physical universe served to highlight what was distinctive about the realm of the spirit: see Anne Bäumer, "Christian Aristotelianism and atomism in embryology: William Harvey, Kenelm Digby, Nathaniel Highmore," in *Science and religion,* ed. Anne Bäumer and Manfred Büttner, Proceedings of a symposium held at the XVIIIth International Congress of History of Science, Hamburg–Munich 1–9 August 1989 (Abhandlungen zur Geschichte Geowissenschaften und Religion/Umweltforschung, Bd. 3), Bochum, 1989. The centrality of theological motives in Mersenne's adoption of a mechanistic physics is argued by Robert Lenoble, *Mersenne ou la naissance du mécanisme,* 2d ed. (Paris, 1971), and in large measure accepted in the more recent study by Peter Dear, *Mersenne and the learning of the schools* (Ithaca, N.Y., 1988).

Descartes's mechanical philosophy, as applied to living organisms, is discussed by Norman Kemp Smith, *New studies in the philosophy of Descartes* (London, 1953), pp. 125–60. His arguments for the uniqueness and immateriality of the human mind can be read in his "Sixth Meditation," in *Descartes: philosophical writings,* ed. Elizabeth Anscombe and Peter Thomas Geach (London, 1954), pp. 109–24. The incorporation of the beast-machine doctrine in a religious apologia is documented by Leonora C. Rosenfield, *From beast-machine to man-machine: animal soul in French letters from Descartes to La Mettrie* (New York, 1968). For a twentieth-century reaffirmation of the view that the indivisibility of the human ego may require nonphysical explanation, see Sir John Eccles, *The brain and the unity of conscious experience* (Cambridge, 1965). That Decartes was, however, guilty of a category mistake in treating the human mind as a separate substance is the contention of Gilbert Ryle, *The concept of mind* (London, 1949), pp. 11–24. The strength of the case for differentiation between humans and animals is considered in Mary Midgley, *Beast and man* (London, 1980). Respects in which Cartesian doctrines could lead to an exploitative attitude toward both animals and the rest of nature are placed in broader perspective by John Passmore, *Man's responsibility for nature* (London, 1974). One should, however, be wary of general theses linking exploitative mentalities with some supposedly dominant theology of creation, whether Christian or otherwise. Criticism of Passmore will be found in R. Attfield, "Christian attitudes to nature," *Journal of the History of Ideas* 44 (1983), 369–86. The ambivalence of Boyle's position with respect to animal suffering is documented by Malcolm R. Oster, "The 'beaume of diuinity': animal suffering in the early thought of

Robert Boyle," *British Journal for the History of Science* 22 (1989), 151–79. See also Michael Hunter, "Alchemy, magic and moralism in the thought of Robert Boyle," *British Journal for the History of Science* 23 (1990), 387–410.

Formative influences on the development of Boyle's natural philosophy and religious sensibilities are currently the subject of reinvestigation. The contextual studies that provide the main basis for discussion are James R. Jacob, *Robert Boyle and the English Revolution: a study in social and intellectual change* (New York, 1977), and Shapin and Schaffer, *Leviathan and the air-pump.* For current issues in Hobbes scholarship, see G. A. J. Rogers and Alan Ryan, eds., *Perspectives on Thomas Hobbes* (Oxford, 1989). Primary sources permitting the study of Boyle's matter theory, mechanical philosophy, and natural theology are available in Marie Boas Hall, *Robert Boyle on natural philosophy* (Bloomington, Ind., 1965); Goodman, *Science and religious belief 1600–1900;* and M. A. Stewart, ed., *Selected philosophical papers of Robert Boyle* (Manchester, 1979). Boyle's voluntarist theology is discussed by J. E. McGuire, "Boyle's conception of nature," *Journal of the History of Ideas* 33 (1972), 523–42.

Probably the best *introduction* to the many facets of Newton's life and work is now Fauvel et al., *Let Newton be!* There is, however, no substitute for the fine biography by Richard S. Westfall, *Never at rest* (Cambridge, 1984). A succinct account of the development of Newton's celestial dynamics is given by D. T. Whiteside, "Before the *Principia:* the maturing of Newton's thoughts on dynamical astronomy, 1664–1684," *Journal for the History of Astronomy* 1 (1970), 5–19. Whiteside's edition of *The mathematical papers of Isaac Newton,* 8 vols. (Cambridge, 1967–81) constitutes a definitive account of Newton the mathematician. On Newton the physical theorist, see Richard S. Westfall, *Force in Newton's physics* (London, 1971). For Newton the historian, see Frank E. Manuel, *Isaac Newton, historian* (Cambridge, 1963). Newton the alchemist is the subject of Betty Jo Dobbs, *The foundations of Newton's alchemy* (Cambridge, 1975); Richard S. Westfall, "Newton and alchemy," in Vickers, *Occult and scientific mentalities in the Renaissance,* pp. 315–35; and Jan Golinski, "The secret life of an alchemist," in Fauvel et al., *Let Newton be!,* pp. 147–67. Newton's religious preoccupations are discussed by Frank E. Manuel, *The religion of Isaac Newton* (Oxford, 1974). The suggestion that Newton's theology was so radical as to be verging on deism is made by Richard S. Westfall, "Isaac Newton's 'Theologiae Gentilis Origines Philosophiae,' " in *The secular mind: transformations of faith in modern Europe,* ed. W. Warren Wagar (New York, 1982), pp. 15–34. Some doubts about this interpretation have been expressed by John H. Brooke, "The God of Isaac Newton," in Fauvel et al., *Let Newton be!,* pp. 168–83, and by James E. Force, "The Newtonians and deism," in *Essays on the context, nature and influence of Isaac Newton's theology,* ed. James E. Force and Richard H. Popkin (Dordrecht, 1989), chap. 4. The voluntarist element in Newton's theology is explored by Henry Guerlac, "Theological voluntarism and biological analogies in Newton's physical thought," *Journal of the History of Ideas* 44 (1983), 219–29. The theological connotations of his concept of space are discussed by B. P. Copenhaver, "Jewish theologies of space in the scientific revolution: Henry More, Joseph Raphson, Isaac Newton and their predecessors," *Annals of Science* 37 (1980), 489–548; and by J. E. McGuire, "Newton on place, space, time and God: an unpublished source," *British Journal for the History of Science* 11 (1978), 114–29.

References to religious apprehension occasioned by the inroads of Cartesian mechanism can be found in Gerald Aylmer, "Unbelief in seventeenth-century England," in *Puritans and revolutionaries: essays in seventeenth-century history presented to Christopher Hill,* ed. Donald H. Pennington and Keith Thomas (Oxford, 1978), pp. 22–46, and in Michael Hunter, *Science and society in Restoration England* (Cambridge, 1981). On the particular problems raised for doctrines of transubstantiation, see Trevor McClaughlin, "Censorship and defenders of the Cartesian faith in mid-seventeenth century France," *Journal of the History of Ideas* 40 (1979), 563–81; Richard A. Watson, "Transubstantiation among the Cartesians," in Thomas M. Lennon, ed., *Problems of Cartesianism* (Kingston, Ont., 1982), pp. 127–48; and Pietro Redondi, *Galileo heretic* (Princeton, 1988). The perception (as in Ray's response to Boyle) of deistic tendencies in theologies celebrating mechanism is noted in H. R. McAdoo, *The spirit of Anglicanism* (London, 1965). The problems raised by Newton's attempt to heighten a sense of divine involvement in the world are discussed by David H. Kubrin, "Newton and the cyclical cosmos: Providence and the mechanical philosophy," *Journal of the History of Ideas* 28 (1967), 325–46; and, as an example of a more general problem, by John H. Brooke, "Science and the fortunes of natural theology: some historical perspectives," *Zygon* 24 (1989), 3–22.

The various meanings that Newton attached to his gravitational force are analyzed by Ernan McMullin, *Newton on matter and activity* (Notre Dame, Ind., 1978). Newton's association of providence with the stability of the stars is examined by Michael Hoskin, "Newton, Providence and the universe of stars," *Journal for the History of Astronomy* 8 (1977), 77–101. That Newton did not reduce the moral and theological function of comets is argued by Simon Schaffer, "Newton's comets and the transformation of astrology," in *Astrology science and society: historical essays,* ed. Patrick Curry (Woodbridge, 1987), pp. 219–43. The subsequent "correction" of Newton by Lagrange and Laplace features in Herbert H. Odom, "The estrangement of celestial mechanics and religion," *Journal of the History of Ideas* 27 (1966), 533–48; and in Roger Hahn, "Laplace and the mechanistic universe," in Lindberg and Numbers, *God and nature,* pp. 256–95. The earlier view, as expressed by William Whiston, that Newton's science gave exciting support both for providence and a biblical literalism, is amplified in James E. Force, *William Whiston: honest Newtonian* (Cambridge, 1985).

V. Science and Religion in the Enlightenment

There is a valuable introduction to the place of science in Enlightenment culture and a concomitant bibliographic essay in Thomas L. Hankins, *Science and the Enlightenment* (Cambridge, 1985). Some of the more persistent problems of historical interpretation are addressed in G. S. Rousseau and Roy Porter, eds., *The ferment of knowledge: studies in the historiography of Enlightenment science* (Cambridge, 1980). That there was no monolithic "Enlightenment," but different manifestations in different national contexts is clear from Roy Porter and Mikulas Teich, eds., *The Enlightenment in national context* (Cambridge, 1981). Political aspects of eighteenth-century movements for reform are considered by F. Venturi, *Utopia and reform in the Enlightenment* (Cambridge, 1971). The development of philosophy during the eighteenth century is discussed from a neo-Kantian standpoint by Ernst Cassirer, *Philosophy of the Enlightenment,* trans. Fritz C. A. Koelln and James P. Pettegrove (Boston, 1951). A substantial overview, characterized by an emphasis on the liberating effects of a knowledge of classical cultures, is that of Peter Gay, *The Enlightenment, an interpretation,* 2 vols. (New York, 1969). An alternative approach to the Enlightenment in France, informed by the study of popular culture, can be seen in Robert Darnton, "In search of the Enlightenment: recent attempts to create a social history of ideas," *Journal of Modern History* 43 (1971), 113–32, and "The High Enlightenment and the low-life of literature in pre-Revolutionary France," *Past and Present* 51 (1971), 81–115. The extent to which the natural sciences in the eighteenth century may be regarded as agents of "secularization" is considered in John H. Brooke, "Science and the secularisation of knowledge: perspectives on some eighteenth-century transformations," *Nuncius* 4 (1989), 43–65. The rhetoric of utility, which accompanied the promotion of the sciences, operated on many levels: J. G. Burke, ed., *The uses of science in the age of Newton* (Berkeley, 1983). Its character in the case of that most utilitarian of sciences – chemistry – is well discussed by Jan V. Golinski, "Utility and audience in eighteenth-century chemistry: case studies of William Cullen and Joseph Priestley," *British Journal for the History of Science* 21 (1988), 1–31. For the organization of the sciences in the eighteenth century and the expansion of scientific societies, see James E. McClellan. *Science reorganized: scientific societies in the eighteenth century* (New York, 1985). Although a little dated, Gerald R. Cragg, *The Church and the Age of Reason 1648–1789,* 2d ed. (Harmondsworth, 1966), has the merit of covering the state of the Christian churches in a range of national contexts. How the teaching of natural philosophy in French colleges of higher education could help (albeit unwittingly) to create an audience for the philosophes is one theme in the important study by L. W. B. Brockliss, *French higher education in the seventeenth and eighteenth centuries* (Oxford, 1987). The intimate connections between political and religious controversy in early eighteenth-century England are exposed by Geoffrey Holmes, *Politics, religion and society in England 1679–1742* (London, 1986). A recent survey, particularly useful for attitudes within the Anglican Church toward Newton and his popularizers, is John Gascoigne, *Cambridge in the age of the Enlightenment: science, religion and politics from the Restoration to the French Revolution* (Cambridge, 1989). For a review of the literature that looks at the Enlightenment from a political angle, see Lester G.

Crocker, "Interpreting the Enlightenment: a political approach," *Journal of the History of Ideas* 46 (1985), 211–30.

The manner in which Newton and Locke were perceived to have set new methods and standards of rigorous knowledge has been the subject of an extensive literature. A useful introduction, with illustrative texts, is Gerd Buchdahl, *The image of Newton and Locke in the Age of Reason* (London, 1961). Two studies by Henry Guerlac provide a fuller picture: *Newton on the Continent* (Ithaca, N.Y., 1981) and "Newton's changing reputation in the eighteenth century," in *Carl Becker's heavenly city revisited*, ed. Raymond O. Rockwood (Ithaca, N.Y., 1958), pp. 3–26. Further bibliographic information on this theme can be obtained from Simon Schaffer, "Newtonianism," in Olby et al., *Companion to the history of modern science,* pp. 610–26. The political and intellectual contexts in which Locke had forged his views on religious toleration are analyzed by Richard Ashcraft, *Revolutionary politics and Locke's "Two Treatises of Government"* (Princeton, 1986). The political implications of Locke's epistemology are considered by Neal Wood, *The politics of Locke's philosophy: a social study of "An Essay Concerning Human Understanding"* (Berkeley, 1983).

Changing sensibilities during the eighteenth century to concepts of divine intervention are considered in the context of the physical sciences by Roger Hahn, "Laplace and the mechanistic universe," in Lindberg and Numbers, *God and nature,* pp. 256–76; in the context of the earth sciences by Roy Porter, "Creation and credence: the career of theories of the Earth in Britain 1660–1820," in *Natural order,* ed. Barry Barnes and Steven Shapin (Beverly Hills, Calif., 1979), pp. 97–124, and by Paolo Rossi, *The dark abyss of time* (Chicago, 1984); and in medical contexts by Roy Porter, *Disease, medicine and society in England 1550–1860* (London, 1987). See also Roy Porter and Dorothy Porter, *In sickness and in health: the British experience 1650–1850* (London, 1988). On the perception of religious enthusiasm as a form of madness, see Roy Porter, "The rage of party: a glorious revolution in English psychiatry?" *Medical History* 27 (1983), 35–50.

On the development in Britain of a rhetoric for dependence on the practical utility of science, see Robert E. Schofield, *The Lunar society of Birmingham* (Oxford, 1963). That "utility" could, however, include religious utility is clear from Arnold Thackray, "Natural knowledge in cultural context: the Manchester model," *American Historical Review* 79 (1974),

672–709, and Derek Orange, "Rational dissent and provincial science: William Turner and the Newcastle Literary and Philosophical Society," in *Metropolis and province: science in British culture, 1780–1850,* ed. Ian Inkster and Jack Morrell (London, 1983), pp. 205–30. For eighteenth-century displays of God's latent powers in nature, see Simon Schaffer, "Natural philosophy and public spectacle in the eighteenth century," *History of Science* 21 (1983), 1–43. The role of "latitudinarian" Anglican clergy in popularizing Newton's science, and their political motives for so doing, are emphasized by Margaret C. Jacob, *The Newtonians and the English Revolution, 1689–1720* (Ithaca, N.Y., 1976). The arguments deployed by Bentley, the first Boyle lecturer, can be studied in I. Bernard Cohen, ed., *Isaac Newton's papers and letters on natural philosophy and related documents,* 2d ed. (Cambridge, Mass., 1978), pp. 271–394. Newton's resistance in Cambridge to the wishes of James II is discussed by Westfall, *Never at rest,* pp. 474–9.

The attempt of Margaret Jacob to correlate the promotion of Newtonian matter theory with a distinct set of socioeconomic and political interests has engendered many rejoinders. There are critical remarks in Geoffrey Holmes, "Science, reason and religion in the age of Newton," *British Journal for the History of Science* 11 (1978), 164–71, and Colin A. Russell, *Science and social change 1700–1900* (London, 1983), pp. 52–68. The role of Scotsmen in making the *Principia* accessible is stressed by Gascoigne, *Cambridge in the age of the Enlightenment,* pp. 145–9. That it was possible to be a Newtonian and a tory with high Church leanings is shown by A. Guerrini, "The tory Newtonians: Gregory, Pitcairne and their circle," *Journal of British Studies* 25 (1986), 288–311. Proponents of a correlation between "Newtonianism" and "latitudinarian" theology must also take into account the definitional problems discussed by John Spurr, " 'Latitudinarians' and the Restoration Church," *The Historical Journal* 31 (1988), 61–82. For reasons why such Newtonian spokesmen as Bentley and Clarke might be perceived as anything but orthodox Anglicans, see Eamonn Duffy, " 'Whiston's affair': the trials of a primitive Christian 1709–14," *Journal of Ecclesiastical History* 27 (1976), 129–50.

The objections of Leibniz to the natural philosophy of Newton can be studied in H. G. Alexander, ed., *The Leibniz–Clarke correspondence* (Manchester, 1956). They are placed in wider context by Steven Shapin, "Of Gods and Kings: natural philosophy and politics in the Leibniz–Clarke disputes," *Isis* 72 (1981), 187–215. For an accessible introduction to the philosophy of Leibniz, see Stuart C. Brown, *Leibniz* (Hassocks, 1984). Another recent study is that of E. J. Aiton, *Leibniz: a biography* (Bristol, 1985). The animosity between Newton and Leibniz over priority for the calculus is documented by A. Rupert Hall, *Philosophers at war: the quarrel between Newton and Leibniz* (Cambridge, 1980). Voltaire's enthusiasm for Newton's achievement is discussed by P. M. Rattansi, "Voltaire and the Enlightenment image of Newton," in *History and imagination,* ed. H. Lloyd-Jones (London, 1981), pp. 218–31. The passionate concern of Voltaire for greater religious toleration is explained by J. Brumfitt, *The French Enlightenment* (London, 1972). On demands for greater religious toleration in England during the eighteenth century, see Martin Fitzpatrick, "Joseph Priestley and the cause of universal toleration," *The Price–Priestley Newsletter* 1 (Aberystwyth, 1977), 3–30, and "Toleration and truth," *Enlightenment and Dissent* 1 (1982), 3–31.

That the Christian Churches in the early eighteenth century had more to fear from ridicule than from reason is argued by John Redwood, *Reason, ridicule and religion: the age of Enlightenment in England 1660–1750* (London, 1976). The "deist" threat to institutionalized Christianity can also be approached through Peter Gay, ed., *Deism: an anthology* (Princeton, 1968). The controversy precipitated by Toland's *Christianity not mysterious* and the manifold contemporary meanings of "deism" as a term of abuse are discussed by Robert E. Sullivan, *John Toland and the deist controversy* (Cambridge, Mass., 1982), pp. 109–40 and 205–34. See also Stephen Daniel, *John Toland: his mind, manners and thought* (Toronto, 1984). That Toland explicitly reinterpreted Newton's conception of matter to favor a pantheist creed is shown by Jacob, *The Newtonians and the English Revolution,* chap. 6. My discussion of Tindal in this section of the text is simply based on the content of his *Christianity as old as the creation.* For further analysis of his religious position, see Stephen N. Williams, "Matthew Tindal on perfection, positivity and the life divine," *Enlightenment and Dissent* 5 (1986), 51–69. That the deists' use of Locke was "flashy and superficial" is also shown by John Yolton, *John Locke and the way of ideas* (Oxford, 1956). Joseph Butler's rejoinder to the deists is discussed by Anders Jeffner, *Butler and Hume on religion: a comparative analysis* (Stockholm, 1966).

On the threat to Christianity from materialism, the arguments of Jean Meslier are presented by J. S. Spink, *French free thought from Gassendi to Voltaire* (London, 1960), pp. 277–8, and their Cartesian inspiration by Jean Deprun, "Jean Meslier et l'héritage cartésien," *Studies on Voltaire and the Eighteenth Century* 24 (1963), 443–55. The long shadow of Descartes over eighteenth-century French materialism is given special prominence by Aram Vartanian, *Diderot and Descartes: a study of scientific naturalism in the Enlightenment* (Princeton, 1953). Access to a more recent literature on eighteenth-century materialism can be gained from John W. Yolton, *Thinking matter* (Oxford, 1984), and the essay review by Geoffrey Cantor in *History of Science* 23 (1985), 201–6. The "discoveries" of John Needham, Albrecht von Haller, and Abraham Trembley and their appropriation by apologists for materialism are discussed by David C. Goodman, *Scientific progress and religious dissent* (Milton Keynes: Open University, 1974). On the issues surrounding spontaneous generation, see John Farley, *The spontaneous generation controversy from Descartes to Oparin* (Baltimore, 1977). Trembley's sensational results are also discussed by Aram Vartanian, "Trembley's polyp, La Mettrie and eighteenth-century French materialism," in *Roots of scientific thought,* ed. P. P. Wiener and A. Noland (New York, 1957), pp. 497–516. Diderot's passage into an atheistic materialism is the focal point of Michael J. Buckley, *At the origins of modern atheism* (New Haven, 1987), pp. 194–250. The terms of the dispute between Haller and Wolff on the nature of embryonic development are discussed by Shirley A. Roe, *Matter, life and generation: eighteenth-century embryology and the Haller–Wolff debate* (Cambridge, 1981). The theological dimensions of earlier embryological debates are considered by Anne Bäumer, "Das Ei als Instrumentum Dei. Religion und Embryologie im 17. und 18. Jahrhundert," *Annali dell'Istituto Storico Italico-germanico* 11 (1985 [1986]), 79–102, and "Zum Verhaltnis von Religion und Zoologie im 17. Jahrhundert (William Harvey, Nathaniel Highmore, Jan Swammerdam)," *Berichte zur Wissenschaftsgeschichte* 10 (1987), 69–81. The "conservative" construction that could be placed on Trembley's polyp as a missing link in an ordered chain is noted by Arthur O. Lovejoy, *The great chain of being* (1936; reprint, New York, 1960), p. 233. Additional insight into contemporary reactions can be gained from Virginia P. Dawson, *Nature's enigma: the problem of the polyp in the letters of Bonnet, Trembley, and Réaumur* (Philadelphia, 1987).

That the philosophes, by negating every doctrine of traditional Christianity, came up with a surrogate creed of their own was the celebrated thesis of Carl Becker's *The heavenly city of the eighteenth-century philosophers* (New Haven, 1932). An introduction to the religious polemics and scientific work of Priestley can be gained from the collection of essays in Christopher Lawrence and R. G. W. Anderson, eds., *Science, medicine and dissent: Joseph Priestley 1733–1804* (London, 1987). Priestley's theology is discussed in relation to his scientific interests by Brooke, " 'A sower went forth': Joseph Priestley and the ministry of reform." See also Robert E. Schofield, "Joseph Priestley: theology, physics and metaphysics," *Enlightenment and Dissent* 2 (1983), 69–81. The most sustained attempt to integrate the many facets of Priestley's intellectual life is that of John G. McEvoy and J. E. McGuire, "God and nature: Priestley's way of rational dissent," *Historical Studies in the Physical Sciences* 6 (1975), 325–404. McEvoy also assimilates Priestley's experimental work on airs [gases] to his synoptic scheme in "Joseph Priestley, 'aerial philosopher': metaphysics and methodology in Priestley's chemical thought, from 1762 to 1781," *Ambix* 25 (1978), 1–55, 93–116, 153–75; 26 (1979), 16–38. More recently, attention has been paid to the various audiences for whom Priestley wrote, with the consequent objection that it is possible to impose too unified a scheme on his multifarious practices. For a critique of approaches that make metaphysical coherence the yardstick by which to judge Priestley's activities, see Simon Schaffer, "Priestley's questions," *History of Science* 22 (1984), 151–83; and also Jan V. Golinski and J. R. R. Christie, "The spreading of the word: new directions in historiography of chemistry, 1600–1800," *History of Science* 20 (1982), 235–66. The case of Priestley raises again the question of the role of Unitarians in English provincial scientific societies. For a recent assessment, see Jean Raymond and John V. Pickstone, "The natural sciences and the learning of the English Unitarians," in *Truth, liberty, religion: essays celebrating two hundred years of Manchester College*, ed. Barbara Smith (Oxford: Manchester College, 1986), pp. 127–64.

The Scottish context in which Hume developed his skeptical philosophy is discussed by Nicholas Phillipson, "The Scottish Enlightenment," in Porter and Teich, *The Enlightenment in national context*, pp. 19–40. Phillipson's recent study *Hume* (London, 1989) emphasizes the political predilections that informed Hume's work as a historian. Reference may also be made to James K. Cameron, "Theological controversy: a factor in the origins of the Scottish Enlightenment," in *The origin and nature of the Scottish Enlightenment*, ed. R. H. Campbell and A. S. Skinner (Edinburgh, 1982), pp. 116–30. That many Scots in the eighteenth century came to expect more from natural philosophy and the sciences than they could provide is suggested by Roger L. Emerson, "Natural philosophy and the problem of the Scottish Enlightenment," *Studies on Voltaire and the Eighteenth Century* 242 (1986), 243–91. The specific development of chemistry in Scotland is the subject of Arthur L. Donovan, *Philosophical chemistry in the Scottish Enlightenment* (Edinburgh, 1975).

Hume's critique of natural theology is discussed by Robert H. Hurlbutt, *Hume, Newton and the design argument* (Lincoln, Neb., 1965) and by Jeffner, *Butler and Hume on religion*. There is a useful anthology of Hume's writings on religion in Richard Wollheim, ed., *Hume on religion* (London, 1963). The relevance of his presbyterian background is explored by Stewart R. Sutherland, "The presbyterian inheritance of Hume and Reid," in Campbell and Skinner, *The origin and nature of the Scottish Enlightenment*, pp. 131–49. There is a comprehensive treatment of Hume's philosophy of religion in J. C. A. Gaskin, *Hume's philosophy of religion* (London, 1978). Access to issues currently under debate among Hume scholars may be gained from M. A. Stewart, ed., *Studies in the philosophy of the Scottish Enlightenment*, Oxford Studies in the History of Philosophy, vol. 1 (Oxford, 1990), in which the essays by David Wootton, "Hume's 'Of miracles': probabililty and irreligion" (pp. 191–229), and by Michael Barfoot, "Hume and the culture of science in the early eighteenth century" (pp. 151–90), are of special interest. For Hume's exploitation of Newton's ether, see E. C. Mossner and John V. Price, *Hume's Letter from a gentleman to his friend in Edinburgh* (Edinburgh, 1967), pp. 28–9. My example of Marcellin Berthelot to show how a religious agnosticism could sometimes be congruous with skepticism toward scientific entities is placed in context by Harry W. Paul, *The edge of contingency* (Gainesville, Fla. 1979), pp. 10–12.

John Wesley's attitude to the natural sciences is currently under reinvestigation, but I have based my account on Robert E. Schofield, "John Wesley and science in eighteenth-century England," *Isis* 44 (1953), 331–40, which may be read in conjunction with T. Q. Campbell, "John Wesley, Conyers Middleton and divine intervention in history," *Church History* 55 (1986), 39–59. On Hutchinson's natural philosophy, there are useful remarks in Geoffrey Cantor, "Light and Enlightenment: an exploration of mid-eighteenth-century modes of discourse," in *The discourse of light from the Middle Ages to the Enlightenment,* by David C. Lindberg and Geoffrey Cantor, (Los Angeles, 1985), pp. 67–106. Cantor also discusses one of Hutchinson's followers, Samuel Pike, in "The theological significance of ethers," in *Conceptions of ether: studies in the history of ether theories 1740–1900,* ed. G. N. Cantor and M. J. S. Hodge (Cambridge, 1981), pp. 135–55. On the appeal of Hutchinson's system to High Church tories, see Christopher Wilde, "Hutchinsonianism, natural philosophy and religious controversy in eighteenth-century Britain," *History of Science* 18 (1980), 1–24. The diversity of beliefs about providence to be found in eighteenth-century America, and the consequent inadmissibility of theses suggesting a general decline, are stressed by Henry F. May, "The decline of Providence?" *Studies on Voltaire and the 18th-century* 154 (1976), 1401–16.

VI. The Fortunes and Functions of Natural Theology

There is a general introduction to eighteenth-century natural theology in John H. Brooke, "Natural theology in Britain from Boyle to Paley," in *New interactions between theology and natural science,* by John H. Brooke and R. Hooykaas (Milton Keynes: Open University, 1974), pp. 5–54. The British natural theology literature is also discussed by Basil Willey, *The eighteenth-century background* (London, 1940); David M. Knight, *Natural science books in English 1600–1900* (London, 1972); and Lynn Barber, *The heyday of natural history* (Garden City, N.Y., 1980). A wider European focus, though coupled with an unsympathetic assessment, is provided by John Dillenberger, *Protestant thought and natural science* (London, 1961), pp. 133–90. Some weaknesses of physico-theology are also identified by Clarence J. Glacken, *Traces on the Rho-*

dian shore: nature and culture in Western thought from ancient times to the end of the eighteenth century (Berkeley, 1967), pp. 504–50. For a philosophical analysis of the different forms of the design argument, see Thomas McPherson, *The argument from design* (London, 1972).

A sympathetic account of Paley's version of the argument is given by Frederick Ferré in his introduction to *Paley: Natural theology, Selections,* Bobbs-Merrill Library of Liberal Arts (Indianapolis, 1963). That Paley capitalized on a current vogue for industrial innovation by stressing the identity (not merely the analogy) of organisms and machines is argued by Neal C. Gillespie, "Divine design and the industrial revolution," 81 (1990), 214–29. Further information on Paley can be gained from D. L. Mahieu, *The mind of William Paley: a philosopher and his age* (Lincoln, Neb., 1976). The contrasting use of reason, as illustrated by Tom Paine, in attacks on the Christian revelation is outlined by John C. Greene, *Darwin and the modern world view* (New York, 1963), pp. 13–15. The ambiguities arising from the use of design arguments by both Christians and their opponents are discussed by John H. Brooke, "The natural theology of the geologists: some theological strata," in Jordanova and Porter, *Images of the earth,* pp. 39–64. The dialectical relationship between theistic proofs and atheistic denial, together with its implications for an assessment of Newtonian physico-theology, is fully explored in Michael J. Buckley, *At the origins of modern atheism* (New Haven, 1987). The contrast between the scope accorded to natural theology in eighteenth-century physico-theology and that permitted by Calvin can be substantiated by reference to E. A. Dowey, *The knowledge of God in Calvin's theology* (New York, 1952); the contrast with the theology of Aquinas by reference to Frederick Copleston, *Thomas Aquinas* (1955; reprint, London, 1976), pp. 130–41. Connections between empiricism, the celebration of design and liberation from scholasticism are revealed in the Spanish context by David C. Goodman, "Science and the clergy in the Spanish Enlightenment," *History of Science* 21 (1983), 111–40. That an empirically based physico-theology was not merely a quirk of the British is abundantly clear from the proliferation of texts in the Netherlands, of which the best known was the apologetic of Bernard Nieuwentijt (1654–1718). See Ben Vermeulen, "Theology and science: the case of Bernard Nieuwentijt's theological positivism," in Sergio Rossi, ed., *Science and imag-*

ination in eighteenth-century British culture (Milan, 1987), pp. 379–90. French texts, including that of Pluche, are discussed by Norman Hampson, *The Enlightenment* (Harmondsworth, 1968), pp. 73–96.

The historical circumstances that encouraged the apologetic use of the argument from design have been identified in four recent studies: John H. Brooke, "Why did the English mix their science and their religion?" in Rossi *Science and imaginaton in eighteenth-century British culture,* pp. 57–78; John Gascoigne, "From Bentley to the Victorians: the rise and fall of British Newtonian natural theology," *Science in Context* 2 (1988), 219–56; Neal C. Gillespie, "Natural history, natural theology and social order: John Ray and the 'Newtonian ideology,' " *Journal of the History of Biology* 20 (1987), 1–49; and Richard Olson, "On the nature of God's existence, wisdom and power: the interplay between organic and mechanistic imagery in Anglican natural theology 1640–1740," in *Approaches to organic form,* ed. Frederick Burwick (Dordrecht, 1987), pp. 1–48. For the divine right of kings as an underground loyalty in England, after 1714, see J. C. D. Clark, *English society 1688–1832: ideology, social structure and political practice during the ancien régime* (Cambridge, 1985), pp. 141–61. My contrast in the text between France and England is particularly indebted to the essays by Roy Porter and Norman Hampson in Porter and Teich, *The Enlightenment in national context,* pp. 1–18 and 41–53. The use of natural theology by William Buckland in promoting the science of geology at Oxford is placed in context by Nicolaas A. Rupke, *The great chain of history: William Buckland and the English school of geology 1814–1849* (Oxford, 1983).

An introduction to the philosophy of Kant can be gained from S. Körner, *Kant* (Harmondsworth, 1955), and Ralph C. S. Walker, *Kant* (London, 1978). Kant's objections to physico-theology are considered by John L. Mackie, *The miracle of theism: arguments for and against the existence of God* (Oxford, 1982). My discussion in the text is particularly indebted in Michel Despland, *Kant on history and religion* (Montreal, 1973); F. E. England, *Kant's conception of God* (New York, 1968), pp. 143–68; and John P. Clayton, "Gottesbeweise," in *Theologische Realenzyklopädie,* ed. Gerhard Krause and Gerhard Müller (Berlin, 1984), pp. 724–84. Of the relevance of Kant's analysis of teleological explanation to the study of living systems, there is an excellent discussion in Timothy

Lenoir, *The strategy of life* (Dordrecht, 1982), pp. 17–53. For a comparison between Kant and Charles Darwin on organic teleology, see John F. Cornell, "Newton of the grassblade? Darwin and the problem of organic teleology," *Isis* 77 (1986), 405–21.

In specifying the continuing functions of natural theology, I have drawn on two studies of my own: John H. Brooke, "The natural theology of the geologists: some theological strata," in Jordanova and Porter, *Images of the earth*, pp. 39–64, and "Indications of a Creator: Whewell as apologist and priest," in *William Whewell: a composite portrait*, ed. Menachem Fisch and Simon Schaffer (Oxford, 1991), pp. 149–73. Whewell's natural theology is also discussed by Richard R. Yeo, "William Whewell, natural theology and the philosophy of science in mid nineteenth century Britain," *Annals of Science* 36 (1979), 493–516. Allusions in the text to Chalmers and Buckland can be pursued by reference to Crosbie Smith, "From design to dissolution: Thomas Chalmers' debt to John Robison," *British Journal for the History of Science* 12 (1979), 59–70; David Cairns, "Thomas Chalmers's *Astronomical discourses:* a study in natural theology," *Scottish Journal of Theology* 9 (1956), 410–21; and Stephen J. Gould, *Hen's teeth and horse's toes* (New York, 1983), pp. 32–45 and 81–3. The manifestation of natural theology within the British Association for the Advancement of Science is described by Jack Morrell and Arnold Thackray, *Gentlemen of science: early years of the British Association for the Advancement of Science* (Oxford, 1981), pp. 224–45. The reaction to the works of Erasmus Darwin as indicative of changing sensibilities toward radical science is outlined by N. Garfinkle, "Science and religion in England, 1790–1800: the critical response to the works of Erasmus Darwin," *Journal of the History of Ideas* 16 (1955), 376–88. See also Maurice Crosland, "The image of science as a threat: Burke versus Priestley and the 'philosophic revolution,'" *British Journal for the History of Science* 20 (1987), 277–307.

On the relevance of natural theology to science, the reference to eighteenth-century German anatomy is amplified in Andreas-Holger Maehle, " 'Est Deus ossa probant' – human anatomy and physico-theology in 17th and 18th century Germany," in Bäumer and Büttner *Science and religion*. An anatomical difficulty for natural theology is identified by William F. Bynum, "The anatomical method, natural theology, and the functions of the brain," *Isis* 64 (1973), 445–68. For the regulative role of Hutton's teleology, see Gould, *Hen's teeth and horse's toes*, pp. 79–93, and Ralph Grant, "Hutton's theory of the earth," in Jordanova and Porter, *Images of the earth*, pp. 23–38. Gould has also discussed Hutton in *Time's arrow, time's cycle: myth and metaphor in the discovery of geological time* (Cambridge, Mass., 1987). A possible theological ambivalence in Hutton is noted in Peter Harman [Heimann], "Voluntarism and immanence: conceptions of nature in eighteenth-century thought," *Journal of the History of Ideas* 39 (1978), 271–83. The role of teleological reasoning in the reconstruction of fossil forms is emphasized by Martin J. S. Rudwick, *The meaning of fossils: episodes in the history of palaeontology*, 2d ed. (New York, 1976), pp. 153–6. My earlier account of the natural theology of Richard Owen [John H. Brooke, "Richard Owen, Willam Whewell and the *Vestiges*," *British Journal for the History of Science* 19 (1977), 132–45] has been largely superseded by Evelleen Richards, "A question of property rights: Richard Owen's evolutionism reassessed," *British Journal for the History of Science* 20 (1987), 129–71. There is also a first-rate account of the cultural and political meaning of Owen's natural theology in Adrian Desmond, *Archetypes and ancestors: palaeontology in Victorian London, 1850–1875* (London, 1982). A diminishing relevance of natural theology to the practise of natural history is one theme of Neal C. Gillespie, "Preparing for Darwin: conchology and natural theology in Anglo-American natural history," *Studies in History of Biology* 7 (1983), 93–145. The association of design arguments with creationist presuppositions is inflated into a historiographic principle in Gillespie's *Charles Darwin and the problem of creation* (Chicago, 1979). Compare James R. Moore, "Creation and the problem of Charles Darwin," *British Journal for the History of Science* 14 (1981), 189–200. The plausible case for structural similarities between Darwin's universe and that of the natural theologians made by Walter [Susan] F. Cannon, "The bases of Darwin's achievement: a revaluation," *Victorian Studies* 5 (1961), 109–34, is critically engaged by M. J. S. Hodge, "The history of the earth, life, and man: Whewell's philosophy of palaetiological science," in Fisch and Schaffer, *William Whewell*, pp. 255–88.

On the relevance of scientific innovation to natural theology, further references to the Darwinian debates are given under Chapters VII and VIII. The reaction of Wallace to the inanity of certain claims for design is noted by John R. Durant, "Scientific naturalism and social reform in the thought of Alfred Russel Wallace," *British Journal for the History of Science* 12 (1979), 31–58. The diversification of natural theology earlier in the nineteenth century and the subtlety of its transformation through the assimilation of French science are discussed by John H. Brooke, "Scientific thought and its meaning for religion: the impact of French science on British natural theology, 1827–1859," *Revue de Synthèse* 4 (1989), 33–59. The special case of the use of chemistry by Humphry Davy and the romantic poet Samuel Taylor Coleridge to challenge French materialism is considered by David M. Knight, *The transcendental part of chemistry* (Folkestone, 1978), pp. 61–90, and Trevor H. Levere, *Poetry realized in nature: Samuel Taylor Coleridge and early nineteenth-century science* (Cambridge, 1981). Further insight into the cultural dimensions of scientific inquiry in nineteenth-century Britain can be obtained from Susan F. Cannon, *Science and culture: the early Victorian period* (Folkestone, 1978).

Baden Powell's changing conceptions of natural theology and the social and political pressures shaping them are fully discussed by Pietro Corsi, *Science and religion: Baden Powell and the Anglican debate, 1800–1860* (Cambridge, 1988). Two sources have been the conventional starting points for information on Chambers and *Vestiges:* Milton Millhauser, *Just before Darwin: Robert Chambers and "Vestiges"* (Middletown, Conn., 1959), and Charles C. Gillispie, *Genesis and geology* (New York, 1959), pp. 149–83. Chambers is, however, currently the subject of a major new study by James Secord, a foretaste of which can be obtained from his "Behind the veil: Robert Chambers and *Vestiges,"* in *History, humanity and evolution,* ed. James R. Moore (Cambridge, 1989), pp. 165–94. For the difficult question posed by *Vestiges* concerning the grounds of scientific authority, see Richard R. Yeo, "Science and intellectual authority: Robert Chambers and *Vestiges of the natural history of creation,"* *Victorian Studies* 28 (1984), 5–31. On phrenology as "popular science" there is a sophisticated analysis in Roger Cooter, *The cultural meaning of popular science: phrenology and the organization of consent in nineteenth-century Britain* (Cambridge, 1984). Chambers's mechanism for transformation via protracted embryonic gestation is explained by M. J. S. Hodge, "The universal gestation of nature: Chambers's *Vestiges* and *Explanations,*" *Journal of the History of Biology* 5 (1972), 127–51.

That there were religious objections from within the Christian churches to the kind of physico-theology associated with Paley is made clear by Gascoigne, *Cambridge in the age of the Enlightenment,* pp. 237–69; Corsi, *Science and religion,* pp. 178–93; Morrell and Thackray, *Gentlemen of science,* pp. 229–45. The qualifications to the ultimate value of natural theology, as voiced by Whewell in his sermons, are considered by John H. Brooke, "Indications of a Creator," in Fisch and Schaffer, *William Whewell,* pp. 149–73. Whewell's theology also receives careful consideration in Michael J. Crowe, *The extra-terrestrial life debate 1750–1900: the idea of a plurality of worlds from Kant to Lowell* (Cambridge, 1986), pp. 265–355. On the eventual demise of natural theology in Britain there are important perspectives in Robert M. Young, *Darwin's metaphor: nature's place in Victorian culture* (Cambridge, 1985), pp. 126–63. For preexisting divisions within the discourse of natural theology, exacerbated by the publication of *Vestiges,* see John H. Brooke, "Natural theology and the plurality of worlds: observations on the Brewster–Whewell debate," *Annals of Science* 34 (1977), 221–86.

VII. Visions of the Past: Religious Belief and the Historial Sciences

My introductory remarks on the disturbing effects of the historical sciences may be followed up by reference to David M. Knight, *The age of science* (Oxford, 1986), pp. 109–27; Gillian Beer, *Darwin's plots* (London, 1983), p. 9; Marcia Poynton, "Geology and landscape painting in nineteenth-century England," in Jordanova and Porter, *Images of the earth,* pp. 84–116; Dorinda Outram, *Georges Cuvier: vocation, science and authority in post-Revolutionary France* (Manchester, 1984), p. 156; and Susan Gliserman, "Early Victorian science writers and Tennyson's *In Memoriam,*" *Victorian Studies* 18 (1975), 277–308 and 437–59. For a general introduction to the development of the historical sciences, see John C. Greene, *The death of Adam: evolution and its impact on Western thought* [1959] (New York, 1961), and John H. Brooke, "Precursors of Darwin?" in *The crisis of evolution,* by John H. Brooke and Alan Richardson (Milton Keynes: Open University, 1974). The recognition of extinction as a central precondition of a truly historical account of organic transformation is stressed by Michel Foucault, *The order of things* (London, 1970), chap. 5. The association of organic transformation with political radicalism is explored by Adrian Desmond, "Artisan resistance and evolution in Britain, 1819–1848," *Osiris*, 2d ser., 3 (1987), 77–110. See also Desmond's major study, *The politics of evolution* (Chicago, 1989). There are particularly instructive remarks on how the cosmological meanings attributed to developments in the historical sciences should be interpreted in Martin J. S. Rudwick, "The shape and meaning of earth history," in Lindberg and Numbers, *God and nature*, pp. 296–321.

One of the best discussions of Linnaeus is that of James L. Larson, *Reason and experience: the representation of natural order in the work of Carl von Linné* (Berkeley, 1971). Other informative accounts include Frans A. Stafleu, *Linnaeus and the Linneans: the spreading of their ideas in systematic botany 1735–1789* (Utrecht, 1971) and Tore Frängsmyr (ed.), *Linnaeus: the man and his work* (Berkeley, 1983). Longer-term perspectives on Linnaeus's interpretation of hybridization are provided by Robert Olby, "The emergence of genetics," in Olby et al., *Companion to the history of modern science*, pp. 521–36. His departure from a biblical literalism is noted by Browne, *The secular ark: studies in the history of biogeography*, pp. 16–31. Contrasts between Linnaeus and Buffon are developed by Phillip R. Sloan, "The Buffon–Linnaeus controversy," *Isis* 67 (1976), 356–75. A good general introduction to Buffon and the responses he attracted is provided by David C. Goodman, *Buffon's natural history* (Milton Keynes: Open University, 1980). A further valuable resource is John Lyon and Phillip R. Sloan, eds., *From natural history to the history of nature: readings from Buffon and his critics* (Notre Dame, Ind., 1981). There is also a classic treatment of Buffon's place in the development of French natural history in Jacques Roger, *Les sciences de la vie dans la pensée française du xviii siècle*, 2d ed. (Paris, 1971). For Buffon's concept of species, see Phillip R. Sloan, "Buffon, German biology, and the historical interpretation of biological species," *British Journal for the History of Science* 12 (1979), 109–53. There is a useful caution against too simple a view of the clash between scientific conjectures and biblical authority in Rhoda Rappaport, "Geology and orthodoxy: the case of Noah's flood in eighteenth-century thought," *British Journal for the History of Science* 11 (1978), 1–18.

An introduction to the literature on Laplace can be gained through Roger Hahn, "Laplace and the mechanistic universe," in Lindberg and Numbers, *God and nature*, pp. 256–76; Ronald L. Numbers, *Creation by natural law: Laplace's nebular hypothesis in American thought* (Seattle, 1977), and John H. Brooke, "Nebular contraction and the expansion of naturalism," *British Journal for the History of Science* 12 (1979), 200–211. For images of Laplace and William Herschel in late eighteenth- and early nineteenth-century Britain, see Simon Schaffer, "Herschel in Bedlam: natural history and stellar astronomy," *British Journal for the History of Science* 13 (1980), 211–39, and Maurice Crosland and Crosbie Smith, "The transmission of physics from France to Britain," *Historical studies in the physical sciences* 9 (1978), 1–61. Further bibliographic guidance on the emergence of a historical astronomy can be obtained from Stephen G. Brush, "The nebular hypothesis and the evolutionary world view," *History of Science* 25 (1987), 245–78.

Three studies can be especially recommended for their authoritative treatment of the science and career of Lamarck: Ludmilla Jordanova, *Lamarck* (Oxford, 1984); Richard W. Burkhardt, *The spirit of system: Lamarck and evolutionary biology* (Cambridge, Mass., 1977); and Pietro Corsi, *The age of Lamarck: evolutionary theories in France 1790–1830* (Berkeley, 1988). For contrasts between Lamarck and Paley, see Ludmilla Jordanova, "Nature's powers: a reading of Lamarck's distinction between creation and production," in Moore, *History, humanity and evolution*, pp. 71–98. Extracts in English translation from Lamarck's scientific writings are available in H. L. McKinney, ed., *Lamarck to Darwin: contributions to evolutionary biology 1809–1859* (Lawrence, Kans., 1971). The view that newly learned habits can affect the direction and pace of evolution has had its strenuous exponent in Sir Alister Hardy, *Darwin and the spirit of man* (London, 1984).

The best introduction to the scientific work of Cuvier is William Coleman, *Georges Cuvier: zoologist* (Cambridge, Mass., 1964). His religious reputation and the social and political contexts in which he forged his career are sensitively analyzed by Outram, *Georges Cuvier*. Further assessment of the impact of Cuvier's work in palaeontology can be found in

Martin J. S. Rudwick, *The meaning of fossils: episodes in the history of palaeontology,* 2d ed. (New York, 1976), and a sympathetic account of his "catastrophism" in Gould, *Hen's teeth and horse's toes,* pp. 94–106.

An extensive literature on Lyell may best be approached through a series of essays by Martin J. S. Rudwick: "Lyell on Etna, and the antiquity of the earth," in *Toward a history of geology,* ed. Cecil J. Schneer (Cambridge, Mass., 1969), pp. 288–304; "The strategy of Lyell's *Principles of Geology,*" *Isis* 61 (1970), 4–33; "Poulett Scrope on the volcanoes of Auvergne: Lyellian time and political economy," *British Journal for the History of Science* 7 (1974), 205–42; "Charles Lyell's dream of a statistical palaeontology," *Palaeontology* 21 (1978), 225–44; and "Transposed concepts from the human sciences in the early work of Charles Lyell," in Jordanova and Porter, *Images of the earth,* pp. 67–83. A parallel between Lyell's geology and Niebuhr's history is affirmed in the last essay. On the nature and origins of "historical" and "causal" geology, see Rachel Laudan, *From mineralogy to geology: the foundations of the earth sciences, 1660–1830* (Chicago, 1987). Lyell emerged as a secular champion of geology in Charles C. Gillispie, *Genesis and geology* (New York, 1959). An alternative account of "The uniformitarian-catastrophist debate" is given by Walter F. Cannon in *Isis* 51 (1960), 38–55. That the debate has been wrongly structured by Whewell's terms is argued by Martin J. S. Rudwick, "Uniformity and progression: reflections on the structure of geological theory in the age of Lyell," in *Perspectives in the history of science and technology,* ed. Duane H. D. Roller (Norman, Okla., 1971), pp. 209–27. That a "catastrophist" position can too easily be caricatured is one of the contentions of R. Hooykaas, *Catastrophism in geology, its scientific character in relation to actualism and uniformatarianism* (Amsterdam, 1970). For the effect of Lyell's reading of Lamarck in turning him against a progression in the fossil record, see Michael J. Bartholomew, "Lyell and evolution: an account of Lyell's response to the prospect of an evolutionary ancestry for man," *British Journal for the History of Science* 6 (1973), 261–303. Lyell's arguments against the transformation of species are analyzed by William Coleman, "Lyell and the reality of species," *Isis* 53 (1962), 325–38. Conflicting interpretations of "Didelphis" are discussed by Adrian Desmond, "Interpreting the origin of mammals: new approaches to the history of palaeontology," *Zoological Journal of the Lin-*

nean Society 82 (1984), 7–16. Desmond also brings out the social and political investment in different interpretations in two further studies: "Robert E. Grant: the social predicament of a pre-Darwinian transmutationist," *Journal of the History of Biology* 17 (1984), 189–223, and "Designing the dinosaur: Richard Owen's response to Robert Edmond Grant," *Isis* 70 (1979), 224–34. Nineteenth-century controversies within the science of geology, and the historiographical techniques required to comprehend them, have been much discussed following the publication of Martin J. S. Rudwick, *The great Devonian controversy: the shaping of scientific knowledge by gentlemanly specialists* (Chicago, 1985), and James A. Secord, *Controversy in Victorian geology: the Cambrian–Silurian dispute* (Princeton, 1986).

The most comprehensive guide to the literature on Darwin and the Darwinian revolution is afforded by David Kohn, ed., *The Darwinian heritage* (Princeton, 1985), pp. 1021–99. A first introduction (with bibliographic recommendations) can be obtained from M. J. S. Hodge, "Origins and species before and after Darwin," in Olby et al., *Companion to the history of modern science*, pp. 374–95. Three books by Peter J. Bowler cover many of the major issues: *Fossils and progress: paleontology, and the idea of progressive evolution in the nineteenth century* (New York, 1976); *Evolution: the history of an idea* (Berkeley, 1984); and *The non-Darwinian revolution: reinterpreting a historical myth* (Baltimore, 1988). For the development of Darwin's views on the evolution of the human mind, see Robert J. Richards, *Darwin and the emergence of evolutionary theories of mind and behavior* (Chicago, 1987). The evolution of Wallace's conception of evolution is sympathetically treated by H. L. McKinney, *Wallace and natural selection* (New Haven, 1972).

Access to a highly specialized literature on the development of Darwin's theory can be gained from Phillip R. Sloan, "Darwin's invertebrate program, 1826–1836: preconditions for transformism," in Kohn, *The Darwinian heritage*, pp. 71–120; Dov Ospovat, *The development of Darwin's theory: natural history, natural theology and natural selection, 1838–1859* (Cambridge, 1981); M. J. S. Hodge and David Kohn, "The immediate origins of natural selection," in Kohn, *The Darwinian heritage*, pp. 185–206. There is an inversion of the traditional myth concerning Darwin's inductive inference from the Galapagos finches in Frank J. Sulloway, "Darwin and his finches: the evolution of a legend," *Journal of the History of Biology* 15 (1982), 1–53. The dialectical relationship between the emergence of Darwin's theory and Lyell's inadequate conceptualization of species distribution is fully explored by M. J. S. Hodge, "Darwin and the laws of the animate part of the terrestrial system (1835–1837): on the Lyellian origins of his zoonomical explanatory program," *Studies in the History of Biology* 6 (1983), 1–106. Darwin's early realization that humanity should be embraced by an evolutionary scheme is stressed by Sandra Herbert, "Man in the development of Darwin's theory of transmutation: part 2," *Journal of the History of Biology* 10 (1977), 155–227. On this point see also John R. Durant, "The ascent of nature in Darwin's *Descent of man*," in Kohn, *The Darwinian heritage*, pp. 283–303. That the interaction between Darwin's science and his changing religious sensibilities cannot be treated as a one-way process is argued by John H. Brooke, "The relations between Darwin's science and his religion," in *Darwinism and divinity*, ed. John R. Durant (Oxford, 1985), pp. 40–75. Darwin's theistic pronouncements are also analyzed by Frank Burch Brown, "The evolution of Darwin's theism," *Journal of the History of Biology* 19 (1986), 1–45. For the wider cultural resources on which he drew, see Edward Manier, *The young Darwin and his cultural circle* (Dordrecht, 1978). An extensive literature on the precise nature of his debt to Malthus is reviewed by Antonello La Vergata, "Images of Darwin: a historiographic overview," in Kohn, *The Darwinian heritage*, pp. 901–72, esp. pp. 953–8.

The Darwin inhibited by fears of persecution can be found in Howard E. Gruber and Paul H. Barrett. *Darwin on man: a psychological study of scientific creativity* (New York, 1974). That his theory was shaped by an enduring legacy from natural theology is one of the contentions of Ospovat, *The development of Darwin's theory*. For a critique of Ospovat, which also takes Darwin's anxieties (and their implications) seriously, see David Kohn, "Darwin's ambiguity: the secularization of biological meaning," *British Journal for the History of Science* 22 (1989), 215–39. Kohn has also discussed the relatively late perception on Darwin's part of how natural selection would explain divergence: "On the origin of the principle of divergence," *Science* 213 (1981), 1105–8. Darwin's analogy between organic specialization and division of labor is the subject of S. S. Schweber, "Darwin and the political economists:

divergence of character," *Journal of the History of Biology* 13 (1980), 195–289. The most illuminating account of Darwin's debt to the pigeon fanciers is that of James A. Secord, "Darwin and the breeders: a social history," in Kohn, *The Darwinian heritage,* pp. 519–42.

The development of biblical criticism is surveyed by W. Neil, "The criticism and theological uses of the Bible, 1700–1950," in *The Cambridge history of the Bible,* ed. S. L. Greenslade (Cambridge, 1975), pp. 238–93. There is an introduction to the philosophy of Spinoza in Stuart Hampshire, *Spinoza,* 2d ed. (Harmondsworth, 1962), and Roger Scruton, *Spinoza* (Oxford, 1986). His critical analysis of biblical references to divine activity is investigated in greater detail by Leo Strauss, *Spinoza's critique of religion* (New York, 1965), pp. 123–46. For reactions to Spinoza in England and the Netherlands, see Rosalie L. Colie, *Light and Enlightenment: a study of the Cambridge Platonists and the Dutch Arminians* (Cambridge, 1957), and "Spinoza in England, 1665–1730," *Proceedings of the American Philosophical Society* 107 (1963), 183–219. There are translated extracts from Reimarus in Charles H. Talbot, ed., *Reimarus: fragments* (London, 1971). Strauss's *The life of Jesus critically examined* is available with an introduction by Peter C. Hodgson (London, 1973). For a further introduction, see Hans Frei, "David Friedrich Strauss," in *Nineteenth century religious thought in the West,* vol. 1, ed. N. Smart, J. Clayton, S. Katz, and P. Sherry (Cambridge, 1985), pp. 215–60. Additional insight can be gained from Van A. Harvey, "D. F. Strauss's *Life of Jesus* revisited," *Church History* 30 (1961), 191–211, and Richard S. Cromwell, *David Friedrich Strauss and his place in modern thought* (Fair Lawn, N.J., 1974). An excellent analysis of the political context and implications of Strauss's critique is given by Marilyn Chapin Massey, *Christ unmasked: the meaning of the "Life of Jesus" in German politics* (Chapel Hill, N.C., 1983).

The mid-nineteenth-century "crisis" of faith, in which scientific and historical criticism played variable parts, is discussed from different points of view in A. Symondson, ed., *The Victorian crisis of faith* (London, 1970). Some literary manifestations of this crisis are discussed in R. L. Wolff, *Gains and losses: novels of faith and doubt in Victorian England* (New York, 1977). Moral revulsion against the presentation of certain Christian doctrines is documented by H. R. Murphy, "The ethical revolt against Christian orthodoxy in early Victorian

England," *American Historical Review* 60 (1955), 800–817. See also James R. Moore, "1859 and all that: remaking the story of evolution-and-religion," in *Charles Darwin, 1809–1882: a centennial commemorative*, ed. R. G. Chapman and C. T. Duval (Wellington, New Zealand, 1982), pp. 167–94. For unbelief in America, see James Turner, *Without God, without creed: the origins of unbelief in America* (Baltimore, 1985), which may be read in conjunction with Herbert Hovenkamp, *Science and religion in America, 1800–1860* (Philadelphia, 1978). Further perspectives on the association of science with unbelief can be gained from Owen Chadwick, *The secularization of the European mind in the nineteenth century* (Cambridge, 1975), and Bernard Lightman, *The origins of agnosticism: Victorian unbelief and the limts of knowledge* (Baltimore, 1987). It should be stressed that the reasons given by atheists and agnostics for their loss of faith often made little or no reference to advances in science: see Susan Budd, *Varieties of unbelief: atheists and agnostics in English society 1850–1960* (London, 1977), pp. 104–23.

The controversy surrounding publication of *Essays and reviews* is explored by Ieuan Ellis, *Seven against Christ: a study of "Essays and Reviews"* (Leiden, 1980). The harmonization schemes, rejected by Goodwin, are discussed in greater detail in Francis C. Haber, *The age of the world: Moses to Darwin* (Baltimore, 1959). The particular contribution of Baden Powell is placed in context by Corsi, *Science and religion*, pp. 209–24. The stand taken by Temple is spotlighted by John Durant, "Darwinism and divinity: a century of debate," in Durant, *Darwinism and divinity*, pp. 9–39.

VIII. Evolutionary Theory and Religious Belief

The colossal compilation by Sydney Eisen and Bernard V. Lightman, *Victorian science and religion: a bibliography with emphasis on evolution, belief and unbelief* (Hamden, Conn., 1984) incorporates secondary literature published between 1900 and 1975. A wealth of information on the reception of Darwin's theory in different constituencies can be obtained from David L. Hull, *Darwin and his critics: the reception of Darwin's theory by the scientific community* (Cambridge, Mass., 1973), and Alvar Ellegard, *Darwin and the general reader* (Göteborg, 1958). The nature of the response within the English churches is examined by Owen Chadwick, *The Victorian Church*, 2 vols.

(London, 1966–70), and among Protestant intellectuals in America by Jon H. Roberts, *Darwinism and the Divine in America: Protestant intellectuals and organic evolution, 1859–1900* (Madison, Wis., 1988). The relevance of a residual theism to Darwin's own outlook is explored, against a background of encroaching positivism, by Neal C. Gillespie, *Charles Darwin and the problem of creation* (Chicago, 1979), pp. 134–45. For perspectives on evolutionary progress as a surrogate religion, see John W. Burrow, *Evolution and society: a study in Victorian social theory* (Cambridge, 1966). Some primary sources germane to the Victorian debate are collated in Tess Cosslett, ed., *Science and religion in the nineteenth century* (Cambridge, 1984). The most elaborate study of theological responses to Darwin in Britain and America is James R. Moore, *The post-Darwinian controversies: a study of the Protestant struggle to come to terms with Darwin in Great Britain and America 1870–1900* (Cambridge, 1979). The relationship between Darwin's *Descent of man* and other theories of human evolution is discussed by Peter J. Bowler, *Theories of human evolution* (Oxford, 1986). The significance of Darwin's challenge to faith in moral absolutes can be appreciated from H. G. Wood, *Belief and unbelief since 1850* (Cambridge, 1955), chap. 6, and Walter E. Houghton, *The Victorian frame of mind, 1830–1870* (New Haven, 1957), pp. 218–42. On the attractions of Darwinism for secular moralists and political liberals, such as Leslie Stephen, see Greta Jones, *Social Darwinism and English thought: the interaction between biological and social theory* (Brighton, 1980), pp. 35–53.

T. H. Huxley's retrospective account of the success of Darwin's theory remains an instructive source: "On the reception of the 'Origin of species,'" in *The life and letters of Charles Darwin*, vol. 2, ed. Francis Darwin (London, 1887), pp. 179–204. That Huxley himself had modified Darwin's mechanism is shown by Michael J. Bartholomew, "Huxley's defence of Darwin," *Annals of Science* 32 (1975), 525–35. The resistance of Agassiz is discussed by Edward Lurie, "Louis Agassiz and the idea of evolution," *Victorian Studies* 3 (1959), 87–108. Mivart's model of evolutionary convergence is elucidated by Adrian Desmond, *Archetypes and ancestors* (London, 1982), pp. 175–201. Mivart's critique of Darwin is also assessed by Peter J. Vorzimmer, *Charles Darwin: the years of controversy. The "Origin of species" and its critics, 1859–1882* (Philadelphia, 1970), a study that is also useful in explaining how Darwin dealt with criticism based on the "swamping" of advantageous variation. The objections leveled by Lord Kelvin are given prominence in Joe D. Burchfield, *Lord Kelvin and the age of the earth* (New York, 1975). Kelvin's scientific career receives handsome treatment in C. W. Smith and M. N. Wise, *Energy and empire: a biographical study of Lord Kelvin* (Cambridge, 1989).

On the philosophical structure of Darwin's theorizing and the objections to which it gave rise, there is useful discussion in Michael Ruse, *The Darwinian revolution* (Chicago, 1979), pp. 174–80, and David R. Oldroyd, *Darwinian impacts: an introduction to the Darwinian revolution* (Milton Keynes: Open University, 1980), pp. 114–38. These can be supplemented by reference to R. Curtis, "Darwin as an epistemologist," *Annals of Science* 44 (1987), 379–407, and M. T. Ghiselin, *The triumph of the Darwinian method* (Berkeley, 1969). The growing respect for a hypotheticodeductive methodology in nineteenth-century science is a major theme in Laurens L. Laudan, *Science and hypothesis* (Dordrecht, 1981). The immediate applicability of Darwin's theory to such technical issues as the mimicry of one species by another is discussed by Muriel Blaisdell, "Natural theology and nature's disguises," *Journal of the History of Biology* 15 (1982), 163–89. Haeckel's evolutionary phylogenies and their attractions for Huxley are discussed by Mario Di Gregorio, *T. H. Huxley's place in natural science* (New Haven, 1984), and Desmond, *Archetypes and ancestors,* pp. 147–74.

Darwin's use of language and its susceptibility to often contrary extensions of meaning are analyzed by Gillian Beer, *Darwin's plots* (London, 1983), esp. pp. 117–28. The sheer variety even of English "social Darwinism" is made apparent in Jones, *Social Darwinism and English thought.* For Herbert Spencer and the accolades he enjoyed in America, see James R. Moore, "Herbert Spencer's henchmen: the evolution of Protestant liberals in late nineteenth-century America," in Durant, *Darwinism and divinity,* pp. 76–100, and R. Hofstadter, *Social Darwinism in American thought,* 2d ed. (Boston, 1955). To the question whether Darwin was a "social Darwinist," an answer has been given by John C. Greene, "Darwin as a social evolutionist," *Journal of the History of Biology* 10 (1977), 1–27, reproduced in John C. Greene, *Science, ideology and worldview* (Berkeley, 1981), pp. 95–127. See also the controversial essay by Robert Young, "Darwinism *is* social," in Kohn, *The Darwinian heritage,* pp. 609–38. For connections between biological and social theory in the perceptions of Wallace, see Roger Smith, "Alfred Russel Wallace: philosophy of nature and man," *British Journal for the History of Science* 6 (1972), 177–99; John R. Durant, "Scientific naturalism and social reform in the thought of Alfred Russel Wallace," *British Journal for the History of Science* 12 (1979), 31–58; and M. J. Kottler, "Alfred Russel Wallace, the origin of man, and spiritualism," *Isis* 65 (1974), 145–92.

For the dissemination of Darwin's science in different national contexts, see Thomas F. Glick, ed., *The comparative reception of Darwinism* (Austin, 1974). The changing character of the response in France has been discussed by Yvette Conry, *L'introduction du darwinisme en France au XIXe siècle* (Paris, 1974), and by Harry W. Paul, *The edge of contingency: French Catholic reaction to scientific change from Darwin to Duhem* (Gainesville, Fla., 1979). Reference should also be made to the excellent review of the literature on "Darwinism in Germany, France, and Italy" by Pietro Corsi and Paul J. Weindling, in Kohn, *The Darwinian heritage,* pp. 683–729. The relative isolation of Catholic scientists during the Third Republic is evaluated by Harry W. Paul, "The crucifix and the crucible: catholic scientists in the Third Republic," *Catholic Historical Review* 58 (1972), 195–219. Mary Jo Nye discusses the productive isolation of the Catholic chemist Paul Sabatier in "Nonconformity and creativity: a study of Paul Sabatier, chemical theory, and the French scientific community," *Isis* 68 (1977), 375–91, and explains the motivation behind the *Revue des Questions Scientifiques* in "The moral freedom of man and the determinism of nature: the Catholic synthesis of science and history in the *Revue des Questions Scientifiques,*" *British Journal for the History of Science* 9 (1976), 243–73. Perspectives on the reception of Darwinism in Germany can be gained from Alfred Kelly, *The descent of Darwin: the popularization of Darwinism in Germany, 1860–1914* (Chapel Hill, N.C., 1981), and from two studies by Frederick Gregory: *Scientific materialism in nineteenth century Germany* (Dordrecht, 1977), and "The impact of Darwinian evolution on Protestant theology in the nineteenth century," in Lindberg and Numbers, *God and nature,* pp. 369–90. A contrast between the cultural connotations of Darwinism in Germany in the 1860s and 1890s is skillfully drawn by Paul Weindling, "Ernst Haeckel, Darwinismus and the secularization of nature," in Moore, *History, humanity and evolution,* pp. 311–27. For Darwinism in Russia, see Daniel P. Todes, *Darwin without Malthus: the struggle for existence in Russian evolutionary thought* (Oxford, 1989), and Alexander Vucinich, *Darwin in Russian thought* (Berkeley, 1989).

For the section on the quasi-religious character of scientific naturalism, there are further insights into Tyndall in Tess Cosslett, *The scientific movement and Victorian literature* (Brighton, 1982). The considerate, though ultimately negative, response of Charles Hodge is discussed by David N. Livingstone, "The idea of design: the vicissitudes of a key concept in the Princeton response to Darwin," *Scottish Journal of Theology* 37 (1984), 329–57. On the role of "chance" in the evolutionary process, there is a good modern discussion in Elliot Sober, *The nature of selection: evolutionary theory in philosophical focus* (Cambridge, Mass., 1984), pp. 103–34. For Darwin's perception of natural theology as incurably anthropocentric, see Maurice Mandelbaum, "Darwin's religious views," *Journal of the History of Ideas* 19 (1958), 363–78; and for his stand on suffering as symbol of modernity, Donald Fleming, "Charles Darwin, the anaesthetic man," *Victorian studies* 4 (1961), 219–36. Late nineteenth-century evolutionary naturalism is well described by Frank M. Turner, *Between science and religion: the reaction to scientific naturalism in late Victorian England* (New Haven, 1974), pp. 1–37. Turner's study is especially valuable because it discusses six eminent thinkers (including Alfred Russel Wallace) who refused to accept a reduction to two opposing creeds. Further information on Galton's stance can be obtained from Ruth S. Cowan, "Nature and nurture in the work of Francis Galton," *Studies in History of Biology* 1 (1977), 133–208. A special issue of the *British Journal for the History of Science,* vol. 22 (1989), pp. 257–375, addresses the current literature on *Genetics, eugenics and evolution.* The concern of Stokes that evolutionary theory was winning adherents for the wrong reasons is evident from David B. Wilson, "A physicist's alternative to materialism: the religious thought of George Gabriel Stokes," *Victorian Studies* 28 (1984), 69–96. Mivart's changing relations with the Catholic Church and with the "Church scientific" are analyzed by J. W. Gruber, *A conscience in conflict: the life of St. George Jackson Mivart* (New York, 1960).

On the theme of evolution within religion, the contrasting

responses of Hodge and McCosh are discussed by David N. Livingstone, *Darwin's forgotten defenders* (Edinburgh, 1987). A sense in which theologians might perceive themselves as completing Darwin's work is noted by Alan Richardson, *The Bible in the age of science* (London, 1961), pp. 69–71. For further light on Troeltsch, see John P. Clayton, ed., *Ernst Troeltsch and the future of theology* (Cambridge, 1976). Lyman Abbott receives special attention in Moore, *The post-Darwinian controversies*. A receptivity toward theistic evolution on the part of theologians wishing to stress the Incarnation and immanence of God is noted by Arthur Peacocke, "Biological evolution and Christian theology – yesterday and today," in Durant, *Darwinism and divinity,* pp. 101–30. William Carpenter's use of *foraminifera* to argue for a mind behind evolution features in Peter J. Bowler, "Darwinism and the argument from design: suggestions for a re-evaluation," *Journal of the History of Biology* 10 (1977), 29–43. The view of C. S. Peirce that the role of chance in the evolutionary process is supportive of divine personality can be found in the extracts in D. Browning, ed., *Philosophers of process* (New York, 1965), pp. 57–109. There is a biography of Asa Gray, who used a Darwinian teleology to illuminate the problem of suffering, in A. Hunter Dupree, *Asa Gray 1810–88* (Cambridge, Mass., 1959). Temple's argument from the unity of the evolutionary process is noted by Greene, *Darwin and the modern world view,* p. 66.

For longer-term perspectives, there is an emphasis on the resurgence of neo-Lamarckian evolution in Peter J. Bowler, *The eclipse of Darwinism: anti-Darwinian evolution theories in the decades around 1900* (Baltimore, 1983). For the development of the modern "evolutionary synthesis," see Ernst Mayr and William B. Provine, eds., *The evolutionary synthesis: perspectives on the unification of biology* (Cambridge, Mass., 1980). The issues as perceived by a Roman Catholic biologist in the period following the encyclical *Humani generis.*(1950) can be studied in P. G. Fothergill, *Evolution and Christians* (London, 1961), pp. 303–40. Present perspectives from liberal Roman Catholic and other standpoints can be found in Ernan McMullin, ed., *Evolution and creation* (Notre Dame, Ind., 1986). Illustrative of a large contemporary literature on the integration of evolutionary theory with Protestant theology would be Arthur Peacocke, *Creation and the world of science* (Oxford, 1979).

The shattering impact of Darwinian naturalism among Boston intellectuals and the development of a philosophical "pragmatism" are described by Bruce Kuklick, *The rise of American philosophy 1860–1930* (New Haven, 1977). Recent studies of William James include Graham Bird, *William James* (London, 1986), and Gerald E. Myers, *William James: his life and thought* (New Haven, 1986). Both Peirce and James are also discussed by Cynthia E. Russett, *Darwin in America: the intellectual response* (San Francisco, 1976). The development of psychical research in which James took so keen an interest is assessed by John J. Cerullo, *The secularization of the soul* (Philadelphia, 1982).

Postscript: Science and Religion in the Twentieth Century

With the ever increasing specialization of both scientific and theological research, there are relatively few contextual surveys of twentieth-century dialogues. There is, however, useful material in several of the general texts mentioned under the Introduction and Chapter I, notably Ian G. Barbour, *Issues in science and religion;* Robert John Russell et al., *Physics, philosophy and theology: a common quest for understanding;* and William H. Austin, *The relevance of natural science to theology.* One may also turn to Arthur Peacocke, ed., *The sciences and theology in the twentieth century* (London, 1981), especially for the essay by Martin J. S. Rudwick, "Senses of the natural world and senses of God: another look at the historical relation of science and religion," pp. 241–61. The periodical *Zygon,* subtitled "Journal of religion and science," offers a variety of opinion on contemporary points of contact. Twentieth-century innovations in Christian doctrine are included in Jaroslav Pelikan, *The Christian tradition: a history of the development of doctrine,* vol. 5: *Christian doctrine and modern culture since 1700* (Chicago, 1989). Fundamentalist opposition in America to evolutionary biology is discussed in its cultural aspects by George M. Marsden, *Fundamentalism and American culture: the shaping of twentieth-century evangelicalism* (New York, 1980), and in its legal aspects by Edward J. Larson, *Trial and error: the American controversy over creation and evolution* (New York, 1985). For the intensification of individualism in twentieth-century Western culture and the common recourse to Freud for the rationalization of unbelief, see Philip Rieff, *The triumph of the therapeutic* (London, 1966).

On Freud, there is an introductory essay (coupled with a substantial bibliography) by Raymond E. Fancher in Olby et al., *Companion to the history of modern science,* pp. 425–41. In discussing Freud's analysis of religious belief, I have been indebted to J. N. Isbister, *Freud: an introduction to his life and work* (Oxford, 1985), and Clarence J. Karier, *Scientists of the mind* (Urbana, Ill., 1986), which also contains valuable essays on William James and Carl Jung. Further access to a massive literature can be gained from David Bakan, *Sigmund Freud and the Jewish mystical tradition* (New York, 1969); Peter Gay, *A godless Jew: Freud, atheism, and the making of psycho-analysis* (New Haven, 1987); and *Freud: a life for our time* (New York, 1988). For the impact of Freud in America, see N. G. Hale, *Freud and the Americans: the beginnings of psycho-analysis in the United States 1876–1917* (New York, 1971). On the philosophical foundations of psychoanalysis the review by John Forrester in *Isis* 77 (1986), 670–4, makes the important point that Freud did not believe his science to be a science like physics. My reference to Roberto Assagioli as illustrative of more eclectic forms of psychotherapy can be followed up in Roberto Assagioli, *Psychosynthesis* (Wellingborough, 1965). Positivist strictures against religious discourse, based on criteria of verifiability and falsifiability have been much debated in theological circles. Representative of the kind of debate that ensues would be A. G. Flew and A. C. MacIntyre, eds., *New essays in philosophical theology* (London, 1955), and Derek Stanesby, *Science, reason and religion* (London, 1988).

A clear introduction to the revolution in physics is provided by the Open University text: *Modern physics and problems of knowledge,* for the course *Science and belief from Darwin to Einstein* (Milton Keynes: Open University, 1981). There are also introductory essays, with bibliographic suggestions, by John Stachel, "The theory of relativity," and Michael Redhead, "Quantum theory," in Olby et al., *Companion to the history of modern science,* pp. 442–78. The different contexts in which statistical reasoning entered the sciences are explored in Gerd Gigerenzer, Zeno Swijtink, Theodore Porter, Lorraine Daston, John Beatty, and Lorenz Kruger, *The empire of chance: how probability changed science and everyday life* (Cambridge, 1989). For what the "new physics" has come to mean today, see the collection of essays edited by Paul Davies. *The new physics* (Cambridge, 1989). On the philosophical issues raised by quantum theory there is an accessible discussion by Jonathan Powers, *Philosophy and the new physics* (London, 1982). Two primary sources deserve special mention: Niels Bohr, *Atomic physics and human knowledge* (New York, 1961), and Werner Heisenberg, *The physicist's conception of nature* (London, 1958). For a survey of the theological ramifications, see Erwin N. Hiebert, "Modern physics and Christian faith," in Lindberg and Numbers, *God and nature,* pp. 424–47. More detailed studies of Einstein, Bohr, and Schrodinger include Abraham Pais, *"Subtle is the Lord . . .": the science and the life of Albert Einstein* (Oxford, 1982); Henry J. Folse, *The philosophy of Niels Bohr: the framework of complementarity* (Amsterdam, 1985); and Walter Moore, *Schrodinger: life and thought* (Cambridge, 1989). On the cultural context in which a nondeterministic physics took shape, there is the influential essay by Paul Forman, "Weimar culture, causality, and quantum theory, 1918–1927: adaptation by German physicists and mathematicians to a hostile intellectual environment," *Historical Studies in the Physical Sciences* 3 (1971), 1–115. Critical commentaries on Forman's thesis can be found in John Hendry, "Weimar culture and quantum causality," *History of Science* 18 (1980), 115–80, and in P. Kraft and P. Kroes, "Adaptation of scientific knowledge to an intellectual environment. Paul Forman's 'Weimar culture . . .': analysis and criticism," *Centaurus* 27 (1984), 76–99. The significance of Bohr's familiarity with the works of Kierkegaard is assessed by Gerald Holton, *Thematic origins of modern science: Kepler to Einstein* (Cambridge, Mass., 1973), pp. 115–61.

The problems posed by quantum physics for scientific reductionism are noted by Powers, *Philosophy and the new physics,* and given popular, if sometimes questionable, emphasis by Fritjof Capra, *The Tao of physics* (1975; reprint, London, 1983), and *The turning point* (London, 1983); and also by Gary Zukav, *The dancing wu-li masters: an overview of the new physics* (London, 1979). The status of reductionist programs in different disciplines is addressed in Arthur Peacocke, ed., *Reductionism in academic disciplines* (Guildford Surrey, 1985). An intellectually demanding defense of the universe as an "unbroken whole" is given by David Bohm, *Wholeness and the implicate order* (London, 1980). There is a lucid exposition of process theology by one of its most respected exponents in Charles Hartshorne, *A natural theology for our time* (La Salle,

Ill., 1967). See also David A. Pailin, *God and the processes of reality* (London, 1989). That process theology could assist the reconstruction of a Christian philosophy of nature is suggested by Norman Pittenger, *Process thought and Christian faith* (Welwyn Herts, 1968). That a more traditional concept of providence can still be defended in the light of modern physics is claimed by John Polkinghorne, *Science and providence* (London, 1989). Claims for the applicability of process philosophy to ecological and other political issues are made by John B. Cobb, *Process theology as political theology* (Manchester, 1982).

There is a rapidly expanding literature on the interrelation of human values and science. A good place to start would be Jerome K. Ravetz, *Scientific knowledge and its social problems* (Oxford, 1971). Some basic issues as perceived from within the philosophy of science are raised in Laurens L. Laudan, ed., *Science and values: the aims of science and their role in scientific debate* (Berkeley, 1984). The wide-ranging and historically informed treatment by Loren Graham, *Between science and values* (New York, 1981), can be particularly recommended. My retrospective remarks in the text are also indebted to Leonard M. Marsak, "Bernard de Fontenelle: in defense of science," in *The rise of science in relation to society*, ed. Leonard M. Marsak (New York, 1964), pp. 67–79. The place of theological considerations in assessing the problems raised by the biological and medical sciences is considered in Earl E. Shelp, ed., *Theology and bioethics: explaining the foundations and frontiers* (Dordrecht, 1985). See also David Brain and Harry Lesser, eds., *Ethics, technology and medicine* (Aldershot Hants., 1988), and Anthony Dyson and John Harris, eds., *Experiments on embryos* (London, 1989). The protest over fetal research in America in the 1970s is placed in perspective by Steven Maynard-Moody, "The fetal research dispute," in *Controversy*, ed. Dorothy Nelkin (Beverly Hills, Calif., 1979), chap. 10. The contrast between a secular and a distinctively Catholic bioethics is affirmed in Edmund D. Pellegrino, John P. Langan, and John Collins Harvey, eds., *Catholic perspectives on medical morals* (Dordrecht, 1989).

Whether there is a distinctive set of values peculiar to Islam that could find application in the orientation of scientific research is one of the questions raised in Ziauddin Sardar, ed., *The touch of Midas: science, values and environment in Islam and the West* (Manchester, 1984). My general remarks about the subtlety of the relations between science and "secularization" can be read in conjunction with James A. Beckford and Thomas Luckmann, eds., *The changing face of religion* (London, 1989). On "scientific creationism" as an oppositional stance, the literature that I have found most helpful includes Eileen Barker, "In the Beginning: the battle of creationist science against evolutionism," in *On the margins of science: the social construction of rejected knowledge,* ed. Roy Wallis (Keele, 1979), pp. 179–200, and also by Barker, "Let there be light: scientific creationism in the twentieth century," in Durant, *Darwinism and divinity,* pp. 181–204; James R. Moore, "Interpreting the new creationism," *Michigan Quarterly Review* 22 (1983), 321–34; Dorothy Nelkin, *The creation controversy: science or scripture in the schools* (New York, 1983); Ronald L. Numbers, "The Creationists," in Lindberg and Numbers, *God and nature,* pp. 391–423; J. Patrick Gray and Linda D. Wolfe, "Sociobiology and creationism: two ethnosociologies of American culture," *American Anthropologist* 84 (1982), 580–94. For anthropological perspectives on fundamentalism as a feature of diverse religious traditions, see Lionel Caplan, ed., *Studies in religious fundamentalism* (London, 1988). Where Darwin needs defending, the task has been most authoritatively accomplished by Philip Kitcher, *Abusing science: the case against Creationism* (Cambridge, Mass., 1982).

A lively exchange concerning the claims of sociobiology can be followed in Michael Ruse, *Taking Darwin seriously* (Oxford, 1986), and James R. Moore, "Born-again social Darwinism," *Annals of Science* 44 (1987), 409–17. Critical reactions to sociobiology can also be found in Philip Kitcher, *Vaulting ambition: sociobiology and the quest for human nature* (Cambridge, Mass., 1985), and Roger Trigg, *The shaping of man: philosophical aspects of sociobiology* (Oxford, 1982). Whether contemporary philosophy can meet the unprecedented demands of the modern world for guidance on ethical matters is considered by Bernard Williams, *Ethics and the limits of philosophy* (London, 1985). For what might still be said of a "sense of God" after the human sciences have sought to explain it, see the sensitive discussion by John Bowker, *The sense of God: sociological, anthropological and psychological approaches to the origin of the sense of God* (Oxford, 1973).

Sources of Quotations

I. Interaction between Science and Religion

1 Paracelsus, in Jolande Jacobi, ed., *Paracelsus: selected writings* (London, 1951), pp. 196–7.

II. Science and Religion in the Scientific Revolution

1 Wallis, quoted in C. J. Scriba, "The autobiography of John Wallis," *Notes and Records of the Royal Society* 25 (1970), 40.
2 Luther, quoted in Stanton J. Linden, "Alchemy and eschatology in seventeenth century poetry," *Ambix* 31 (1984), 102.
3 Descartes, quoted in D. C. Goodman, ed., *Science and religious belief 1600–1900: a selection of primary sources* (Dorchester, 1973), pp. 54–5.

III. The Parallel between Scientific and Religious Reform

1 Bacon, quoted in Eugene M. Klaaren, *Religious origins of modern science* (Grand Rapids, Mich., 1977), p. 92.
2 Ross, quoted in Grant McColley, "The Ross–Wilkins Controversy," *Annals of Science* 3 (1938), 158.
3 Alexander Ross, *The new planet no planet or the earth no wandring star except in the wandring heads of Galileans* (London, 1646), p. 101.
4 Milton, quoted in McColley, "The Ross–Wilkins Controversy," p. 172.
5 Beale, quoted in Charles Webster, *The Great Instauration: science, medicine and reform 1626–1660* (London, 1975), p. 12.
6 Robert K. Merton, *Science, technology and society in seventeenth century England*, preface to the 1970 edition (New York, 1970), p. xxviii.

7 Lotte Mulligan, "Civil War politics, religion and the Royal Society," *Past and Present* 59 (1973), 108.

IV. Divine Activity in a Mechanical Universe

1 Leonardo da Vinci, quoted in Peter Burke, *Tradition and innovation in Renaissance Italy* (London, 1974), p. 208.
2 Kepler, quoted in Robert S. Westman, "Magical reform and astronomical reform: the Yates thesis reconsidered," in *Hermeticism and the scientific revolution,* by Robert S. Westman and J. E. McGuire (Los Angeles, 1977), p. 41.
3 Isaac Newton, Query 31 from *Opticks,* 4th ed. (London, 1730), in H. S. Thayer, ed., *Newton's philosophy of nature: selections from his writings* (New York, 1953), p. 172.
4 Francis Bacon, "Of atheism," in *Essays by Francis Bacon* (London, 1937), p. 66.
5 Robert Boyle, "Some considerations touching the usefulness of experimental natural philosophy," in *Robert Boyle: the works,* ed. T. Birch (London, 1772), 2:58.
6 Newton, quoted in Richard S. Westfall, *Force in Newton's physics* (London, 1971), p. 340.
7 Newton, quoted in ibid., p. 397.
8 Newton, in the third of four letters to Richard Bentley (1692–3), in Thayer, *Newton's philosophy of nature* p. 54.

V. Science and Religion in the Enlightenment

1 Voltaire, "Sermon of the fifty," in *Deism: an anthology,* ed. Peter Gay (Princeton, 1968), pp. 152–3.
2 D'Alembert, quoted in R. Grimsley, *Jean d'Alembert 1717–1783* (Oxford, 1963), p. 199.
3 Leibniz, "Fourth paper, being an answer to Dr. Clarke's third reply," in *The Leibniz–Clarke correspondence,* ed. H. G. Alexander (Manchester, 1956), p. 44.
4 Voltaire, in Gay, *Deism* pp. 157–8.
5 A. A. Cooper, earl of Shaftesbury, quoted in John Redwood, *Reason, ridicule and religion* (London, 1976), pp. 85–6.
6 John Toland, *Christianity not mysterious* (London, 1702), p. 44.
7 Matthew Tindal, *Christianity as old as the creation: or the Gospel a republication of the religion of nature* (London, 1732), p. 102.
8 Priestley, quoted in Derek Orange, "Oxygen and one God," *History Today* 24 (1974), 781.
9 David Hume, "Of miracles," in *Hume on religion,* ed. R. Wollheim (London, 1963), p. 208.
10 Ibid., p. 223.
11 Hume, quoted in J. C. A. Gaskin, *Hume's philosophy of religion* (London, 1978), p. 153.

VI. The Fortunes and Functions of Natural Theology

1 C. Linnaeus, *Reflections on the study of nature* (1754), trans. J. E. Smith (1786), quoted in D. C. Goodman, *Buffon's natural history* (Milton Keynes: Open University, 1980), p. 18.
2 Voltaire, quoted in Roy Porter, "The Enlightenment in England," in *The Enlightenment in national context,* ed. R. S. Porter and M. Teich (Cambridge, 1981), p. 14.
3 Porter, ibid., pp. 7–8 and 13.
4 Owen, quoted in William Whewell, *History of the inductive sciences,* 3d ed. (London, 1857; reprint, London, 1967), 3:559.
5 Coleridge, quoted in R. F. Brinkley, ed., *Coleridge on the seventeenth century* (Durham, N.C., 1955), p. 402.

VII. Visions of the Past

1 Wallace, quoted in H. L. McKinney, *Wallace and natural selection* (New Haven, 1972), p. 41.
2 Strauss, quoted in E. Benz, *Evolution and Christian hope,* trans. Heinz G. Frank (London, 1967), p. 74.
3 C. W. Goodwin, "On the Mosaic cosmogony," in *Essays and reviews,* ed. F. Temple, 5th ed. (London, 1861), p. 250.

VIII. Evolutionary Theory and Religious Belief

1 *Punch* 69 (8 April 1871), p. 145, quoted in A. Ellegard, *Darwin and the general reader* (Göteborg, 1958), p. 190.
2 Charles Darwin to Asa Gray, 20 July 1856, in F. Darwin, ed., *The life and letters of Charles Darwin,* 3 vols., 2d ed., vol. 2 (London, 1887), pp. 78–9.
3 Sumner, quoted in R. Hofstadter, *Social Darwinism in American thought* (Boston, 1955), p. 57.
4 F. C. Selous, *Sunshine and storm in Rhodesia* (1896; reprint, New York, 1969), p. 67.
5 John Tyndall, "The Belfast address" to the British Association for the Advancement of Science (1874), in *Science and religion in the nineteenth century,* ed. Tess Coslett (Cambridge, 1984), p. 182.
6 Jousset, quoted in Harry W. Paul, *The edge of contingency* (Gainesville, Fla. 1979), pp. 46–7.
7 Troeltsch, quoted in S. W. Sykes, "Ernst Troeltsch and Christianity's essence," in *Ernst Troeltsch and the future of theology,* ed. John P. Clayton (Cambridge, 1976), pp. 157–8.
8 Moore, quoted in Arthur Peacocke, "Biological evolution and Christian theology – yesterday and today," in *Darwinism and divinity,* ed. John Durant (Oxford, 1985), p. 111.

9 Asa Gray, *Darwiniana,* ed. A. Hunter Dupree (Cambridge, Mass., 1963), pp. 310–11.
10 William James, *The will to believe and other essays in popular philosophy* (New York, 1899), p. 213.

Index